Second Edition

Intelligent Systems *for* Engineers and Scientists

Second Edition

Intelligent Systems *for* Engineers and Scientists

Adrian A. Hopgood

CRC Press
Boca Raton London New York Washington, D.C.

Library of Congress Cataloging-in-Publication Data

Hopgood, Adrian A.
Intelligent systems for engineers and scientists / Adrian A.
Hopgood.--2nd ed.
p. cm.
Includes bibliographical references and index.
ISBN 0-8493-0456-3
1. Expert systems (Computer science) 2. Computer-aided engineering.
I. Title.
QA76.76.E95 H675 2000
006.3'3'02462--dc21 00-010341

Direct all inquiries to CRC Press LLC, 2000 N.W. Corporate Blvd., Boca Raton, Florida 33431.

International Standard Book Number 0-8493-0456-3
Library of Congress Card Number 00-010341
Printed in the United States of America 3 4 5 6 7 8 9 0
Printed on acid-free paper

Preface

"Intelligent systems" is a broad term, covering a range of computing techniques that have emerged from research into artificial intelligence. It includes symbolic approaches — in which knowledge is explicitly expressed in words and symbols — and numerical approaches such as neural networks, genetic algorithms, and fuzzy logic. In fact, many practical intelligent systems are a hybrid of different approaches. Whether any of these systems is really capable of displaying intelligent behavior is a moot point. Nevertheless, they are extremely useful and they have enabled elegant solutions to a wide variety of difficult problems.

There are plenty of other books available on intelligent systems and related technologies, but I hope this one is substantially different. It takes a practical view, showing the issues encountered in the development of applied systems. I have tried to describe a wide range of intelligent systems techniques, with the help of realistic problems in engineering and science. The examples included here have been specifically selected for the details of the techniques that they illustrate, rather than merely to survey current practice.

The book can be roughly divided into two parts. Chapters 1 to 10 describe the techniques of intelligent systems, while Chapters 11 to 14 look at four broad categories of applications. These latter chapters explore in depth the design and implementation issues of applied systems, together with their advantages and difficulties. The four application areas have much in common, as they all concern automated decision making, while making the best use of the available information.

The first edition of this book was published as *Knowledge-Based Systems for Engineers and Scientists*. It was adopted by the Open University for its course *T396: Artificial Intelligence for Technology* and, as a result, I have received a lot of useful feedback. I hope that this new edition addresses the weaknesses of the previous one, while retaining and building upon its strengths. As well as updating the entire book, I have added new chapters on intelligent agents, neural networks, optimization algorithms (especially genetic algorithms), and hybrid systems. A new title was therefore needed to reflect the broader scope of this new edition. *Intelligent Systems for Engineers and*

Scientists seems appropriate, as it embraces both the explicit knowledge-based models that are retained from the first edition and the implicit numerical models represented by neural networks and optimization algorithms.

I hope the book will appeal to a wide readership. In particular, I hope that students will be drawn toward this fascinating area from all scientific and engineering subjects, not just from the computer sciences. Beyond academia, the book will appeal to engineers and scientists who either are building intelligent systems or simply want to know more about them.

The first edition was mostly written while I was working at the Telstra Research Laboratories in Victoria, Australia, and subsequently finished upon my return to the Open University in the UK. I am still at the Open University, where this second edition was written.

Many people have helped me, and I am grateful to them all. The following all helped either directly or indirectly with the first edition (in alphabetical order): Mike Brayshaw, David Carpenter, Nicholas Hallam, David Hopgood, Sue Hopgood, Adam Kowalzyk, Sean Ogden, Phil Picton, Chris Price, Peter Richardson, Philip Sargent, Navin Sullivan, Neil Woodcock, and John Zucker.

I am also indebted to those who have helped in any way with this new edition. I am particularly grateful to Tony Hirst for his detailed suggestions for inclusion and for his thoughtful comments on the drafts. I also extend my thanks to Lars Nolle for his helpful comments and for supplying Figures 7.1, 7.7, and 8.18; to Jon Hall for his comments on Chapter 5; to Sara Parkin and Carole Gustafson for their careful proofreading; and to Dawn Mesa for making the publication arrangements. Finally, I am indebted to Sue and Emily for letting me get on with it. Normal family life can now resume.

Adrian Hopgood
www.adrianhopgood.com

The author

Adrian Hopgood has earned his BSc from Bristol University, PhD from Oxford University, and MBA from the Open University. After completing his PhD in 1984, he spent two years developing applied intelligent systems for Systems Designers PLC. He subsequently joined the academic staff of the Open University, where he has established his research in intelligent systems and their application in engineering and science. Between 1990 and 1992 he worked for Telstra Research Laboratories in Australia, where he contributed to the development of intelligent systems for telecommunications applications. Following his return to the Open University he led the development of the course *T396 – Artificial Intelligence for Technology*. He has further developed his interests in intelligent systems and pioneered the development of the blackboard system, ARBS.

For Sue and Emily

Contents

Chapter one

Introduction

1.1 Intelligent systems

Over many centuries, tools of increasing sophistication have been developed to serve the human race. Physical tools such as chisels, hammers, spears, arrows, guns, carts, cars, and aircraft all have their place in the history of civilization. The human race has also developed tools of communication — spoken language, written language, and the language of mathematics. These tools have not only enabled the exchange and storage of information, but have also allowed the expression of concepts that simply could not exist outside of the language.

The last few decades have seen the arrival of a new tool — the digital computer. Computers are able to perform the same sort of numerical and symbolic manipulations that an ordinary person can, but faster and more reliably. They have therefore been able to remove the tedium from many tasks that were previously performed manually, and have allowed the achievement of new feats. Such feats range from huge scientific "number-crunching" experiments to the more familiar electronic banking facilities.

Although these uses of the computer are impressive, it is actually only performing quite simple operations, albeit rapidly. In such applications, the computer is still only a complex calculating machine. The intriguing idea now is whether we can build a computer (or a computer program) that can *think*. As Penrose [1] has pointed out, most of us are quite happy with machines that enable us to do physical things more easily or more quickly, such as digging a hole or traveling along a freeway. We are also happy to use machines that enable us to do physical things that would otherwise be impossible, such as flying. However, the idea of a machine that can think for us is a huge leap forward in our ambitions, and one which raises many ethical and philosophical questions.

Research in artificial intelligence (or simply AI) is directed toward building such a machine and improving our understanding of intelligence. The ultimate achievement in this field would be to construct a machine that can mimic or exceed human mental capabilities, including reasoning, understanding, imagination, recognition, creativity, and emotions. We are a long way from achieving this, but some successes have been achieved in mimicking specific areas of human mental activity. For instance, machines are now able to play chess at the highest level, to interpret spoken sentences, and to diagnose medical complaints. An objection to these claimed successes might be that the machine does not tackle these problems in the same way that a human would. This objection will not concern us in this book, which is intended as a guide to practical systems and not a philosophical thesis.

In achieving these modest successes, research into artificial intelligence, together with other branches of computer science, has resulted in the development of several useful computing tools that form the basis of this book. These tools have a range of potential applications, but this book emphasizes their use in engineering and science. The tools of particular interest can be roughly divided among knowledge-based systems, computational intelligence, and hybrid systems. Knowledge-based systems include expert and rule-based systems, object-oriented and frame-based systems, and intelligent agents. Computational intelligence includes neural networks, genetic algorithms and other optimization algorithms. Techniques for handling uncertainty, such as fuzzy logic, fit into both categories.

Knowledge-based systems, computational intelligence, and their hybrids are collectively referred to here as *intelligent systems*. Intelligent systems have not solved the problem of building an artificial mind and, indeed, some would argue that they show little, if any, real intelligence. Nevertheless, they have enabled a range of problems to be tackled that were previously considered too difficult, and have enabled a large number of other problems to be tackled more effectively. From a pragmatic point of view, this in itself makes them interesting and useful.

1.2 Knowledge-based systems

The principal difference between a knowledge-based system (KBS) and a conventional program lies in the structure. In a conventional program, domain knowledge is intimately intertwined with software for controlling the application of that knowledge. In a knowledge-based system, the two roles are explicitly separated. In the simplest case there are two modules — the knowledge module is called the *knowledge base*, and the control module is called the *inference engine* (Figure 1.1). In more complex systems, the inference engine itself may be a knowledge-based system containing *meta-knowledge*, i.e., knowledge of how to apply the domain knowledge.

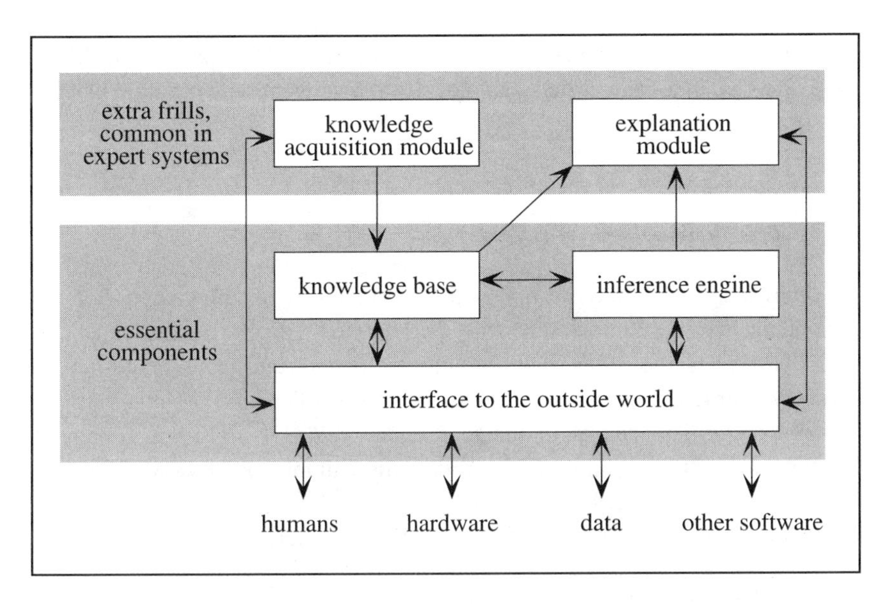

Figure 1.1 The main components of a knowledge-based system

The explicit separation of knowledge from control makes it easier to add new knowledge, either during program development or in the light of experience during the program's lifetime. There is an analogy with the brain, the control processes of which are approximately unchanging in their nature (cf. the inference engine), even though individual behavior is continually modified by new knowledge and experience (cf. updating the knowledge base).

Suppose that a professional engineer uses a conventional program to support his or her everyday work. Altering the behavior of the program would require him or her to become immersed in the details of the program's implementation. Typically this would involve altering control structures of the form:

```
if...then...else...
```

or

```
for x from a to b do...
```

To achieve these changes, the engineer needs to be a proficient programmer. Even if he or she does have this skill, modifications of the kind described are unwieldy and are difficult to make without unwittingly altering some other aspect of the program's behavior.

The knowledge-based system approach is more straightforward. The knowledge is represented *explicitly* in the knowledge base, not *implicitly* within the structure of a program. Thus, the knowledge can be altered with relative ease. The inference engine uses the knowledge base to tackle a particular task in a manner that is analogous to a conventional program using a data file.

1.3 The knowledge base

The knowledge base may be rich with diverse forms of knowledge. For the time being, we will simply state that the knowledge base contains rules and facts. However, the rules may be complex, and the facts may include sequences, structured entities, attributes of such entities, and the relationships between them. The details of the representation used vary from system to system, so the syntax shown in the following examples is chosen arbitrarily. Let us consider a knowledge-based system for dealing with the payroll of ACME, Inc. A fact and a rule in the knowledge base may be:

```
/* Fact 1.1 */
Joe Bloggs works for ACME

/* Rule 1.1 */
IF ?x works for ACME THEN ?x earns a large salary
```

The question marks are used to indicate that `x` is a variable that can be replaced by a constant value, such as `Joe Bloggs` or `Mary Smith`.

Let us now consider how we might represent the fact and the rule in a conventional program. We might start by creating a "record" (a data structure for grouping together different data types) for each employee. The rule could be expressed easily enough as a conditional statement (`IF...THEN...`), but it would need to be carefully positioned within the program so that:

- the statement is applied whenever it is needed;
- all relevant variables are in scope (the scope of a variable is that part of the program to which the declaration of the variable applies);
- any values that are assigned to variables remain active for as long as they are needed; and
- the rest of the program is not disrupted.

In effect, the fact and the rule are "hard-wired," so that they become an intrinsic part of the program. As many systems need hundreds or thousands of facts and rules, slotting them into a conventional program is a difficult task. This can be contrasted with a knowledge-based system in which the rule and the fact are represented explicitly and can be changed at will.

Rules such as Rule 1.1 are a useful way of expressing many types of knowledge, and are discussed in more detail in Chapter 2. It has been assumed so far that we are dealing with certain knowledge. This is not always the case, and Chapter 3 discusses the use of uncertainty in rules. In the case of Rule 1.1, uncertainty may arise from three distinct sources:

- *uncertain evidence*
  (perhaps we are not certain that Joe Bloggs works for ACME)

- *uncertain link between evidence and conclusion*
  (We cannot be certain that an ACME employee earns a large salary, we just know that it is likely)

- *vague rule*
  (what is a "large" salary anyway?)

The first two sources of uncertainty can be handled by Bayesian updating, or variants of this idea. The last source of uncertainty can be handled by fuzzy sets and fuzzy logic.

Let us now consider facts in more detail. Facts may be static, in which case they can be written into the knowledge base. Fact 1.1 falls into this category. Note that static facts need not be permanent, but they change sufficiently infrequently that changes can be accommodated by updating the knowledge base when necessary. In contrast, some facts may be transient. Transient facts (e.g., "Oil pressure is 3000 Pa," "the user of this program is Adrian") apply at a specific instance only, or for a single run of the system. The knowledge base may contain defaults, which can be used as facts in the absence of transient facts to the contrary. Here is a collection of facts about my car:

```
My car is a car                  (static relationship)
A car is a vehicle               (static relationship)
A car has four wheels            (static attribute)
A car's speed is 0mph            (default attribute)
My car is red                    (static attribute)
My car is in my garage           (default relationship)
My garage is a garage            (static relationship)
A garage is a building           (static relationship)
My garage is made from brick     (static attribute)
My car is in the High Street     (transient relationship)
The High Street is a street      (static relationship)
A street is a road               (static relationship)
```

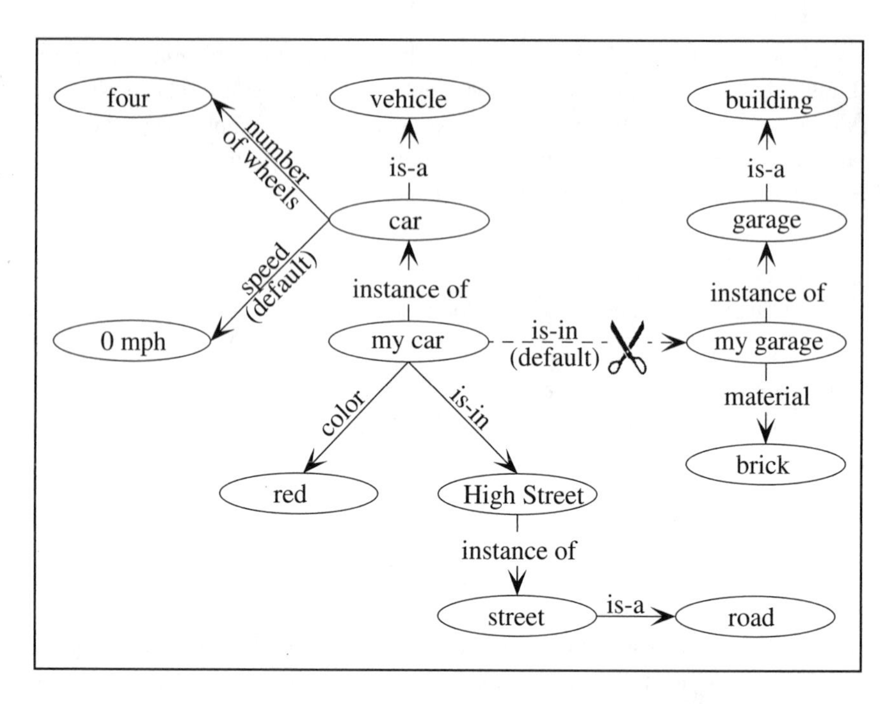

Figure 1.2 A semantic network with an overridden default

Notice that in this list we have distinguished between attributes and relationships. Attributes are properties of object *instances* (such as my car) or object *classes* (such as cars and vehicles). Relationships exist among instances of objects and classes of objects. In this way we can begin to build a model of the subject area of interest, and this reliance on a model will be a recurring topic throughout this book. Attributes and relationships can be represented as a network, known as an *associative* or *semantic network*, as shown in Figure 1.2. In this representation, attributes are treated in the same way as relationships. In Chapter 4 we will explore object-oriented systems, in which relationships and attributes are represented explicitly in a formalized manner. Object-oriented systems offer many other benefits, which will also be discussed.

The facts that have been described so far are all made available to the knowledge-based system either at the outset (static facts) or while the system is running (transient facts). Both may therefore be described as *given* facts. One or more given facts may satisfy the condition of a rule, resulting in the generation of a new fact, known as a *derived* fact. For example, by applying Rule 1.1 to Fact 1.1, we can derive:

```
/* Fact 1.2 */
Joe Bloggs earns a large salary
```

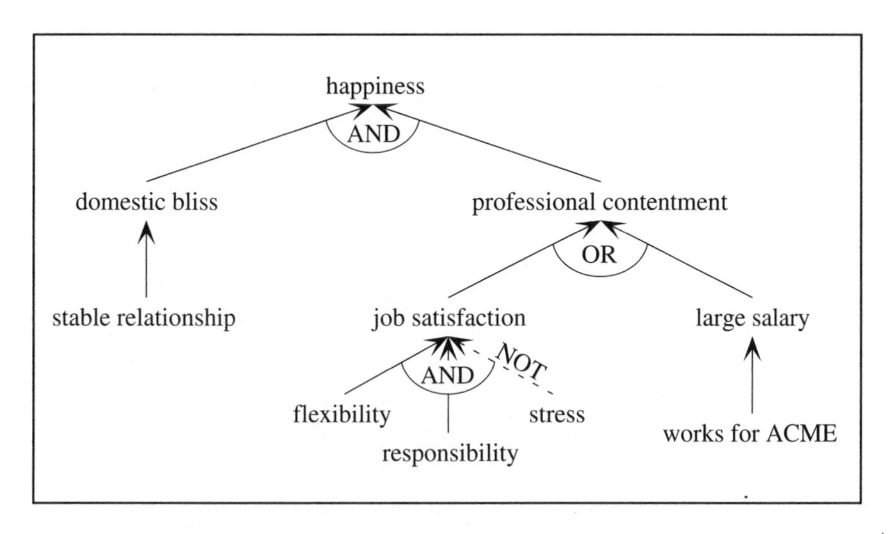

Figure 1.3 An inference network

The derived fact may satisfy, or partially satisfy, another rule, such as:

```
/* Rule 1.2 */
IF ?x earns a large salary OR ?x has job satisfaction
THEN ?x is professionally content
```

This in turn may lead to the generation of a new derived fact. Rules 1.1 and 1.2 are interdependent, since the conclusion of one can satisfy the condition of the other. The interdependencies amongst the rules define a network, as shown in Figure 1.3, known as an *inference network.*

1.4 Deduction, abduction, and induction

The rules that make up the inference network in Figure 1.3, and the network taken as a whole, are used to link cause and effect:

```
IF <cause> THEN <effect>
```

Using the inference network, we can infer that if Joe Bloggs works for ACME and is in a stable relationship (the causes) then he is happy (the effect). This is the process of *deduction*. Many problems, such as diagnosis, involve reasoning in the reverse direction, i.e., we wish to ascertain a cause, given an effect. This is *abduction*. Given the observation that Joe Bloggs is happy, we can infer by abduction that Joe Bloggs enjoys domestic bliss and professional contentment. However, this is only a valid conclusion if the inference network shows *all* of the ways in which a person can find happiness. This is the *closed-world assumption*, the implications of which are discussed in Chapters 2 and 11.

The inference network therefore represents a closed world, where nothing is known beyond its boundaries. As each node represents a possible state of some aspect of the world, a model of the current overall state of the world can be maintained. Such a model is dependent on the extent of the relationships between the nodes in the inference network. In particular, if a change occurs in one aspect of the world, many other nodes could be affected. Determining what else is changed in the world model as a consequence of changing one particular thing is known as the *frame problem.* In the description of Joe Bloggs' world represented in Figure 1.3, this is equivalent to determining the extent of the relationships between the nodes. For example, if Joe Bloggs gets a new job, Figure 1.3 suggests that the only direct change is his salary, which could change his professional contentment and happiness. However, in a more complex model of Joe Bloggs' world, many other nodes could also be affected.

If we have many examples of cause and effect, we can infer the rule (or inference network) that links them. For instance, if every employee of ACME that we have met earns a large salary, then we might infer Rule 1.1:

```
/* Rule 1.1 */
IF ?x works for ACME THEN ?x earns a large salary.
```

Inferring a rule from a set of example cases of cause and effect is termed *induction.*

We can summarize deduction, abduction, and induction as follows:

- *deduction:* cause + rule ⇒ effect
- *abduction:* effect + rule ⇒ cause
- *induction:* cause + effect ⇒ rule

1.5 The inference engine

Inference engines vary greatly according to the type and complexity of knowledge with which they deal. Two important types of inference engines can be distinguished: *forward-chaining* and *backward-chaining*. These may also be known as *data-driven* and *goal-driven*, respectively. A knowledge-based system working in data-driven mode takes the available information (the "given" facts) and generates as many derived facts as it can. The output is therefore unpredictable. This may have either the advantage of leading to novel or innovative solutions to a problem or the disadvantage of wasting time generating irrelevant information. The data-driven approach might typically be used for problems of interpretation, where we wish to know whatever the system can tell us about some data. A goal-driven strategy is appropriate when a more tightly focused solution is required. For instance, a planning system may be required to generate a plan for manufacturing a consumer product. Any other plans are irrelevant. A backward-chaining system might be presented with the proposition: a plan exists for manufacturing a widget. It will then attempt to ascertain the truth of this proposition by generating the plan, or it may conclude that the proposition is false and no plan is possible. Forward- and backward-chaining are discussed in more detail in Chapter 2. Planning is discussed in Chapter 13.

1.6 Declarative and procedural programming

We have already seen that a distinctive characteristic of a knowledge-based system is that knowledge is separated from reasoning. Within the knowledge base, the programmer expresses information about the problem to be solved. Often this information is declarative, i.e., the programmer states some facts, rules, or relationships without having to be concerned with the detail of *how* and *when* that information is applied. The following are all examples of declarative programming:

```
/* Rule 1.3 */
IF pressure is above threshold THEN close valve

/* Fact 1.3 */
valve A is shut                                  /* a simple fact */

/* Fact 1.4 */
valve B is connected to tank 3                   /* a relation */
```

Each example represents a piece of knowledge that could form part of a knowledge base. The declarative programmer does not necessarily need to state explicitly how, when, and if the knowledge should be used. These details are implicit in the inference engine. An inference engine is normally programmed procedurally — a set of sequential commands is obeyed, which involves extracting and using information from the knowledge base. This task can be made explicit by using *metaknowledge* (knowledge about knowledge), e.g.:

```
/* Metarule 1.4 */
Examine rules about valves before rules about pipes
```

Most conventional programming is procedural. Consider, for example, the following C program:

```
/* A program in C to read 10 integers from a file and */
/* print them out */
#include <stdio.h>
FILE *openfile;
main()
{ int j, mynumber;
  openfile = fopen("myfile.dat", "r");
  if (openfile == NULL)
    printf("error opening file");
  else
  {
    for (j=1; j<=10; j=j+1)
    {
      fscanf(openfile,"%d",&mynumber);
      printf("Number %d is %d\n", j, mynumber);
    }
    fclose(openfile);
  }
}
```

This program contains explicit step-by-step instructions telling the computer to perform the following actions:

(i) open a data file;
(ii) print a message if it cannot open the file, otherwise perform the remaining steps;
(iii) set the value of `j` to 1;
(iv) read an integer from the file and store it in the variable `mynumber`;
(v) print out the value of `mynumber`;
(vi) add 1 to `j`;
(vii) if `j`≤10 repeat steps (iv)–(vi), otherwise move on to step (viii);
(viii) close the data file.

The data file, on the other hand, contains no instructions for the computer at all, just information in the form of a set of integers. The procedural instructions for determining what the computer should do with the integers resides in the program. The data file is therefore declarative, while the program is procedural. The data file is analogous to a trivial knowledge base, and the program is analogous to the corresponding inference engine. Of course, a proper knowledge base would be richer in content, perhaps containing a combination of rules, facts, relations, and data. The corresponding inference engine would be expected to interpret the knowledge, to combine it to form an overall view, to apply the knowledge to data, and to make decisions.

It would be an oversimplification to think that all knowledge bases are written declaratively and all inference engines are written procedurally. In real systems, a collection of declarative information, such as a rule set, often needs to be embellished by some procedural information. Similarly, there may be some inference engines that have been programmed declaratively, notably those implemented in Prolog (described in Chapter 10). Nevertheless, the declarative instructions must eventually be translated into procedural ones, as the computer can only understand procedural instructions at the machine-code level.

1.7 Expert systems

Expert systems are a type of knowledge-based system designed to embody expertise in a particular specialized domain. Example domains might be configuring computer networks, diagnosing faults in telephones, or mineral prospecting. An expert system is intended to act as a human expert who can be consulted on a range of problems that fall within his or her domain of expertise. Typically, the user of an expert system will enter into a dialogue in which he or she describes the problem (such as the symptoms of a fault) and the expert system offers advice, suggestions, or recommendations. The dialogue may be led by the expert system, so that the user responds to a series of questions or enters information into a spreadsheet. Alternatively, the expert system may allow the user to take the initiative in the consultation by allowing him or her to supply information without necessarily being asked for it.

Since an expert system is a knowledge-based system that acts as a specialist consultant, it is often proposed that an expert system must offer certain capabilities that mirror those of a human consultant. In particular, it is often claimed that an expert system must be capable of justifying its current line of inquiry and explaining its reasoning in arriving at a conclusion. This is the purpose of the explanation module in Figure 1.1. However, the best that most expert systems can achieve is to produce a trace of the facts and rules that have been used. This is equivalent to a trace of the execution path for a conventional program, a capability that is normally regarded as standard and not particularly noteworthy.

An expert system shell is an expert system with an empty knowledge base. These are sold as software packages of varying complexity. In principle, it should be possible to buy an expert system shell, build up a knowledge base, and thereby produce an expert system. However, all domains are different and it is difficult for a software supplier to build a shell that adequately handles them all. The best shells are flexible in their ability to represent and apply knowledge. Without this flexibility, it may be necessary to generate rules and facts in a convoluted style in order to fit the syntax or to force a certain kind of behavior from the system. This situation is scarcely better than building a conventional program.

1.8 Knowledge acquisition

The representation of knowledge in a knowledge base can only be addressed once the knowledge is known. There are three distinct approaches to acquiring the relevant knowledge for a particular domain:

- the knowledge is teased out of a domain expert;
- the builder of the knowledge-based system *is* a domain expert;
- the system learns automatically from examples.

The first approach is commonly used, but is fraught with difficulties. The person who extracts the knowledge from the expert and encodes it in the knowledge base is termed the *knowledge engineer*. Typically, the knowledge engineer interviews one or more domain experts and tries to make them articulate their in-depth knowledge in a manner that the knowledge engineer can understand. The inevitable communication difficulties can be avoided by the second approach, in which the domain expert becomes a knowledge engineer or the knowledge engineer becomes a domain expert.

Finally, there are many circumstances in which the knowledge is either unknown or cannot be expressed explicitly. In these circumstances it may be preferable to have the system generate its own knowledge base from a set of examples. Techniques for automatic learning are discussed in Chapters 6 to 9.

1.9 Search

Search is the key to practically all problem-solving tasks, and has been a major focus of research in intelligent systems. Problem solving concerns the search for a solution. The detailed engineering applications discussed in this book include the search for a design, plan, control action, or diagnosis of a fault. All of these applications involve searching through the possible solutions (the *search space*) to find one or more that are optimal or satisfactory. Search is also a key issue for the internal workings of a knowledge-based system. The

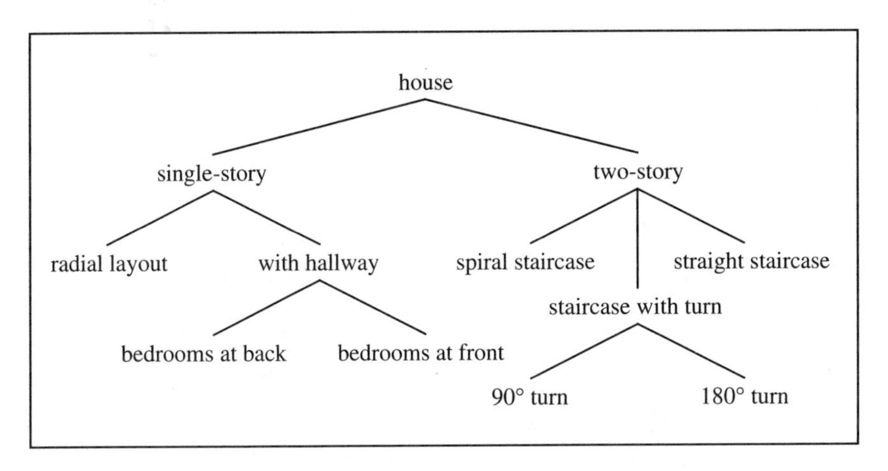

Figure 1.4 A search tree for house designs

knowledge base may contain hundreds or thousands of rules and facts. The principal role of the inference engine is to search for the most appropriate item of knowledge to apply at any given moment (see Chapter 2).

In the case of searching the knowledge base, it is feasible (although not very efficient) to test all of the alternatives before selecting one. This is known as *exhaustive search*. In the search space of an application such as design or diagnosis, exhaustive search is likely to be impractical since the number of candidate solutions is so vast. A practical search therefore has to be selective. To this end, the candidate solutions that make up the search space can be organized as a *search tree*.

The expression *search tree* indicates that solutions can be categorized in some fashion, so that similar solutions are clustered together on the same branch of the tree. Figure 1.4 shows a possible search tree for designing a house. Unlike a real tree, the tree shown here has its root at the top and its leaves at the bottom. As we progress toward the leaves, the differences between the designs become less significant. Each alternative design is either generated automatically or found in a database, and then tested for suitability. This is described as a *generate and test* strategy. If a solution passes the test, the search may continue in order to find further acceptable solutions, or it may stop.

Two alternative strategies for systematically searching the tree are depth-first and breadth-first searches. An example of *depth-first* search is shown in Figure 1.5. In this example, a progressively more detailed description of a single-story house is built up and tested before other classifications, such as a two-story house, are considered. When a node fails the test, the search resumes at the previous node where a branch was selected. This process of *backtracking*

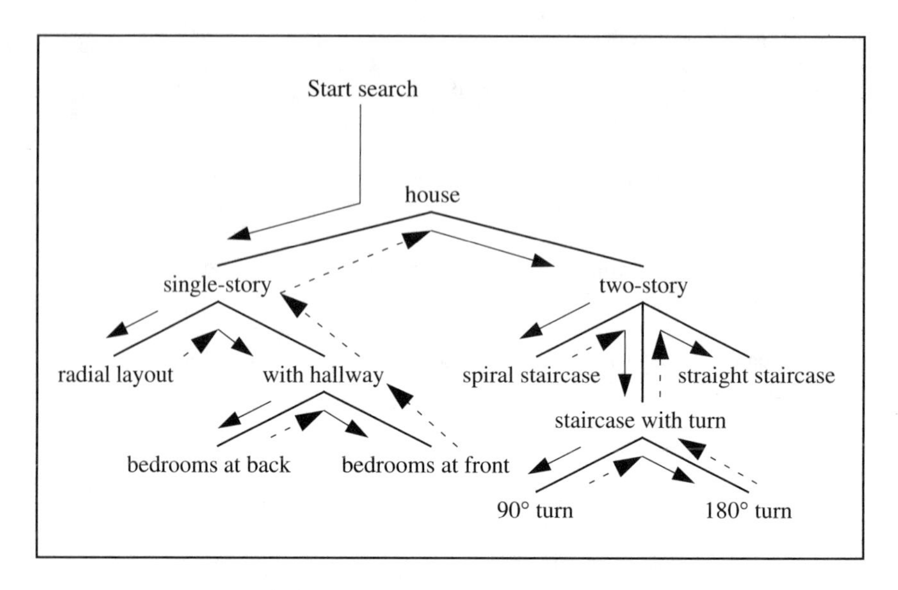

Figure 1.5 Depth-first search (the broken arrows indicate backtracking)

(see Chapters 2 and 10) is indicated in Figure 1.5 by the broken arrows, which are directed toward the root rather than the leaves of the tree.

In contrast, the *breadth-first* approach (Figure 1.6) involves examining all nodes at a given level in the tree before progressing to the next level. Each node is a generalization of the nodes on its subbranches. Therefore, if a node fails the test, all subcategories are assumed to fail the test and are eliminated from the search.

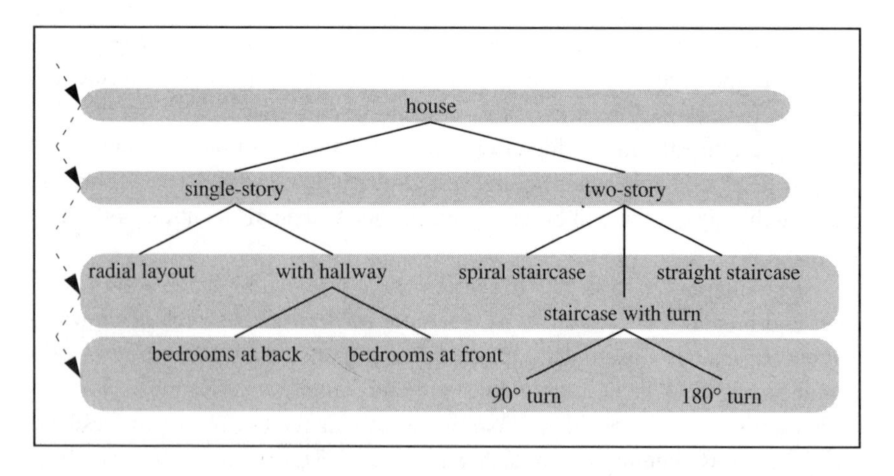

Figure 1.6 Breadth-first search

The systematic search strategies described above are examples of *blind search*. The search can be made more efficient either by eliminating unfeasible categories ("pruning the search tree") or by ensuring that the most likely alternatives are tested before less likely ones. To achieve either of these, we need to apply heuristics to the search process. Blind search is thereby modified to *heuristic search*. Barr and Feigenbaum [2] have surveyed the use of the word *heuristic* and produced this general description:

> *A heuristic is a rule of thumb, strategy, trick, simplification, or any other kind of device which drastically limits search for solutions in large search spaces. Heuristics do not guarantee optimal solutions; in fact they do not guarantee any solution at all; all that can be said for a useful heuristic is that it offers solutions which are good enough most of the time.*

In a diagnostic system for plumbing, a useful heuristic may be that pipes are more likely to leak at joints than along lengths. This heuristic defines a search strategy, namely to look for leaking joints first. In a design system for a house, we might use a heuristic to eliminate all designs that would require us to walk through a bedroom to reach the bathroom. In a short-term planning system, a heuristic might be used to ensure that no plans looked ahead more than a week. In a control system for a nuclear reactor, a heuristic may ensure that the control rods are never raised while the coolant supply is turned off.

1.10 Computational intelligence

The discussion so far has concentrated on types of knowledge-based systems. They are all symbolic representations, in which knowledge is explicitly represented in words and symbols that are combined to form rules, facts, relations, or other forms of knowledge representation. As the knowledge is explicitly written, it can be read and understood by a human. These symbolic techniques contrast with numerical techniques such as genetic algorithms (Chapter 7) and neural networks (Chapter 8). Here the knowledge is not explicitly stated but is represented by numbers which are adjusted as the system improves its accuracy. These techniques are collectively known as *computational intelligence* (CI) or *soft computing*.

Chapter 3 describes three techniques for handling uncertainty: Bayesian updating, certainty factors, and fuzzy logic. These techniques all use a mixture of rules and associated numerical values, and they can therefore be considered as both computational intelligence tools and knowledge-based system tools. In summary, computational intelligence embraces:

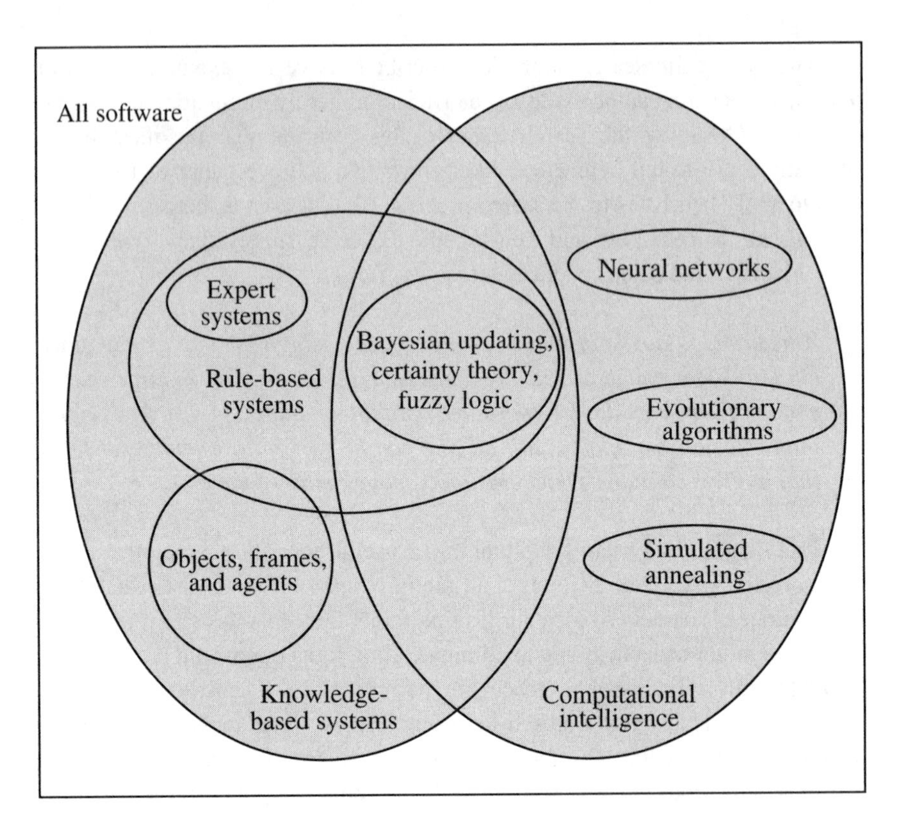

Figure 1.7 Categories of intelligent system software

- neural networks;
- genetic algorithms or, more generally, evolutionary algorithms;
- probabilistic methods such as Bayesian updating and certainty factors;
- fuzzy logic;
- combinations of these techniques with each other and with KBSs.

1.11 Integration with other software

This book will take a broad view of intelligent systems, and will stress the interrelationship of the various KBS and CI techniques with each other and with conventional programming (Figure 1.7). These techniques do not necessarily represent exclusive alternatives, but can often be used cooperatively. For example, a designer may already have an excellent conventional program for simulating the aerodynamic properties of a car. In this case, a knowledge-based system might be used as an interface to the program, allowing the program to be used to its full potential.

Chapter 9 describes some of the ways in which intelligent systems techniques can be used cooperatively within hybrid systems. Chapters 11 to 14 describe some practical applications, most of which are hybrids of some form. If a problem can be broken down into subtasks, a blackboard system (described in Chapter 9) might provide a suitable way of tackling it. Blackboard systems allow each subtask to be handled using an appropriate technique, thereby contributing most effectively to the overall solution.

As with any other technique, KBSs and CI are not suitable for all types of problems. Each problem calls for the most appropriate tool, but KBSs and CI can be used for many problems that would be impracticable by other means.

References

1. Penrose, R., *The Emperor's New Mind*, Oxford University Press, 1989.
2. Barr, A. and Feigenbaum, E. A., *The Handbook of Artificial Intelligence*, vol. 1, Addison-Wesley, 1986.

Further reading

- Finlay, J. and Dix, A., *An Introduction to Artificial Intelligence*, UCL Press, 1996.
- Pedrycz, W., *Computational Intelligence: an introduction*, CRC Press, 1997.
- Sriram, R. D., *Intelligent Systems for Engineering: a knowledge-based approach*, Springer-Verlag Telos, 1997.
- Torsun, I. S., *Foundations of Intelligent Knowledge-Based Systems*, Academic Press, 1995.
- Winston, P. H., *Artificial Intelligence*, 3rd ed., Addison-Wesley, 1992.

Chapter two

Rule-based systems

2.1 Rules and facts

A rule-based system is a knowledge-based system where the knowledge base is represented in the form of a set, or sets, of *rules*. Rules are an elegant, expressive, straightforward, and flexible means of expressing knowledge. The simplest type of rule is called a *production rule* and takes the form:

```
IF <condition> THEN <conclusion>
```

An example of a production rule might be:

```
IF the tap is open THEN water flows
```

Part of the attraction of using production rules is that they can often be written in a form that closely resembles natural language, as opposed to a computer language. A simple rule like the one above is intelligible to anyone who understands English. Although rules can be considerably more complex than this, their explicit nature still makes them more intelligible than conventional computer code.

In order for rules to be applied, and hence for a rule-based system to be of any use, the system will need to have access to *facts*. Facts are unconditional statements which are assumed to be correct at the time that they are used. F example, `the tap is open` is a fact. Facts can be:

- looked up from a database;
- already stored in computer memory;
- determined from sensors connected to the computer;
- obtained by prompting the user for information;
- derived by applying rules to other facts.

Facts can be thought of as special rules, where the condition part is always true. Therefore, the fact `the tap is open` could also be thought of as a rule:

```
IF TRUE THEN the tap is open
```

Given the rule `IF the tap is open THEN water flows` and the fact `the tap is open`, the derived fact `water flows` can be generated. The new fact is stored in computer memory and can be used to satisfy the conditions of other rules, thereby leading to further derived facts. The collection of facts which are known to the system at any given time is called the *fact base*.

Rule-writing is a type of declarative programming (see Section 1.6), because rules represent knowledge that can be used by the computer, without specifying how and when to apply that knowledge. The ordering of rules in a program should ideally be unimportant, and it should be possible to add new rules or modify existing ones without fear of side effects. We will see by reference to some simple examples that these ideals cannot always be taken for granted.

For the declared rules and facts to be useful, an inference engine for interpreting and applying them is required (see Section 1.5). Inference engines are incorporated into a range of software tools, discussed in Chapter 10, that includes expert system shells, artificial intelligence toolkits, software libraries, and the Prolog language.

2.2 A rule-based system for boiler control

Whereas the above discussion describes rule-based systems in an abstract fashion, a physical example is introduced in this section. We will consider a rule-based system to monitor the state of a power station boiler and to advise appropriate actions. The boiler in our example (Figure 2.1) is used to produce steam to drive a turbine and generator. Water is heated in the boiler tubes to produce a steam and water mixture that rises to the steam drum, which is a cylindrical vessel mounted horizontally near the top of the boiler. The purpose of the drum is to separate the steam from the water. Steam is taken from the drum, passed through the superheater and applied to the turbine that turns the generator. Sensors are fitted to the drum in order to monitor:

- the temperature of the steam in the drum;
- the voltage output from a transducer, which in turn monitors the level of water in the drum;
- the status of pressure release valve (i.e., open or closed);
- the rate of flow of water through the control valve.

The following rules have been written for controlling the boiler:

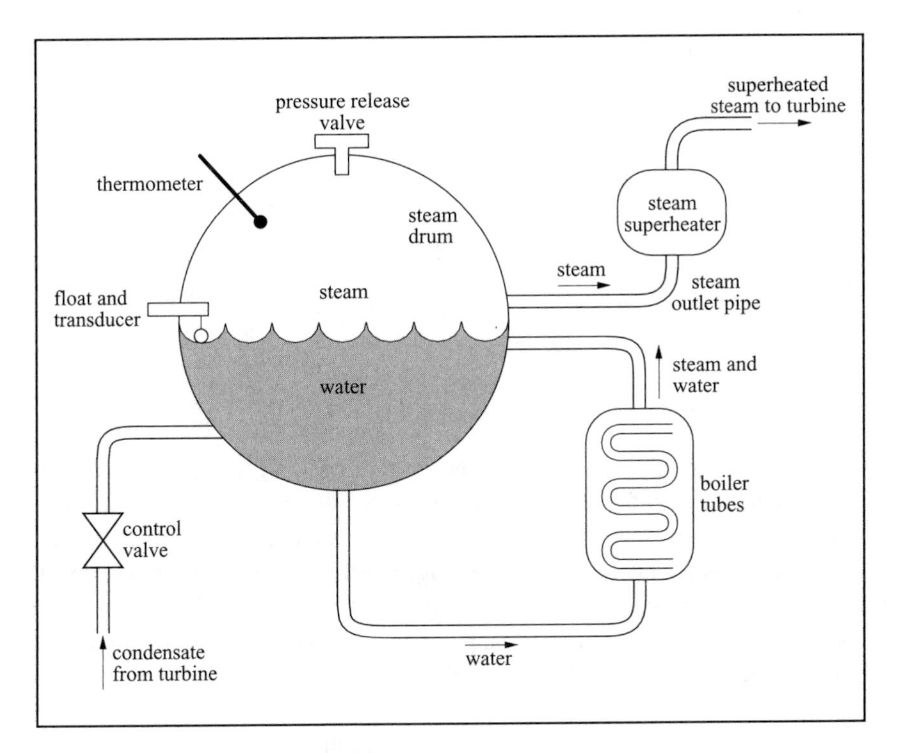

Figure 2.1 A power station boiler

```
/* Rule 2.1 */
IF water level low THEN open control valve

/* Rule 2.2 */
IF temperature high AND water level low
THEN open control valve AND shut down boiler tubes

/* Rule 2.3 */
IF steam outlet blocked THEN replace outlet pipe

/* Rule 2.4 */
IF release valve stuck THEN steam outlet blocked

/* Rule 2.5 */
IF pressure high AND release valve closed
THEN release valve stuck

/* Rule 2.6 */
IF steam escaping THEN steam outlet blocked
```

```
/* Rule 2.7 */
IF temperature high AND NOT(water level low)
THEN pressure high

/* Rule 2.8 */
IF transducer output low THEN water level low

/* Rule 2.9 */
IF release valve open AND flow rate high
THEN steam escaping

/* Rule 2.10 */
IF flow rate low THEN control valve closed
```

The conclusions of three of the above rules (2.1, 2.2, and 2.3) consist of recommendations to the boiler operators. In a fully automated system, such rules would be able to perform their recommended actions rather than simply making a recommendation. The remaining rules all involve taking a low-level fact, such as a transducer reading, and deriving a higher-level fact, such as the quantity of water in the drum. The input data to the system (sensor readings in our example) are low-level facts; higher-level facts are facts derived from them.

Most of the rules in our rule base are specific to one particular boiler arrangement and would not apply to other situations. These rules could be described as *shallow*, because they represent shallow knowledge. On the other hand, Rule 2.7 expresses a fundamental rule of physics, namely that the boiling temperature of a liquid increases with increasing applied pressure. This is valid under any circumstances and is not specific to the boiler shown in Figure 2.1. It is an example of a *deep* rule expressing deep knowledge.

The distinction between deep and shallow rules should not be confused with the distinction between *high-level* and *low-level* rules. Low-level rules are those that depend on low-level facts. Rule 2.8 is a low-level rule since it is dependent on a transducer reading. High-level rules make use of more abstract information, such as Rule 2.3 which relates the occurrence of a steam outlet blockage to a recommendation to replace a pipe. Higher-level rules are those which are closest to providing a solution to a problem, while lower-level rules represent the first stages toward reaching a conclusion.

2.3 Rule examination and rule firing

In Section 2.2, a rule base for boiler control was described without mention of how the rules would be applied. The task of interpreting and applying the rules belongs to the inference engine (see Chapter 1). The application of rules can be broken down as follows:

(i) selecting rules to examine — these are the *available rules*;
(ii) determining which of these are applicable — these are the *triggered* rules; they make up the *conflict set*;
(iii) selecting a rule to *fire* (described below).

The distinction between examination and firing of rules is best explained by example. Suppose the rule-based system has access to the transducer output and to the temperature readings. A sensible set of rules to *examine* would be 2.2, 2.7, and 2.8, as these rules are conditional on the boiler temperature and transducer output. If the transducer level is found to be low, then Rule 2.8 is applicable. If it is selected and used to make the deduction `water level low`, then the rule is said to have *fired*. If the rule is examined but cannot fire (because the transducer reading is not low), the rule is said to *fail*.

The condition part of Rule 2.2 can be satisfied only if Rule 2.8 has been fired. For this reason, it makes sense to examine Rule 2.8 before Rule 2.2. If Rule 2.8 fails, then Rule 2.2 need not be examined as it too will fail. The interdependence between rules is discussed further in Sections 2.4 and 2.5.

The method for rule examination and firing described so far is a form of forward-chaining. This strategy and others are discussed in more detail in Sections 2.7 through 2.10.

2.4 Maintaining consistency

A key advantage of rule-based systems is their flexibility. New rules can be added at will, but only if each rule is written with care and without assuming the behavior of other rules. Consider Rule 2.4:

```
/* Rule 2.4 */
IF release valve stuck THEN steam outlet blocked
```

Given the current rule base, the fact `release valve stuck` could only be established by first firing Rule 2.5:

```
/* Rule 2.5 */
IF pressure high AND release valve closed
THEN release valve stuck
```

Rule 2.5 is sensible, since the purpose of the release valve is to open itself automatically if the pressure becomes high, thereby releasing the excess pressure. Rule 2.4, however, is less sensible. The fact `release valve stuck` is not in itself sufficient evidence to deduce that the steam outlet is blocked. The other necessary evidence is that the pressure in the drum must be high. The reason that the rule base works in its current form is that in order for the system to believe the release valve to be stuck, the pressure *must* be high. Although the rule base works, it is not robust and is not tolerant of new knowledge being added. Consider, for instance, the effect of adding the following rule:

```
/* Rule 2.11 */
IF pressure low AND release valve open THEN release valve stuck
```

This rule is in itself sensible. However, the addition of the rule has an unwanted effect on Rule 2.4. Because of the unintended interaction between Rules 2.4 and 2.11, low pressure in the drum and the observation that the release valve is open result in the erroneous conclusion that the steam outlet is blocked. Problems of this sort can be avoided by making each rule an accurate statement in its own right. Thus in our example, Rule 2.4 should be written as:

```
/* Rule 2.4a */
IF pressure high AND release valve stuck
THEN steam outlet blocked
```

A typical rule firing order, given that the drum pressure is high and the release valve closed, might be:

```
/* Rule 2.5 */
IF pressure high AND release valve closed
THEN release valve stuck
        ↓
/* Rule 2.4a */
IF pressure high AND release valve stuck
THEN steam outlet blocked
        ↓
/* Rule 2.3 */
IF steam outlet blocked THEN replace outlet pipe
```

The modification that has been introduced in Rule 2.4a means that the conditions of both Rules 2.5 and 2.4a involve checking to see whether the drum pressure is high. This source of inefficiency can be justified through the improved robustness of the rule base. In fact, the Rete algorithm, described in Section 2.7.2, allows rule conditions to be duplicated in this way with minimal loss of efficiency.

In general, rules can be considerably more complex than the ones we have considered so far. For instance, rules can contain combinations of conditions and conclusions, exemplified by combining Rules 2.3, 2.4, and 2.6 to form a new rule:

```
/* Rule 2.12 */
IF (pressure high AND release valve stuck) OR steam escaping
THEN steam outlet blocked AND outlet pipe needs replacing
```

2.5 The closed-world assumption

If we do not know that a given proposition is true, then in many rule-based systems the proposition is assumed to be false. This assumption, known as the *closed-world assumption*, simplifies the logic required as all propositions are either TRUE or FALSE. If the closed-world assumption is not made, then a third category, namely UNKNOWN, has to be introduced. To illustrate the closed-world assumption, let us return to the boiler control example. If `steam outlet blocked` is not known, then `NOT (steam outlet blocked)` is assumed to be true. Similarly if `water level low` is not known, then `NOT (water level low)` is assumed to be true. The latter example of the closed-world assumption affects the interaction between Rules 2.7 and 2.8:

```
/* Rule 2.7 */
IF temperature high AND NOT(water level low) THEN pressure high

/* Rule 2.8 */
IF transducer output low THEN water level low
```

Consider the case where the temperature reading is high and the transducer output is low. Whether or not the pressure is assumed to be high will depend on the order in which the rules are selected for firing. If we fire Rule 2.7 followed by 2.8, the following deductions will be made:

`temperature high` — *TRUE*
`NOT(water level low)` — *TRUE by closed-world assumption*
therefore `pressure is high` (Rule 2.7)

`transducer output low` — *TRUE*
therefore `water level low` (Rule 2.8)

Alternatively, we could examine Rule 2.8 first:

`transducer output low` — *TRUE*
therefore `water level low` (Rule 2.8)

`temperature high` — *TRUE*
`NOT(water level low)` — *FALSE*
Rule 2.7 fails

It is most likely that the second outcome was intended by the rule-writer. There are two measures that could be taken to avoid this ambiguity, namely, to modify the rules or to modify the inference engine. The latter approach would aim to ensure that Rule 2.8 is examined before 2.7, and a method for achieving this is described in Section 2.10. The former solution could be achieved by altering the rules so that they do not contain any negative conditions, as shown below:

```
/* Rule 2.7a */
IF temperature high AND water level not_low THEN pressure high

/* Rule 2.8 */
IF transducer output low THEN water level low

/* Rule 2.8a */
IF transducer output not_low THEN water level not_low
```

2.6 Use of variables within rules

The boiler shown in Figure 2.1 is a simplified view of a real system, and the accompanying rule set is much smaller than those associated with most real-world problems. In real-world systems, variables can be used to make rules more general, thereby reducing the number of rules needed and keeping the rule set manageable. The sort of rule that is often required is of the form:

```
For all x, IF <condition about x> THEN <conclusion about x>
```

To illustrate this idea, let us imagine a more complex boiler. This boiler may, for instance, have many water supply pipes, each with its own control valve. For each pipe, the flow rate will be related to whether or not the control valve is open. So some possible rules might be of the form:

```
/* Rule 2.13 */
IF control valve 1 is open THEN flow rate in tube 1 is high
```

```
/* Rule 2.14 */
IF control valve 2 is open THEN flow rate in tube 2 is high

/* Rule 2.15 */
IF control valve 3 is open THEN flow rate in tube 3 is high

/* Rule 2.16 */
IF control valve 4 is open THEN flow rate in tube 4 is high

/* Rule 2.17 */
IF control valve 5 is open THEN flow rate in tube 5 is high
```

A much more compact, elegant and flexible representation of these rules would be:

```
/* Rule 2.18 */
IF control valve ?x is open THEN flow rate in tube ?x is high
```

Here we have used a question mark ('?') to denote that `x` is a variable. Now if the sensors detect that any control valve is open, the identity of the control valve is substituted for `x` when the rule is fired. The variable `x` is said to be *instantiated* with a value, in this case an identity number. Thus, if control valve 3 is open, `x` is instantiated with the value 3 and the deduction `flow rate in tube 3 is high` is made.

In this example the possible values of `x` were limited, and so the use of a variable was convenient rather than necessary. Where the possible values of a variable cannot be anticipated in advance, the use of variables becomes essential. This is the case when values are being looked up, perhaps from a database or from a sensor. As an example, consider the following rule:

```
/* Rule 2.19 */
IF (drum pressure is ?p) AND (?p > threshold) THEN
tube pressure is (?p/10)
```

Without the use of the variable name `p`, it would not be possible to generate a derived fact which states explicitly a pressure value. Suppose that the drum pressure sensor is reading a value of 300MNm^{-2} and `threshold` is a variable currently set to 100MNm^{-2}. Rule 2.19 can therefore be fired, and the derived fact `tube pressure is 30` (MNm^{-2}) is generated. Note that in this example the value of the variable `p` has been manipulated, i.e., divided by 10. More sophisticated rule-based systems allow values represented as variables to be manipulated in this way or passed as parameters to procedures and functions.

The association of a specific value (say, 300MNm^{-2}) with a variable name (such as `p`) is sometimes referred to as *unification*. The term applies not only to numerical examples but to any form of data. The word arises because, from the computer's perspective, the following are contradictory pieces of information:

`pressure` is `p`;
`pressure` is `300` (units of MNm^{-2} assumed).

This conflict can be resolved by recognizing that one of the values is a variable name (because in our syntax it is preceded by a question mark) and by making the following assignment or unification:

```
p:=300
```

The use of variable names within rules is integral to the Prolog language (see Chapter 10). In Prolog, rather than using a question mark, variables are distinguished from constants by having an underscore or upper case letter as their first character. Other languages would require the user to program this facility or to purchase suitable software.

2.7 Forward-chaining (a data-driven strategy)

As noted in Section 2.3, the inference engine applies a strategy for deciding which rules to apply and when to apply them. Forward-chaining is the name given to a data-driven strategy, i.e., rules are selected and applied in response to the current fact base. The fact base comprises all facts known by the system, whether derived by rules or supplied directly (see Section 2.1).

A schematic representation of the cyclic selection, examination and firing of rules is shown in Figure 2.2. The cycle of events shown in Figure 2.2 is just one version of forward-chaining, and variations in the strategy are possible. The key points of the scheme shown in Figure 2.2 are as follows:

- rules are examined and fired on the basis of the current fact base, independently of any predetermined goals;
- the set of rules available for examination may comprise *all* of the rules or a subset;
- of the available rules, those whose conditions are satisfied are said to have been *triggered*. These rules make up the *conflict set*, and the method of selecting a rule from the conflict set is *conflict resolution* (Section 2.8);

- although the conflict set may contain many rules, only one rule is fired on a given cycle. This is because once a rule has fired, the stored deductions have potentially changed, and so it cannot be guaranteed that the other rules in the conflict set still have their condition parts satisfied.

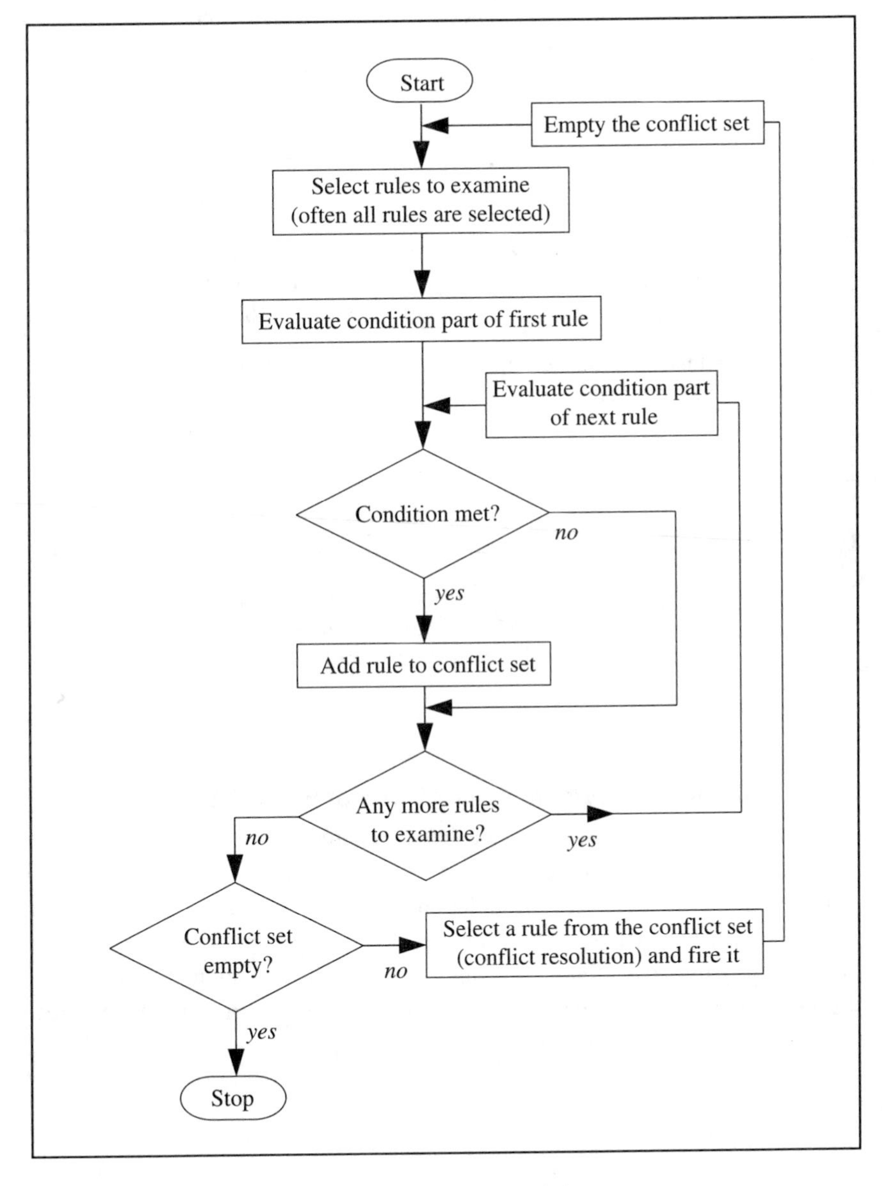

Figure 2.2 Forward-chaining

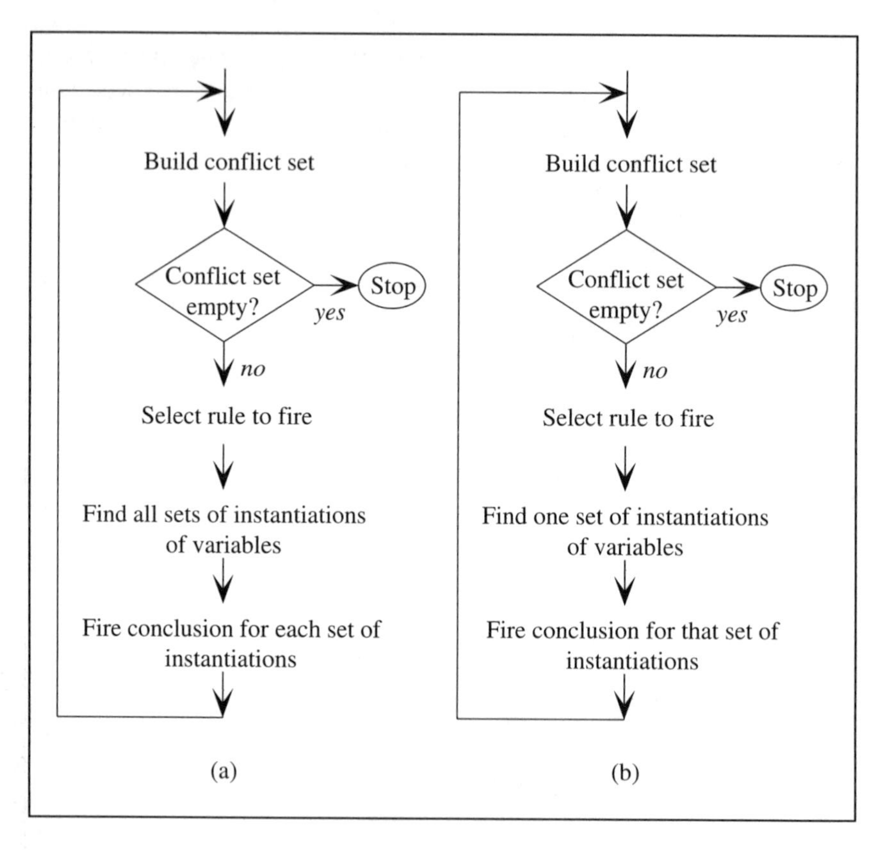

Figure 2.3 Alternative forms of forward-chaining:
(a) multiple instantiation of variables
(b) single instantiation of variables

2.7.1 *Single and multiple instantiation of variables*

As noted above, variations on the basic scheme for forward-chaining are possible. Where variables are used in rules, the conclusions may be performed using just the first set of instantiations that are found — this is single instantiation. Alternatively, the conclusions may be performed repeatedly using all possible instantiations — this is multiple instantiation. The difference between the two approaches is shown in Figure 2.3. As an example, consider the following pair of rules:

```
/* Rule 2.20 */
IF control valve ?x is open THEN flow rate in tube ?x is high
```

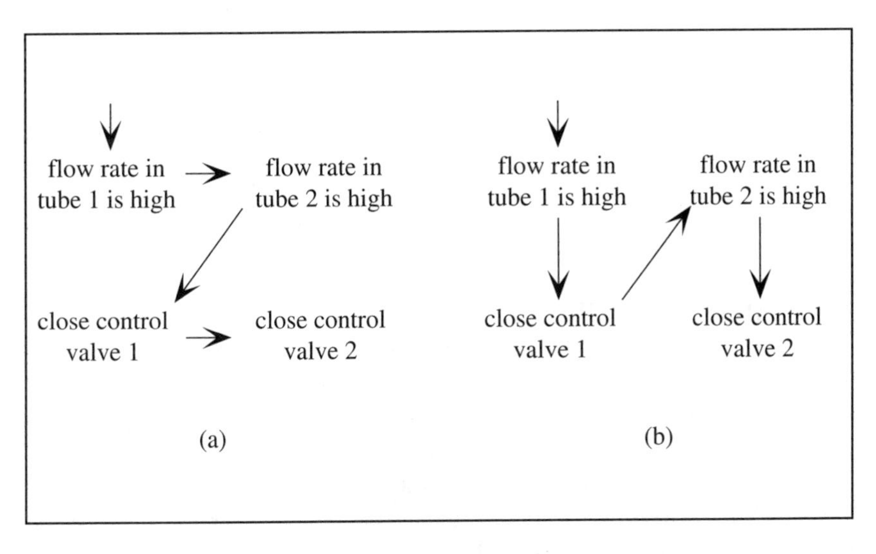

Figure 2.4 Applying Rules 2.20 and 2.21:
(a) multiple instantiation is a breadth-first process
(b) single instantiation is a depth-first process

```
/* Rule 2.21 */
IF flow rate in tube ?x is high THEN close control valve ?x
```

Suppose that we start with two facts:

```
control valve 1 is open
control valve 2 is open
```

Under multiple instantiation, each rule would fire once, generating conclusions in the following order:

```
flow rate in tube 1 is high
flow rate in tube 2 is high
close control valve 1
close control valve 2
```

If the conflict resolution strategy gives preference to Rule 2.21 over Rule 2.20, a different firing order would occur under single instantiation. Each cycle of the inference engine would result in a rule firing on a single instantiation of the variable `x`. After four cycles, conclusions would have been generated in the following order:

```
flow rate in tube 1 is high
close control valve 1
```

```
flow rate in tube 2 is high
close control valve 2
```

Multiple instantiation is a breadth-first approach to problem solving and single instantiation is a depth-first approach, as illustrated in Figure 2.4. The practical implications of the two approaches are discussed in Chapters 11 and 14.

2.7.2 Rete algorithm

The scheme for forward-chaining shown in Figure 2.2 contains at least one source of inefficiency. Once a rule has been selected from the conflict set and fired, the conflict set is thrown away and the process starts all over again. This is because firing a rule alters the fact base, so that a different set of rules may qualify for the conflict set. A new conflict set is therefore drawn up by re-examining the condition parts of all the available rules. In most applications, the firing of a single rule makes only slight changes to the fact base and hence to the membership of the conflict set. Therefore, a more efficient approach would be to examine only those rules whose condition is affected by changes made to the fact base on the previous cycle. The Rete (pronounced "ree-tee") algorithm [1, 2] is one way of achieving this.

The principle of the Rete algorithm can be shown by a simple example, using the following rule:

```
/* Rule 2.22*/
IF ?p is a pipe of bore ?b AND ?v is a valve of bore ?b
THEN ?p and ?v are compatible
```

Prior to running the system, the condition parts of all the rules are assembled into a *Rete network*, where each node represents an atomic condition, i.e., one that contains a simple test. There are two types of nodes — alpha nodes can be satisfied by a single fact, whereas beta nodes can only be satisfied by a pair of facts. The condition part of Rule 2.22 would be broken down into two alpha nodes and one beta node:

```
α1: find a pipe
α2: find a valve
β1: the bore of each must be equal
```

The Rete network for this example is shown in Figure 2.5. Suppose that, initially, the only relevant fact is:

```
p1 is a pipe of bore 100mm
```

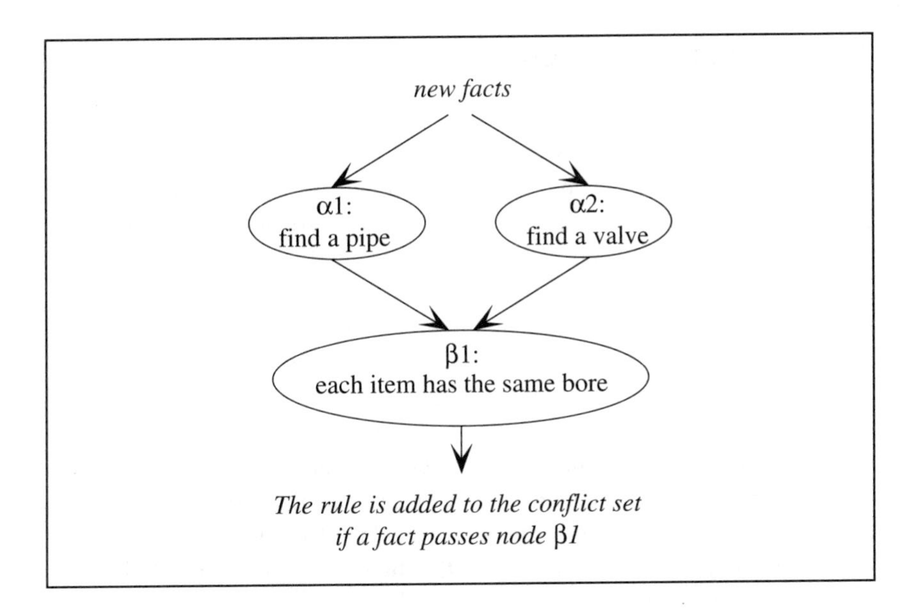

Figure 2.5 A Rete network for Rule 2.22

Node α1 would be satisfied, and so the fact would be passed on to node β1. However, node β1 would not be satisfied as it has received no information from node α2. The fact that there is a pipe of bore 100mm would remain stored at node β1. Imagine now that, as a result of firing other rules, the following fact is derived:

```
v1 is a valve of bore 100mm
```

This fact satisfies node α2 and is passed on to node β1. Node β1 is satisfied by the combination of the new fact and the one that was already stored there. Thus, Rule 2.22 can be added to the conflict set without having to find a pipe again (the task of node α1).

A full Rete network would contain nodes representing the subconditions of all the rules in the rule base. Every time a rule is fired, the altered facts would be fed into the network and the changes to the conflict set generated. Where rules contain identical subconditions, nodes can be shared, thereby avoiding duplicated testing of the conditions. In an evaluation of some commercially available artificial intelligence toolkits that use forward-chaining, those that incorporated the Rete algorithm were found to offer substantial improvements in performance [3].

2.8 Conflict resolution

2.8.1 First come, first served

As noted above, conflict resolution is the method of choosing one rule to fire from those that are able to fire, i.e., from the set of triggered rules, known as the conflict set. In Figure 2.2, the complete conflict set is found before choosing a rule to fire. Since only one rule from the conflict set can actually fire on a given cycle, the time spent evaluating the condition parts of the other rules is wasted unless the result is saved by using a Rete algorithm or similar technique. A strategy which overcomes this inefficiency is to fire immediately the first rule to be found that qualifies for the conflict set (Figure 2.6). In this scheme, the conflict set is not assembled at all, and the order in which rules are selected for examination determines the resolution of conflict. The order of rule examination is often simply the order in which the rules appear in the rule base. If the rule-writer is aware of this, the rules can be ordered in accordance with their perceived priority.

2.8.2 Priority values

Rather than relying on rule ordering as a means of determining rule priorities, rules can be written so that each has an explicitly stated priority value. Where more than one rule is able to fire, the one chosen is the one having the highest priority. The two rules below would be available for firing if the water level had been found to be low (i.e., Rule 2.8 had fired) and if the temperature were high:

```
/* Rule 2.1a */
IF water level low THEN open control valve PRIORITY 4.0

/* Rule 2.2a */
IF temperature high and water level low THEN
open control valve AND shut down boiler tubes PRIORITY 9.0
```

In the scheme shown here, Rule 2.2a would be selected for firing as it has the higher priority value. This scheme arbitrarily uses a scale of priorities from 1 to 10.

As the examination of rules that are not fired represents wasted effort, an efficient use of priorities would be to select rules for examination in order of their priority. Once a rule has been found which is fireable, it could be fired immediately. This scheme is identical to the "first come, first served" strategy (Figure 2.6), except that rules are selected for examination according to their priority value rather than their position in the rule base.

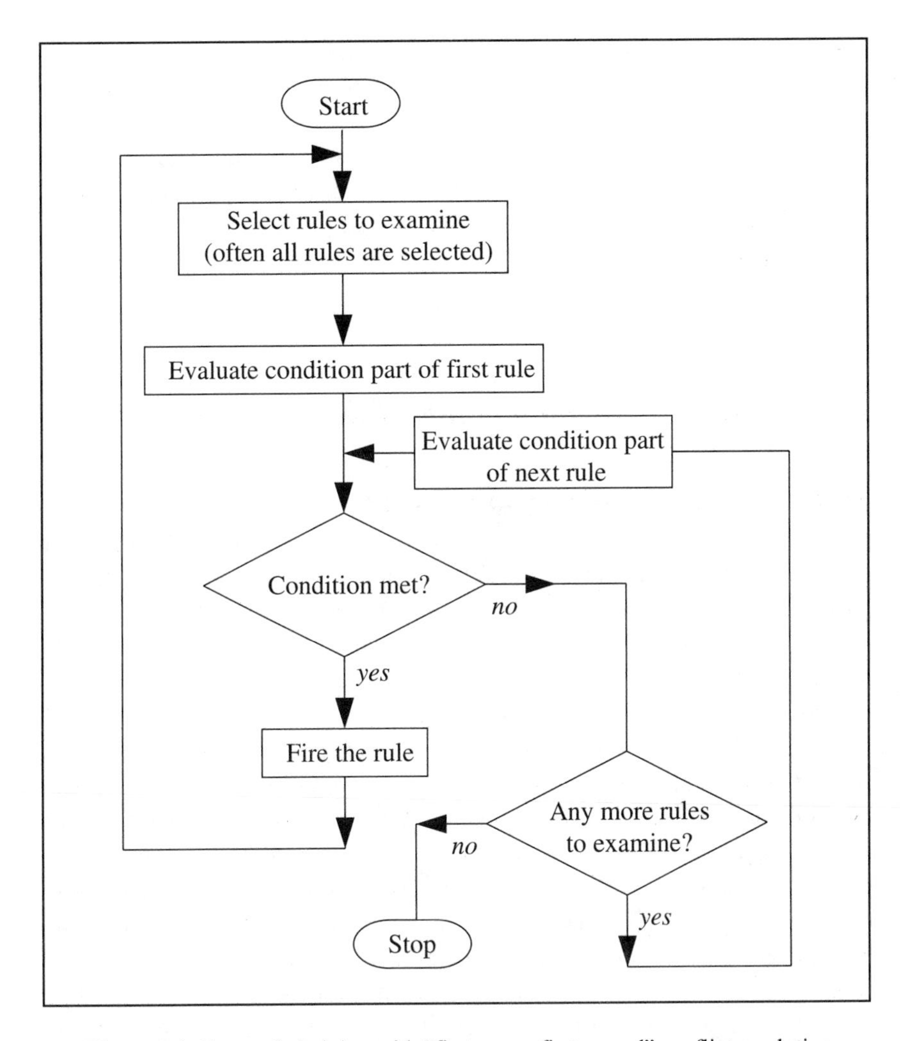

Figure 2.6 Forward-chaining with "first come, first served" conflict resolution

2.8.3 Metarules

Metarules are rules which are not specifically concerned with knowledge about the application at hand, but rather with knowledge about how that knowledge should be applied. Metarules are therefore "rules about rules" (or more generally, "rules about knowledge"). Some examples of metarules might be:

```
/* Metarule 2.23 */
PREFER rules about shut-down TO rules about control valves
```

```
/* Metarule 2.24 */
PREFER high-level rules TO low-level rules
```

If Rules 2.1 and 2.2 are both in the conflict set, Metarule 2.23 will be fired, with the result that Rule 2.2 is then fired. If a conflict arises for which no metarule can be applied, then a default method such as "first come, first served" can be used.

2.9 Backward-chaining (a goal-driven strategy)

2.9.1 The backward-chaining mechanism

Backward-chaining is an inference strategy that assumes the existence of a goal that needs to be established or refuted. In the boiler control example, our goal might be to establish whether it is appropriate to replace the outlet pipe, and we may not be interested in any other deductions that the system is capable of making. Backward-chaining provides the means for achieving this. Initially, only those rules that can lead directly to the fulfillment of the goal are selected for examination. In our case, the only rule that can achieve the goal is Rule 2.3, since it is the only rule whose conclusion is `replace outlet pipe`. The condition part of Rule 2.3 is examined but, since there is no information about a steam outlet blockage in the fact base, Rule 2.3 cannot be fired yet. A new goal is then produced, namely `steam outlet blocked`, corresponding to the condition part of Rule 2.3. Two rules, 2.4 and 2.6, are capable of fulfilling this goal and are therefore *antecedents* of Rule 2.3. What happens next depends on whether a depth-first or breadth-first search strategy is used. These two methods for exploring a search tree were introduced in Chapter 1, but now the nodes of the search tree are *rules*.

For the moment we will assume the use of a depth-first search strategy, as this is normally adopted. The use of a breadth-first search is discussed in Section 2.9.3. One of the two relevant rules (2.4 or 2.6) is selected for examination. Let us suppose that Rule 2.6 is chosen. Rule 2.6 can fire only if steam is escaping from the drum. This information is not in the fact base, so `steam escaping` becomes the new goal. The system searches the rule base for a rule which can satisfy this goal. Rule 2.9 can satisfy the goal, if its condition is met. The condition part of Rule 2.9 relies only on the status of the release valve. If the valve were found to be open, then 2.9 would be able to fire, the goal `steam escaping` could be satisfied, Rule 2.6 would be fireable and the original goal thus fulfilled.

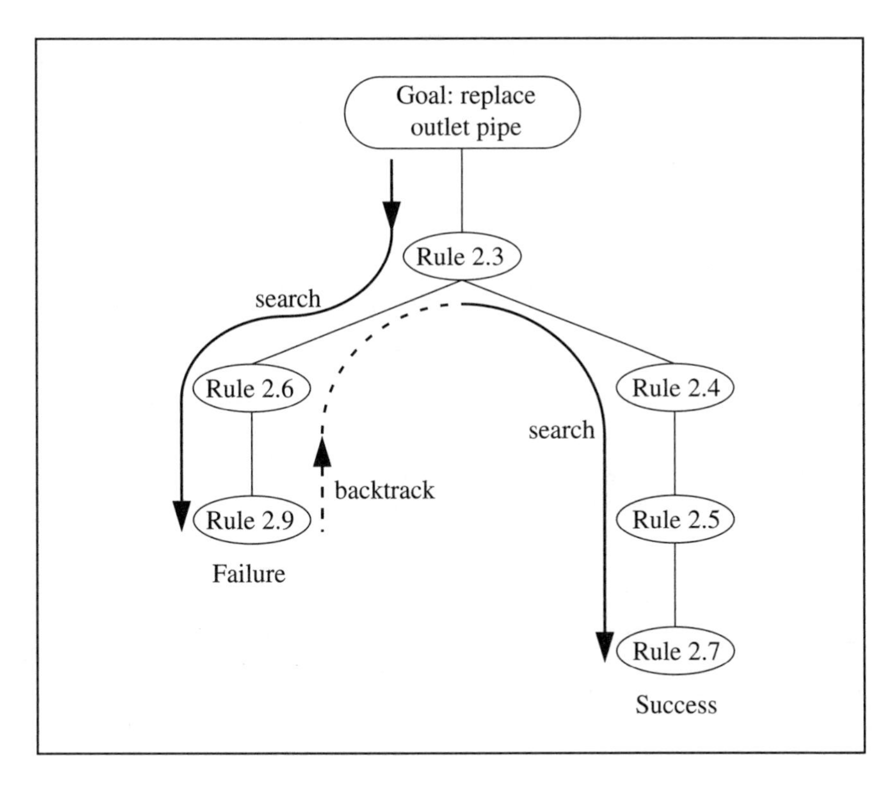

Figure 2.7 Backward-chaining applied to the boiler control rules: the search for rules proceeds in a depth-first manner

Let us suppose, on the other hand, that the release valve is found to be closed. Rule 2.9 therefore fails, with the result that Rule 2.6 also fails. The system *backtracks* to the last place where a choice between possible rules was made and will try an alternative, as shown in Figure 2.7. In the example shown here, this means that Rule 2.4 is examined next. It can fire only if the release valve is stuck but, because this information is not yet known, it becomes the new goal. This process continues until a goal is satisfied by the information in the fact base. When this happens, the original goal is fulfilled, and the chosen path through the rules is the solution. If all possible ways of achieving the overall goal have been explored and failed, then the overall goal fails.

The backward-chaining mechanism described so far has assumed a depth-first search for rules. This means that whenever a choice between rules exists, just one is selected, the others being examined only if backtracking occurs.

The inference mechanism built into the Prolog language (see Chapter 10) is a depth-first backward-chainer, like that described here. There are two sets of circumstances under which backtracking takes place:

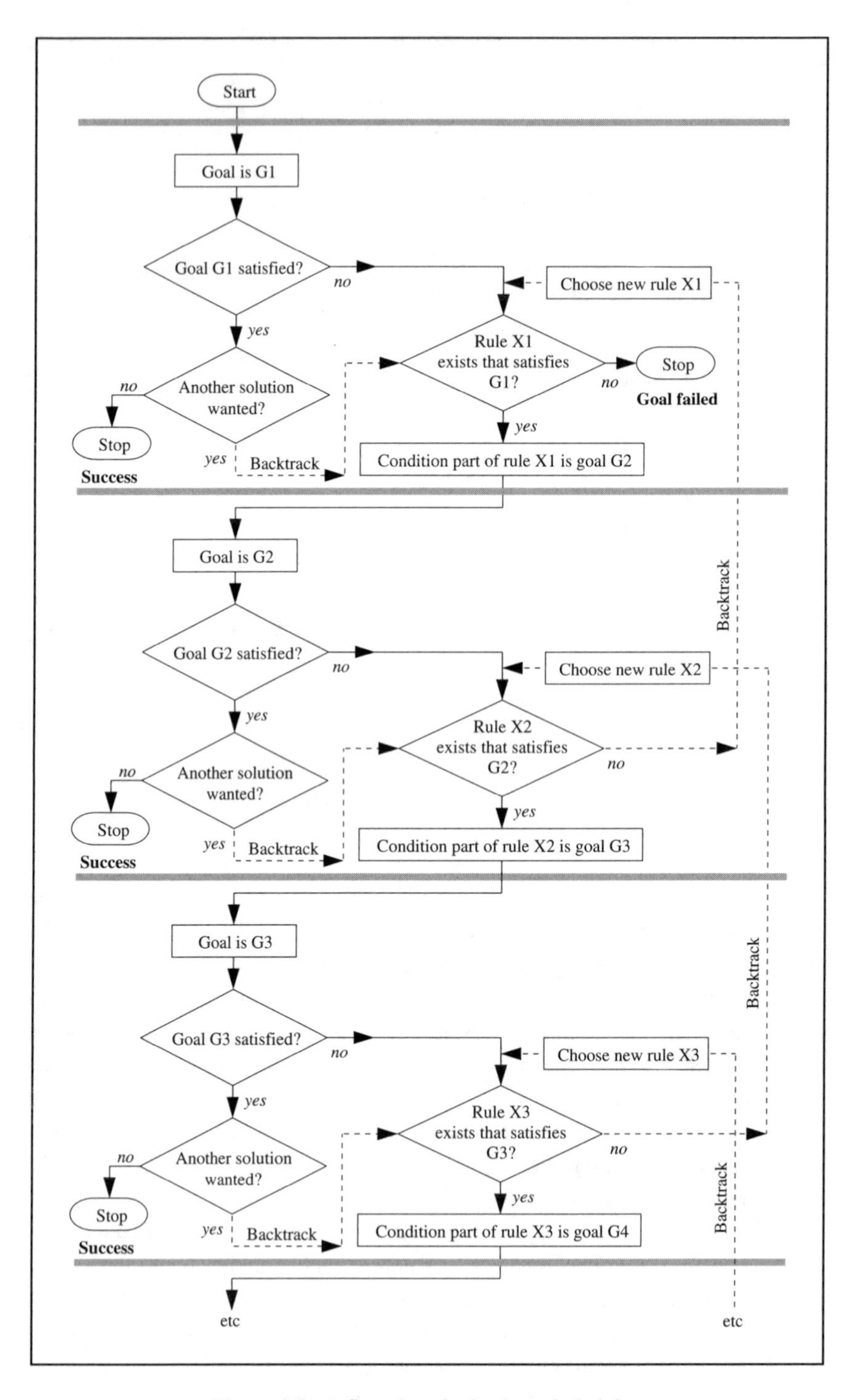

Figure 2.8 A flowchart for backward-chaining

(i) when a goal cannot be satisfied by the set of rules currently under consideration; or
(ii) when a goal has been satisfied and the user wants to investigate other ways of achieving the goal (i.e., to find other solutions).

2.9.2 Implementation of backward-chaining

Figure 2.8 shows a generalized flowchart for backward-chaining from a goal G1. In order to simplify the chart, it has been assumed that each rule has only one condition, so that the satisfaction of a condition can be represented as a single goal. In general, rules have more than one condition. The flowchart in Figure 2.8 is an attempt to represent backward-chaining as an iterative process. This is difficult to achieve, as the length of the chain of rules cannot be predetermined. The flowchart has, of necessity, been left incomplete. The flowchart contains repeating sections that are identical except for the variable names, an indication that while it is difficult to represent the process iteratively, it can be elegantly represented *recursively*. A recursive definition of a function is one that includes the function itself. Recursion is an important aspect of the artificial intelligence languages Lisp and Prolog, discussed in Chapter 10, as well as many other computer languages. Box 2.1 shows a recursive definition of backward-chaining, where it has again been assumed that rules have only one condition. It is not always necessary to write such a function for yourself as backward-chaining forms an integral part of the Prolog language and most expert system shells and artificial intelligence toolkits (see Chapter 10).

2.9.3 Variations of backward-chaining

There are a number of possible variations to the idea of backward-chaining. Some of these are:

(i) depth-first or breadth-first search for rules;
(ii) whether or not to pursue other solutions (i.e., other means of achieving the goal) once a solution has been found;
(iii) different ways of choosing between branches to explore (depth-first search only);
(iv) deciding upon the order in which to examine rules (breadth-first search only);
(v) having found a succession of enabling rules whose lowest-level conditions are satisfied, whether to fire the sequence of rules or simply to conclude that the goal is proven.

```
define function backwardchain(G);
   /* returns a boolean (i.e., true/false) value */
   /* G is the goal being validated */
variable S, X, C;
   result:= false;
   /* ':=' represents assignment of a value to a variable */
   S:= set of rules whose conclusion part matches goal G;
   if S is empty then
      result:= false;
   else
      while (result=false) and (S is not empty) do
         X:= rule selected from S;
         S:= S with X removed;
         C:= condition part of X;
         if C is true then
            result:=true
         elseif C is false then
            result:=false
         elseif (backwardchain(C)=true) then
            result:=true;
         /* note the recursive call of 'backwardchain' */
         /* C is the new goal */
         endif;
      endwhile;
   endif;
return result;
   /* 'result' is the value returned by the function */
   /* 'backwardchain' */
enddefine;
```

Box 2.1 A recursive definition of backward-chaining

Breadth-first backward-chaining is identical to depth-first, except for the mechanism for deciding between alternative rules to examine. In the example described in Section 2.9.1 (and in Figure 2.7), instead of choosing to explore either 2.4 or 2.6, both rules would be examined. Then the preconditions of each (i.e., Rules 2.5 and 2.9) would be examined. If either branch fails, then the system will not need to backtrack, as it will simply carry on down those branches which still have the potential to fulfill the goal. The process is illustrated in Figure 2.9. The first solution to be found will be the one with the shortest (or joint-shortest) chain of rules. A disadvantage with the breadth-first approach is that large amounts of computer memory may be required, as the system needs to keep a record of progress along all branches of the search tree, rather than just one branch.

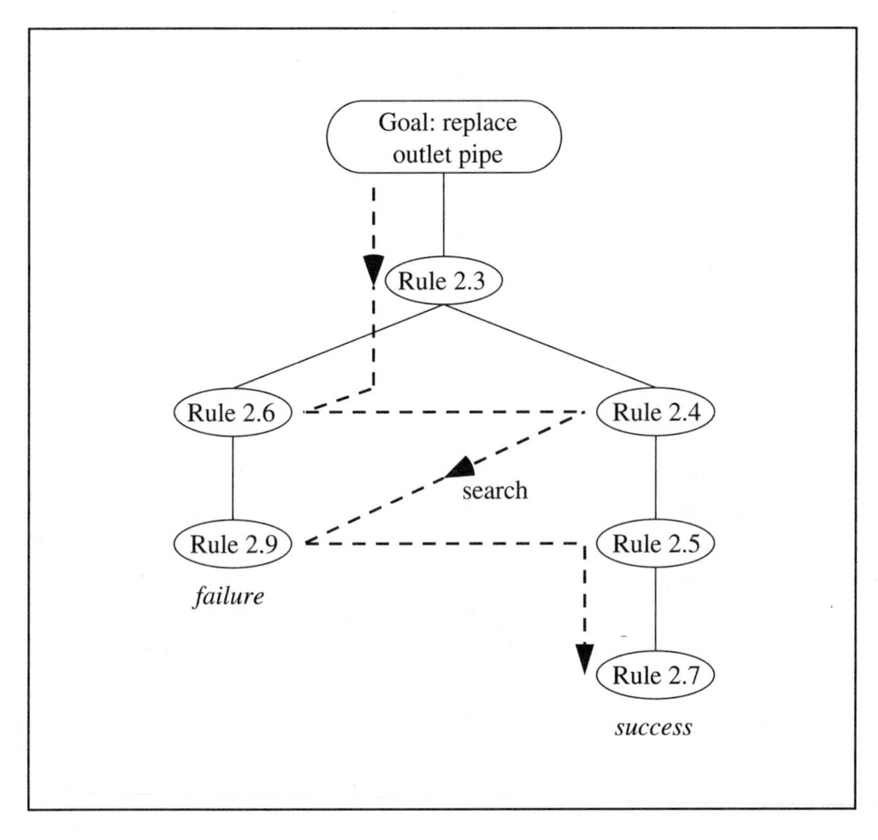

Figure 2.9 A variation of backward-chaining applied to the boiler control rules; here the search for rules proceeds in a breadth-first manner

The last item (v) in the above list of variations of backward-chaining is important but subtle. So far our discussion of backward-chaining has made limited reference to rule firing. If a goal is capable of being satisfied by a succession of rules, the first of which is fireable, then the goal is assumed to be true and *no rules are actually fired*. If all that is required is validation of a goal, then this approach is adequate. However, by actually firing the rules, all of the intermediate deductions such as `water level low` are recorded in the fact base and may be used by other rules, thereby eliminating the need to re-establish them. Therefore, in some implementations of backward-chaining, once the path to a goal has been established, rules are fired until the original goal is fulfilled. In more complex rule-based systems, rules may call up and run a procedurally coded module as part of their conclusion. In such systems, the firing of the rules is essential for the role of the procedural code to be realized.

Finally, it should be noted that in some backward-chaining systems the rule syntax is reversed, compared with the examples that we have discussed so far. A possible syntax would be:

```
DEDUCE <conclusion> IF <condition>
```

The placing of the conclusion before the condition reflects the fact that in backward-chaining systems it is the conclusion part of a rule that is assessed first, and only if the conclusion is relevant is the condition examined.

2.10 A hybrid strategy

A system called ARBS (Algorithmic and Rule-based Blackboard System), described in Chapters 11 and 14, makes use of an inference engine that can be thought of as part forward-chaining and part backward-chaining [4]. Conventional backward-chaining involves initial examination of a rule that achieves the goal. If that cannot fire, its antecedent rules are recursively examined, until rules can be fired and progress made toward the goal. In all problems involving data interpretation (such as the boiler control example), the high-level rules concerning the goal itself can never fire until lower-level rules for data manipulation have been fired. The standard mechanisms for forward- or backward-chaining, therefore, involve a great deal of redundant rule examination. The hybrid strategy is a means of eliminating this source of inefficiency.

Under the hybrid strategy, a *rule dependence network* is built prior to running the system. For each rule, the network shows which other rules may enable it, i.e., its antecedents, and which rules it may enable, i.e., its dependents. The rule dependencies for the boiler control knowledge base are shown in Figure 2.10. In its data-driven mode, known as *directed forward-chaining*, the hybrid strategy achieves improved efficiency by using the dependence network to select rules for examination. Low-level rules concerning the sensor data are initially selected for examination. As shown in Figure 2.10, only Rules 2.8, 2.9, and 2.10 need be examined initially. Then higher-level rules, leading toward a solution, are selected depending on which rules have actually fired. So, if Rules 2.8 and 2.9 fire successfully, the new set of rules to be examined becomes 2.1, 2.2, and 2.6. The technique is an effective way of carrying out the instruction marked "select rules to examine" in the flowchart for forward-chaining (Figure 2.2).

The same mechanism can easily be adapted to provide an efficient goal-driven strategy. Given a particular goal, the control mechanism can select the

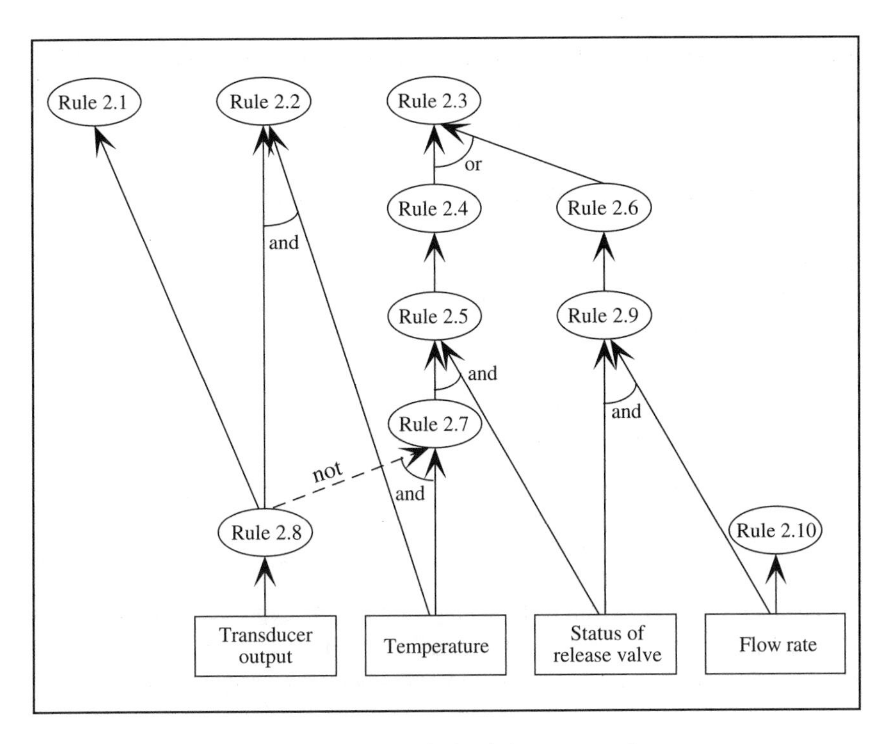

Figure 2.10 A rule dependence network

branch of the dependence network leading to that goal and then backward-chain through the selected rules.

For a given rule base, the dependence network needs to be generated only once, and is then available to the system at run-time. The ordering of rules in the rule base does not affect the system, because the application of rules is dictated by their position in the dependence network rather than in the rule set.

Rule 2.7 has a "negative dependence" on Rule 2.8, meaning that Rule 2.7 is fireable if Rule 2.8 fails to fire:

```
/* Rule 2.7 */
IF temperature high AND NOT(water level low) THEN pressure high

/* Rule 2.8 */
IF transducer output low THEN water level low
```

As discussed in Section 2.5, the closed-world assumption will lead to `NOT(water level low)` being assumed true unless Rule 2.8 is successfully fired. Therefore, for Rule 2.7 to behave as intended, Rule 2.8 must be examined (and either fire or fail) before Rule 2.7 is examined. Using the dependence network to direct rule examination and firing is one way of ensuring this order of events.

The use of dependence networks is more complicated when variables are used within rules because the dependencies between rules are less certain. Consider, for example, the following set of rules that do *not* use variables:

```
/* Rule 2.25 */
IF control valve 1 is open AND pipe 1 is blocked
THEN open release valve

/* Rule 2.26 */
IF flow rate through pipe 1 is high
THEN control valve 1 is open

/* Rule 2.27 */
IF pressure in pipe 1 is high
THEN pipe 1 is blocked
```

A dependence network would show that Rule 2.25 is dependent only on Rules 2.26 and 2.27. Therefore, if 2.26 and 2.27 have fired, 2.25 can definitely fire. Now consider the same rules modified to incorporate the use of variables:

```
/* Rule 2.25a */
IF control valve ?x open AND pipe ?x blocked
THEN open release valve

/* Rule 2.26a */
IF flow rate through pipe ?x is high
THEN control valve ?x open

/* Rule 2.27a */
IF pressure in pipe ?x is high
THEN pipe ?x is blocked
```

Rule 2.25a is dependent on Rules 2.26a and 2.27a. However, it is possible for Rules 2.26a and 2.27a to have fired, but for Rule 2.25a to fail. This is because the condition of Rule 2.25a requires the valve and pipe numbers (represented by `x`) to be identical, whereas Rules 2.26a and 2.27a could each use a different value for `x`. Thus when rules contain variables, a dependence network shows which rules have the *potential* to enable others to fire. Whether or not the dependent rules will actually fire cannot be determined until run-time. The dependence network shows us that Rule 2.25a should be examined if Rules 2.26a and 2.27a have fired, but otherwise it can be ignored.

A similar situation arises when there is a negative dependence between rules containing variables, for example:

```
/* Rule 2.25b */
IF NOT(control valve ?x open) AND pipe ?x blocked
THEN open release valve

/* Rule 2.26b */
IF flow rate through pipe ?x is high
THEN control valve ?x open

/* Rule 2.27b */
IF pressure in pipe ?x is high
THEN pipe ?x is blocked
```

Here 2.25b has a negative dependence on 2.26b and a normal (positive) dependence on 2.27b. Under these circumstances, 2.25b should be examined after both:

(i) 2.27b has fired; *and*
(ii) 2.26b has been examined, *whether or not it fired.*

The first subcondition of Rule 2.25b, `NOT(control valve ?x open)`, will certainly be true if Rule 2.26b fails, owing to the closed-world assumption. However, it may also be true even if Rule 2.26b has fired, since `x` could be instantiated to a different value in each rule. It is assumed that the scope of the variable is the length of the rule, so the value of `x` is the same throughout Rule 2.25b, but may be different in Rule 2.26b.

2.11 Explanation facilities

One of the claims frequently made in support of expert systems is that they are able to explain their reasoning, and that this gives users of such systems confidence in the accuracy or wisdom of the system's decisions. However, as noted in Chapter 1, the explanations offered by many systems are little more than a trace of the firing of rules. While this is an important facility, tracing the flow of a computer program is standard practice and not a particularly special capability.

Explanation facilities can be divided into two categories:

- *how* a conclusion has been derived;
- *why* a particular line of reasoning is being followed.

The first type of explanation would normally be applied when the system has completed its reasoning, whereas the second type is applicable while the system is carrying out its reasoning process. The latter type of explanation is particularly appropriate in an interactive expert system, which involves a dialogue between a user and the computer. During such a dialogue the user will often want to establish why particular questions are being asked. If either type of explanation is incorrect or impenetrable, the user is likely to distrust or ignore the system's findings.

Returning once more to our rule set for boiler control (Section 2.2), the following would be a typical explanation for a recommendation to replace the outlet pipe:

```
Replace outlet pipe
BECAUSE (Rule 2.3) steam outlet blocked

steam outlet blocked
BECAUSE (Rule 2.4) release valve stuck

release valve stuck
BECAUSE (Rule 2.5) pressure high AND release valve closed

pressure high
BECAUSE (Rule 2.7) temperature high AND NOT(water level low)

NOT(water level low)
BECAUSE (Rule 2.8) NOT(transducer output low)

       release valve closed ⎫
           temperature high ⎬  are supplied facts
NOT(transducer output low)  ⎭
```

Explanation facilities are desirable for increasing user confidence in the system, as a teaching aid and as an aid to debugging. However, a simple trace like the one shown is likely to be of little use except for debugging. The quality of explanation can be improved by placing an obligation on the rule-writer to provide an explanatory note for each rule. These notes can then be included in the rule trace, or reproduced at run-time to explain the current line of reasoning. Explanation facilities can also be made more relevant by supplying the user with a level of detail tailored to his or her needs.

2.12 Summary

Rules are an effective way of representing knowledge in many application domains. They are most versatile when variables are used within rules and they can be particularly useful in cooperation with procedural algorithms or object-oriented systems (Chapter 4). The role of interpreting, selecting, and applying rules is fulfilled by the inference engine. Rule-writing should ideally be independent of the details of the inference engine, apart from fulfilling its syntax requirements. In practice, the rule-writer needs to be aware of the strategy for applying rules and any assumptions that are made by the inference engine. For instance, under the closed-world assumption, any facts that have not been supplied or derived are assumed to be false. Forward- and backward-chaining are two distinct strategies for applying rules, but many variations of these strategies are also possible.

References

1. Forgy, C. L., "Rete: a fast algorithm for the many-pattern/many-object-pattern match problem," *Artificial Intelligence*, vol. 19, pp. 17–37, 1982.
2. Graham, P., "Using the Rete algorithm," *AI Expert*, pp. 46–51, December 1990.
3. Mettrey, W., "A comparative evaluation of expert system tools," *IEEE Computer*, vol. 24, issue 2, pp. 19–31, February 1991.
4. Hopgood, A. A., "Rule-based control of a telecommunications network using the blackboard model," *Artificial Intelligence in Engineering*, vol. 9, pp. 29–38, 1994.

Further reading

- Darlington, K. W., *The Essence of Expert Systems*, Prentice Hall, 2000.
- Durkin, J., *Expert Systems: design and development*, MacMillan, 1998.
- Giarratano, J. and Riley, G., *Expert Systems: principles and programming*, 3rd ed., PWS, 1998.
- Jackson, P., *Introduction to Expert Systems*, 3rd ed., Addison-Wesley, 1998.

Chapter three

Dealing with uncertainty

3.1 Sources of uncertainty

The discussion of rule-based systems in Chapter 2 assumed that we live in a clear-cut world, where every hypothesis is either true, false, or unknown. Furthermore, it was pointed out that many systems make use of the *closed-world assumption*, whereby any hypothesis that is unknown is assumed to be false. We were then left with a binary system, where everything is either true or false. While this model of reality is useful in many applications, real reasoning processes are rarely so clear-cut. Referring to the example of the control of a power station boiler, we made use of the following rule:

```
IF transducer output is low THEN water level is low
```

There are three distinct forms of uncertainty that might be associated with this rule:

Uncertainty in the rule itself
A low level of water in the drum is not the only possible explanation for a low transducer output. Another possible cause could be that the float attached to the transducer is stuck. What we really mean by this rule is that if the transducer output is low then the water level is *probably* low.

Uncertainty in the evidence
The evidence upon which the rule is based may be uncertain. There are two possible reasons for this uncertainty. First, the evidence may come from a source that is not totally reliable. For instance, we may not be absolutely certain that the transducer output is low, as this information relies upon a meter to measure the voltage. Second, the evidence itself may have been derived by a rule whose conclusion was probable rather than certain.

Use of vague language
The above rule is based around the notion of a "low" transducer output. Assuming that the output is a voltage, we must consider whether "low" corresponds to 1mV, 1V or 1kV.

It is important to distinguish between these sources of uncertainty, as they need to be handled differently. There are some situations in nature that are truly random and whose outcome, while uncertain, can be anticipated on a statistical basis. For instance, we can anticipate that on average one of six throws of a die will result in a score of four. Some of the techniques that we will be discussing are based upon probability theory. These assume that a statistical approach can be adopted, although this assumption will be only an approximation to the real circumstances unless the problem is truly random.

This chapter will review some of the commonly used techniques for reasoning with uncertainty. Bayesian updating has a rigorous derivation based upon probability theory, but its underlying assumptions, e.g., the statistical independence of multiple pieces of evidence, may not be true in practical situations. Certainty theory does not have a rigorous mathematical basis, but has been devised as a practical way of overcoming some of the limitations of Bayesian updating. Possibility theory, or fuzzy logic, allows the third form of uncertainty, i.e., vague language, to be used in a precise manner. The assumptions and arbitrariness of the various techniques have meant that reasoning under uncertainty remains a controversial issue.

3.2 Bayesian updating

3.2.1 Representing uncertainty by probability

Bayesian updating assumes that it is possible to ascribe a probability to every hypothesis or assertion, and that probabilities can be updated in the light of evidence for or against a hypothesis or assertion. This updating can either use Bayes' theorem directly (Section 3.2.2), or it can be slightly simplified by the calculation of likelihood ratios (Section 3.2.3). One of the earliest successful applications of Bayesian updating to expert systems was PROSPECTOR, a system which assisted mineral prospecting by interpreting geological data [1, 2].

Let us start our discussion by returning to our rule set for control of the power station boiler (see Chapter 2), which included the following two rules:

```
/* Rule 2.4 */
IF release valve stuck THEN steam outlet blocked

/* Rule 2.6 */
IF steam escaping THEN steam outlet blocked
```

We're going to consider the hypothesis that there is a steam outlet blockage. Previously, under the closed-world assumption, we asserted that in the absence of any evidence about a hypothesis, the hypothesis could be treated as false. The Bayesian approach is to ascribe an *a priori* probability (sometimes simply called the *prior* probability) to the hypothesis that the steam outlet is blocked. This is the probability that the steam outlet is blocked, in the absence of any evidence that it is or is not blocked. Bayesian updating is a technique for updating this probability in the light of evidence for or against the hypothesis. So, whereas we had previously assumed that `steam escaping` led to the deduction `steam outlet blockage` with absolute certainty, now we can only say that it supports that deduction. Bayesian updating is cumulative, so that if the probability of a hypothesis has been updated in the light of one piece of evidence, the new probability can then be updated further by a second piece of evidence.

3.2.2 Direct application of Bayes' theorem

Suppose that the prior probability of `steam outlet blockage` is 0.01, which implies that blockages occur only rarely. Our modified version of Rule 2.6 might look like this:

```
IF steam escaping
THEN update P(steam outlet blockage)
```

With this new rule, the observation of steam escaping requires us to update the probability of a steam outlet blockage. This contrasts with Rule 2.6, where the conclusion that there is a steam outlet blockage would be drawn with absolute certainty. In this example, `steam outlet blockage` is considered to be a hypothesis (or assertion), and `steam escaping` is its supporting evidence.

The technique of Bayesian updating provides a mechanism for updating the probability of a hypothesis P(H) in the presence of evidence E. Often the evidence is a symptom and the hypothesis is a diagnosis. The technique is based upon the application of Bayes' theorem (sometimes called Bayes' rule). Bayes' theorem provides an expression for the conditional probability $P(H|E)$ of a hypothesis H given some evidence E, in terms of $P(E|H)$, i.e., the conditional probability of E given H:

$$P(H|E) = \frac{P(H) \times P(E|H)}{P(E)} \tag{3.1}$$

The theorem is easily proved by looking at the definition of dependent probabilities. Of an expected population of events in which E is observed, P(H|E) is the fraction in which H is also observed. Thus:

$$P(H|E) = \frac{P(H \& E)}{P(E)} \tag{3.2}$$

Similarly,

$$P(E|H) = \frac{P(H \& E)}{P(H)} \tag{3.3}$$

The combination of Equations 3.2 and 3.3 yields Equation 3.1. Bayes' theorem can then be expanded as follows:

$$P(H|E) = \frac{P(H) \times P(E|H)}{P(H) \times P(E|H) + P(\sim H) \times P(E|\sim H)} \tag{3.4}$$

where ~H means "not H." The probability of ~H is simply given by:

$$P(\sim H) = 1 - P(H) \tag{3.5}$$

Equation 3.4 provides a mechanism for updating the probability of a hypothesis H in the light of new evidence E. This is done by updating the existing value of P(H) to the value for P(H|E) yielded by Equation 3.4. The application of the equation requires knowledge of the following values:

- P(H), the current probability of the hypothesis. If this is the first update for this hypothesis, then P(H) is the prior probability.
- P(E|H), the conditional probability that the evidence is present, given that the hypothesis is true.
- P(E|~H), the conditional probability that the evidence is present, given that the hypothesis is false.

Thus, to build a system that makes direct use of Bayes' theorem in this way, values are needed in advance for P(H), P(E|H), and P(E|~H) for all the different hypotheses and evidence covered by the rules. Obtaining these values might appear at first glance more formidable than the expression we are hoping to derive, namely P(H|E). However, in the case of diagnosis problems, the conditional probability of evidence, given a hypothesis, is usually more readily available than the conditional probability of a hypothesis, given the evidence. Even if P(E|H) and P(E|~H) are not available as formal statistical observations, they may at least be available as informal estimates. So in our example an expert may have some idea of how often steam is observed escaping when there is an outlet blockage, but is less likely to know how often a steam escape is due to an outlet blockage. Chapter 1 introduced the ideas of deduction, abduction, and induction. Bayes' theorem, in effect, performs abduction (i.e., determining causes) using deductive information (i.e., the likelihood of symptoms, effects, or evidence). The premise that deductive information is more readily available than abductive information is one of the justifications for using Bayesian updating.

3.2.3 Likelihood ratios

Likelihood ratios, defined below, provide an alternative means of representing Bayesian updating. They lead to rules of this general form:

```
IF steam escaping
THEN steam outlet blockage IS X times more likely
```

With a rule like this, if steam is escaping we can update the probability of a steam outlet blockage provided we have an expression for `X`. A value for `X` can be expressed most easily if the hypothesis `steam outlet blockage` is expressed as odds rather than a probability. The odds O(H) of a given hypothesis H are related to its probability P(H) by the relations:

$$O(H) = \frac{P(H)}{P(\sim H)} = \frac{P(H)}{1 - P(H)} \tag{3.6}$$

and

$$P(H) = \frac{O(H)}{O(H) + 1} \tag{3.7}$$

As before, ~H means "not H." Thus a hypothesis with a probability of 0.2 has odds of 0.25 (or "4 to 1 against"). Similarly a hypothesis with a probability of 0.8 has odds of 4 (or "4 to 1 on"). An assertion that is absolutely certain, i.e., has a probability of 1, has infinite odds. In practice, limits are often set on odds values so that, for example, if $O(H)>10^6$ then H is true, and if $O(H)<10^{-6}$ then H is false. Such limits are arbitrary.

In order to derive the updating equations, start by considering the hypothesis "not H," or ~H, in Equation 3.1:

$$P(\sim H|E) = \frac{P(\sim H) \times P(E|\sim H)}{P(E)} \tag{3.8}$$

Division of Equation 3.1 by Equation 3.8 yields:

$$\frac{P(H|E)}{P(\sim H|E)} = \frac{P(H) \times P(E|H)}{P(\sim H) \times P(E|\sim H)} \tag{3.9}$$

By definition, O(H|E), the conditional odds of H given E, is:

$$O(H|E) = \frac{P(H|E)}{P(\sim H|E)} \tag{3.10}$$

Substituting Equations 3.6 and 3.10 into Equation 3.9 yields:

$$O(H|E) = A \times O(H) \tag{3.11}$$

where:

$$A = \frac{P(E|H)}{P(E|\sim H)} \tag{3.12}$$

O(H|E) is the updated odds of H, given the presence of evidence E, and A is the *affirms* weight of evidence E. It is one of two likelihood ratios. The other is the *denies* weight D of evidence E. The *denies* weight can be obtained by considering the absence of evidence, i.e., ~E:

$$O(H|\sim E) = D \times O(H) \tag{3.13}$$

where:

$$D = \frac{P(\sim E|H)}{P(\sim E|\sim H)} = \frac{1 - P(E|H)}{1 - P(E|\sim H)} \tag{3.14}$$

The function represented by Equations 3.11 and 3.13 is shown in Figure 3.1. Rather than displaying odds values, which have an infinite range, the corresponding probabilities have been shown. The weight (A or D) has been shown on a logarithmic scale over the range 0.01 to 100.

3.2.4 Using the likelihood ratios

Equation 3.11 provides a simple way of updating our confidence in hypothesis H in the light of new evidence E, assuming that we have a value for A and for

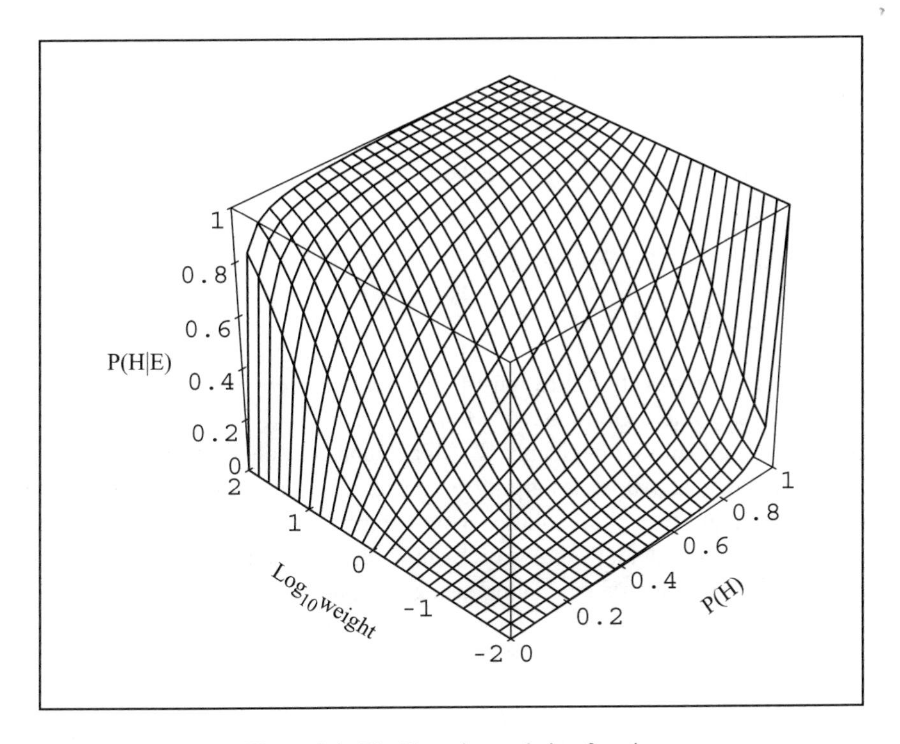

Figure 3.1 The Bayesian updating function

O(H), i.e., the current odds of H. O(H) will be at its *a priori* value if it has not previously been updated by other pieces of evidence. In the case of Rule 2.6, H refers to the hypothesis `steam outlet blockage` and E refers to the evidence `steam escaping`.

In many cases, the absence of a piece of supporting evidence may reduce the likelihood of a certain hypothesis. In other words, the absence of supporting evidence is equivalent to the presence of opposing evidence. The known absence of evidence is distinct from not knowing whether the evidence is present, and can be used to reduce the probability (or odds) of the hypothesis by applying Equation 3.13 using the *denies* weight, D.

If a given piece of evidence E has an *affirms* weight A that is greater than 1, then its *denies* weight must be less than 1 and vice versa:

$A>1$ implies $D<1$,
$A<1$ implies $D>1$.

If $A<1$ and $D>1$, then the absence of evidence is supportive of a hypothesis. Rule 2.7 provides an example of this, where `NOT(water level low)` supports the hypothesis `pressure high` and `water level low` opposes the hypothesis:

```
/* Rule 2.7 */
IF temperature high AND NOT(water level low)
THEN pressure high
```

A Bayesian version of this rule might be:

```
/* Rule 3.1 */
IF temperature high (AFFIRMS 18.0; DENIES 0.11)
AND water level low (AFFIRMS 0.10; DENIES 1.90)
THEN pressure high
```

As with the direct application of Bayes rule, likelihood ratios have the advantage that the definitions of A and D are couched in terms of the conditional probability of evidence, given a hypothesis, rather than the reverse. As pointed out above, it is usually assumed that this information is more readily available than the conditional probability of a hypothesis, given the evidence, at least in an informal way. Even if accurate conditional probabilities are unavailable, Bayesian updating using likelihood ratios is still a useful technique if heuristic values can be attached to A and D.

3.2.5 Dealing with uncertain evidence

So far we have assumed that evidence is either definitely present (i.e., has a probability of 1) or definitely absent (i.e., has a probability of 0). If the probability of the evidence lies between these extremes, then the confidence in the conclusion must be scaled appropriately. There are two reasons why the evidence may be uncertain:

- the evidence could be an assertion generated by another uncertain rule, and which therefore has a probability associated with it;
- the evidence may be in the form of data which are not totally reliable, such as the output from a sensor.

In terms of probabilities, we wish to calculate P(H|E), where E is uncertain. We can handle this problem by assuming that E was asserted by another rule whose evidence was B, where B is certain (has probability 1). Given the evidence B, the probability of E is P(E|B). Our problem then becomes one of calculating P(H|B). An expression for this has been derived by Duda et al. [3]:

$$P(H|B) = P(H|E) \times P(E|B) + P(H|{\sim}E) \times [1 - P(E|B)] \tag{3.15}$$

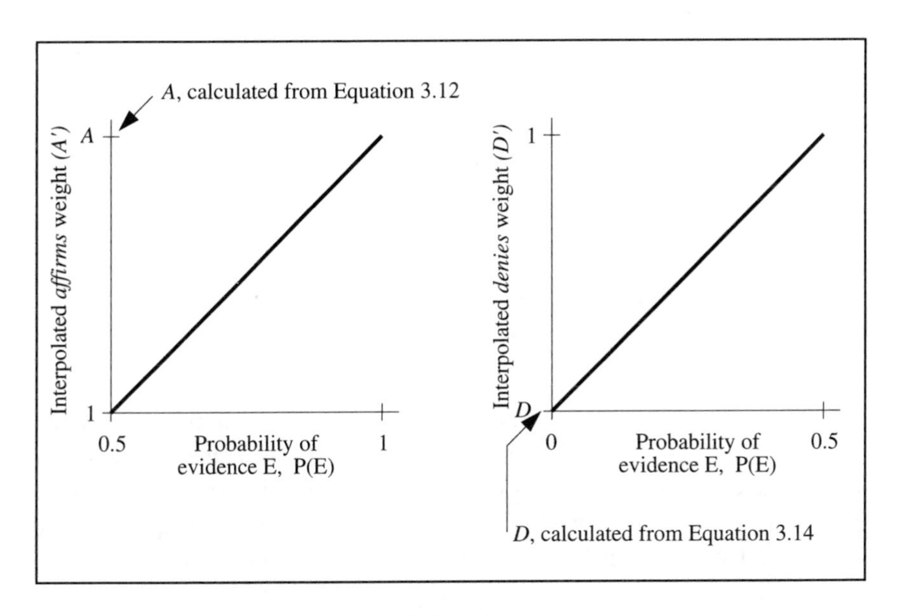

Figure 3.2 Linear interpolation of *affirms* and *denies* weights when the evidence is uncertain

This expression can be useful if Bayes' theorem is being used directly (Section 3.2.2), but an alternative is needed when using likelihood ratios. One technique is to modify the *affirms* and *denies* weights to reflect the uncertainty in E. One means of achieving this is to interpolate the weights linearly as the probability of E varies between 1 and 0. Figure 3.2 illustrates this scaling process, where the interpolated *affirms* and *denies* weights are given the symbols A' and D', respectively. While P(E) is greater than 0.5, the *affirms* weight is used, and when P(E) is less than 0.5, the *denies* weight is used. Over the range of values for P(E), A' and D' vary between 1 (neutral weighting) and A and D, respectively. The interpolation process achieves the right sort of result, but has no rigorous basis. The expressions used to calculate the interpolated values are:

$$A' = [2(A - 1) \times P(E)] + 2 - A \tag{3.16}$$

$$D' = [2(1 - D) \times P(E)] + D \tag{3.17}$$

3.2.6 Combining evidence

Much of the controversy concerning the use of Bayesian updating is centered on the issue of how to combine several pieces of evidence that support the same hypothesis. If n pieces of evidence are found that support a hypothesis H, then the formal restatement of the updating equation is straightforward:

$$\mathrm{O(H|E_1\&E_2\&E_3 \ldots E_n)} = A \times \mathrm{O(H)} \tag{3.18}$$

where

$$A = \frac{\mathrm{P(E_1\&E_2\&E_3 \ldots E_n|H)}}{\mathrm{P(E_1\&E_2\&E_3 \ldots E_n|{\sim}H)}} \tag{3.19}$$

However, the usefulness of this pair of equations is doubtful, since we do not know in advance which pieces of evidence will be available to support the hypothesis H. We would have to write expressions for A covering all possible pieces of evidence E_i, as well as all combinations of the pairs $E_i\&E_j$, of the triples $E_i\&E_j\&E_k$, of quadruples $E_i\&E_j\&E_k\&E_m$, and so on. As this is clearly an unrealistic requirement, especially where the number of possible pieces of evidence (or symptoms in a diagnosis problem) is large, a simplification is normally sought. The problem becomes much more manageable if it is assumed that all pieces of evidence are *statistically independent*. It is this assumption that is one of the most controversial aspects of the use of Bayesian updating in knowledge-based systems, since the assumption is rarely accurate. Statistical independence of two pieces of evidence (E_1 and E_2) means that the probability of observing E_1 given that E_2 has been observed is identical to the probability of observing E_1 given no information about E_2. Stating this more formally, the statistical independence of E_1 and E_2 is defined as:

$$\mathrm{P(E_1|E_2)} = \mathrm{P(E_1)}$$

and (3.20)

$$\mathrm{P(E_2|E_1)} = \mathrm{P(E_2)}$$

If the independence assumption is made, then the rule-writer need only worry about supplying weightings of the form:

$$A_i = \frac{\mathrm{P(E_i|H)}}{\mathrm{P(E_i|{\sim}H)}} \tag{3.21}$$

and

$$D_i = \frac{\mathrm{P({\sim}E_i|H)}}{\mathrm{P({\sim}E_i|{\sim}H)}} \tag{3.22}$$

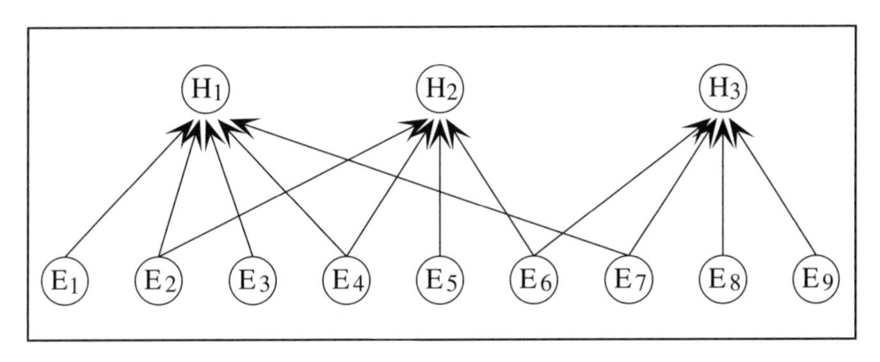

Figure 3.3 A shallow Bayesian inference network (E_i = evidence, H_i = hypothesis)

for each piece of evidence E_i that has the potential to update H. If, in a given run of the system, n pieces of evidence are found that support or oppose H, then the updating equations are simply:

$$O(H|E_1\&E_2\&E_3....E_n) = A_1 \times A_2 \times A_3 \times \ldots \times A_n \times O(H) \quad (3.23)$$

and

$$O(H|\sim E_1\&\sim E_2\&\sim E_3....\sim E_n) = D_1 \times D_2 \times D_3 \times \ldots \times D_n \times O(H) \quad (3.24)$$

Problems arising from the interdependence of pieces of evidence can be avoided if the rule base is properly structured. Where pieces of evidence are known to be dependent on each other, they should not be combined in a single rule. Instead assertions — and the rules that generate them — should be arranged in a hierarchy from low-level input data to high-level conclusions, with many levels of hypotheses between. This does not limit the amount of evidence that is considered in reaching a conclusion, but controls the interactions between the pieces of evidence. Inference networks are a convenient means of representing the levels of assertions from input data, through intermediate deductions to final conclusions. Figures 3.3 and 3.4 show two possible inference networks. Each node represents either a hypothesis or a piece of evidence, and has an associated probability (not shown). In Figure 3.3 the rule-writer has attempted to draw all the evidence that is relevant to particular conclusions together in a single rule for each conclusion. This produces a shallow network, with no intermediate levels between input data and conclusions. Such a system would only be reliable if there was little or no dependence between the input data.

In contrast, the inference network in Figure 3.4 includes several intermediate steps. The probabilities at each node are modified as the

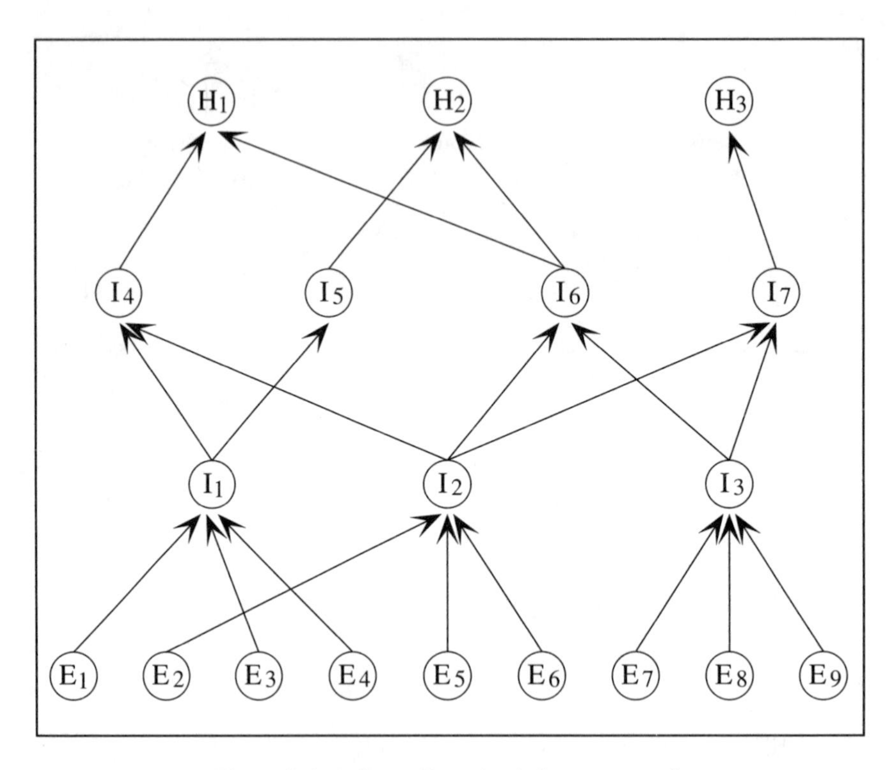

Figure 3.4 A deeper Bayesian inference network
(E_i = evidence, H_i = hypothesis, I_i = intermediate hypothesis)

reasoning process proceeds, until they reach their final values. Note that the rules in the boiler control example made use of several intermediate nodes, which helped to make the rules more understandable and avoided duplication of tests for specific pieces of evidence.

3.2.7 Combining Bayesian rules with production rules

In a practical rule-based system, we may wish to mix uncertain rules with production rules. For instance, we may wish to make use of the production rule:

```
IF release valve is stuck THEN release valve needs cleaning
```

even though the assertion `release valve is stuck` may have been established with a probability less than 1. In this case the hypothesis `release valve needs cleaning` can be asserted with the same probability as the evidence. This avoids the issue of providing a prior probability for the hypothesis or a weighting for the evidence.

If a production rule contains multiple pieces of evidence that are independent from each other, their combined probability can be derived from standard probability theory. Consider, for example, a rule in which two pieces of independent evidence are conjoined (i.e., they are joined by AND):

```
IF evidence E1 AND evidence E2 THEN hypothesis H3
```

The probability of hypothesis H_3 is given by:

$$P(H_3) = P(E_1) \times P(E_2) \tag{3.25}$$

Production rules containing independent evidence that is disjoined (i.e., joined by OR) can be treated in a similar way. So given the rule:

```
IF evidence E1 OR evidence E2 THEN hypothesis H3
```

the probability of hypothesis H_3 is given by:

$$P(H_3) = P(E_1) + P(E_2) - (P(E_1) \times P(E_2)) \tag{3.26}$$

3.2.8 A worked example of Bayesian updating

We will consider the same example that was introduced in Chapter 2, namely control of a power station boiler. Let us start with just four rules:

```
/* Rule 3.1a */
IF release valve is stuck THEN release valve needs cleaning

/* Rule 3.2a */
IF warning light is on THEN release valve is stuck

/* Rule 3.3a */
IF pressure is high THEN release valve is stuck

/* Rule 3.4a */
IF temperature is high AND NOT(water level is low)
THEN pressure is high
```

The conclusion of each of these rules is expressed as an assertion. The four rules contain four assertions (or hypotheses) and three pieces of evidence which are independent of the rules, namely the temperature, the status of the warning light (on or off), and the water level. The various probability estimates for these and their associated *affirms* and *denies* weights are shown in Table 3.1.

H	E	P(H)	O(H)	P(E|H)	P(E|~H)	*A*	*D*
release valve needs cleaning	release valve is stuck	——	——	——	——	——	——
release valve is stuck	warning light is on	0.02	0.02	0.88	0.4	2.20	0.20
release valve is stuck	pressure is high	0.02	0.02	0.85	0.01	85.0	0.15
pressure is high	temperature is high	0.1	0.11	0.90	0.05	18.0	0.11
pressure is high	water level is low	0.1	0.11	0.05	0.5	0.10	1.90

Table 3.1 Values used in the worked example of Bayesian updating

Having calculated the *affirms* and *denies* weights, we can now rewrite our production rules as probabilistic rules. We will leave Rule 3.1a unaltered in order to illustrate the interaction between production rules and probabilistic rules. Our new rule set is therefore as follows:

```
/* Rule 3.1b */
IF release valve is stuck THEN release valve needs cleaning

/* Rule 3.2b */
IF warning light is on        (AFFIRMS 2.20; DENIES 0.20)
THEN release valve is stuck

/* Rule 3.3b */
IF pressure is high           (AFFIRMS 85.0; DENIES 0.15)
THEN release valve is stuck

/* Rule 3.4b */
IF temperature is high        (AFFIRMS 18.0; DENIES 0.11)
AND water level is low        (AFFIRMS 0.10; DENIES 1.90)
THEN pressure is high
```

Rule 3.4b makes use of two pieces of evidence, and it no longer needs a negative condition, as this has been accommodated by the *affirms* and *denies* weights. The requirement that `NOT(water level is low)` be supportive evidence is expressed by the *denies* weight of `water level is low` being greater than 1 while the *affirms* weight is less than 1.

To illustrate how the various weights are used, let us consider how a Bayesian inference engine would use the following set of input data:

- `NOT(water level is low)`
- `warning light is on`
- `temperature is high`

We will assume that the rules fire in the following order:

Rule 3.4b → Rule 3.3b → Rule 3.2b → Rule 3.1b

The resultant rule trace might then appear as follows:

```
Rule 3.4b
H = pressure is high;                         O(H) = 0.11
E1 = temperature is high;                       A1 = 18.0
E2 = water level is low;                        D2 = 1.90
O(H|(E1&~E2)) = O(H) × A1 × D2 = 3.76
/* Updated odds of pressure is high are 3.76 */

Rule 3.3b
H = release valve is stuck;                   O(H) = 0.02
E = pressure is high;                            A = 85.0
Because E is not certain (O(E) = 3.76, P(E) = 0.79), the
inference engine must calculate an interpolated value A' for the
affirms weight of E (see Section 3.2.5).
A'= [2(A-1) × P(E)] + 2 - A = 49.7
O(H|(E)) = O(H) × A' = 0.99
/* Updated odds of release valve is stuck are 0.99, */
/* corresponding to a probability of approximately 0.5 */

Rule 3.2b
H = release valve is stuck;                   O(H) = 0.99
E = warning light is on;                         A = 2.20
O(H|(E)) = O(H) × A = 2.18
/* Updated odds of release valve is stuck are 2.18 */

Rule 3.1b
H = release valve needs cleaning
E = release valve is stuck;
O(E)= 2.18 implies O(H)= 2.18
/* This is a production rule, so the conclusion is asserted with
the same probability as the evidence. */
/* Updated odds of release valve needs cleaning are 2.18 */
```

3.2.9 Discussion of the worked example

The above example serves to illustrate a number of features of Bayesian updating. Our final conclusion that the release valve needs cleaning is reached with a certainty represented as:

```
O(release valve needs cleaning) = 2.18
```

or

```
P(release valve needs cleaning) = 0.69
```

Thus, there is a probability of 0.69 that the valve needs cleaning. In a real-world situation, this is a more realistic outcome than concluding that the valve definitely needs cleaning, which would have been the conclusion had we used the original set of production rules.

The initial three items of evidence were all stated with complete certainty: `NOT(water level is low)`; `warning light is on`; and `temperature is high`. In other words, P(E) = 1 for each of these. Consider the evidence `warning light is on`. A probability of less than 1 might be associated with this evidence if it were generated as an assertion by another probabilistic rule, or if it were supplied as an input to the system, but the user's view of the light was impaired. If P(warning light is on) is 0.8, an interpolated value of the *affirms* weight would be used in Rule 3.2b. Equation 3.16 yields an interpolated value of 1.72 for the *affirms* weight.

However, if P(warning light is on) were less than 0.5, then an interpolated *denies* weighting would be used. If P(warning light is on) were 0.3, an interpolated *denies* weighting of 0.68 is yielded by Equation 3.17.

If P(warning light is on) = 0.5, then the warning light is just as likely to be on as it is to be off. If we try to interpolate either the *affirms* or *denies* weight, a value of 1 will be found. Thus, if each item of evidence for a particular rule has a probability of 0.5, then the rule has no effect whatsoever.

Assuming that the prior probability of a hypothesis is less than 1 and greater than 0, the hypothesis can never be confirmed with complete certainty by the application of likelihood ratios as this would require its odds to become infinite.

While Bayesian updating is a mathematically rigorous technique for updating probabilities, it is important to remember that the results obtained can only be valid if the data supplied are valid. This is the key issue to consider when assessing the virtues of the technique. The probabilities shown in Table 3.1 have not been measured from a series of trials, but instead they are an expert's best guesses. Given that the values upon which the *affirms* and *denies* weights are based are only guesses, then a reasonable alternative to calculating them is to simply take an educated guess at the appropriate weightings. Such an approach is just as valid or invalid as calculating values from unreliable data. If a rule-writer takes such an *ad hoc* approach, the provision of both an *affirms* and *denies* weighting becomes optional. If an *affirms* weight is provided for a piece of evidence E, but not a *denies* weight, then that rule can be ignored when P(E) < 0.5.

As well as relying on the rule-writer's weightings, Bayesian updating is also critically dependent on the values of the prior probabilities. Obtaining accurate estimates for these is also problematic.

Even if we assume that all of the data supplied in the above worked example are accurate, the validity of the final conclusion relies upon the statistical independence from each other of the supporting pieces of evidence. In our example, as with very many real problems, this assumption is dubious. For example, `pressure is high` and `warning light is on` were used as independent pieces of evidence, when in reality there is a cause-and-effect relationship between the two.

3.2.10 Advantages and disadvantages of Bayesian updating

Bayesian updating is a means of handling uncertainty by updating the probability of an assertion when evidence for or against the assertion is provided.

The principal *advantages* of Bayesian updating are:

(i) The technique is based upon a proven statistical theorem.

(ii) Likelihood is expressed as a probability (or odds), which has a clearly defined and familiar meaning.

(iii) The technique requires deductive probabilities, which are generally easier to estimate than abductive ones. The user supplies values for the probability of evidence (the symptoms) given a hypothesis (the cause) rather than the reverse.

(iv) Likelihood ratios and prior probabilities can be replaced by sensible guesses. This is at the expense of advantage (i), as the probabilities subsequently calculated cannot be interpreted literally, but rather as an imprecise measure of likelihood.

(v) Evidence for and against a hypothesis (or the presence and absence of evidence) can be combined in a single rule by using *affirms* and *denies* weights.

(vi) Linear interpolation of the likelihood ratios can be used to take account of any uncertainty in the evidence (i.e., uncertainty about whether the condition part of the rule is satisfied), though this is an *ad hoc* solution.

(vii) The probability of a hypothesis can be updated in response to more than one piece of evidence.

The principal *disadvantages* of Bayesian updating are:

(i) The prior probability of an assertion must be known or guessed at.
(ii) Conditional probabilities must be measured or estimated or, failing those, guesses must be taken at suitable likelihood ratios. Although the conditional probabilities are often easier to judge than the prior probability, they are nevertheless a considerable source of errors. Estimates of likelihood are often clouded by a subjective view of the importance or utility of a piece of information [4].
(iii) The single probability value for the truth of an assertion tells us nothing about its precision.
(iv) Because evidence for and against an assertion are lumped together, no record is kept of how much there is of each.
(v) The addition of a new rule that asserts a new hypothesis often requires alterations to the prior probabilities and weightings of several other rules. This contravenes one of the main advantages of knowledge-based systems.
(vi) The assumption that pieces of evidence are independent is often unfounded. The only alternatives are to calculate *affirms* and *denies* weights for all possible combinations of dependent evidence, or to restructure the rule base so as to minimize these interactions.
(vii) The linear interpolation technique for dealing with uncertain evidence is not mathematically justified.
(viii) Representations based on odds, as required to make use of likelihood ratios, cannot handle absolute truth, i.e., odds = ∞.

3.3 Certainty theory

3.3.1 Introduction

Certainty theory [5] is an adaptation of Bayesian updating that is incorporated into the EMYCIN expert system shell. EMYCIN is based on MYCIN [6], an expert system that assists in the diagnosis of infectious diseases. The name EMYCIN is derived from "essential MYCIN," reflecting the fact that it is not specific to medical diagnosis and that its handling of uncertainty is simplified. Certainty theory represents an attempt to overcome some of the shortcomings of Bayesian updating, although the mathematical rigor of Bayesian updating is lost. As this rigor is rarely justified by the quality of the data, this is not really a problem.

3.3.2 Making uncertain hypotheses

Instead of using probabilities, each assertion in EMYCIN has a certainty value associated with it. Certainty values can range between 1 and –1.

For a given hypothesis H, its certainty value C(H) is given by:

$C(H) = 1.0$ if H is known to be true;
$C(H) = 0.0$ if H is unknown;
$C(H) = -1.0$ if H is known to be false.

There is a similarity between certainty values and probabilities, such that:

$C(H) = 1.0$ corresponds to $P(H)=1.0$;
$C(H) = 0.0$ corresponds to P(H) being at its *a priori* value;
$C(H) = -1.0$ corresponds to $P(H)=0.0$.

Each rule also has a certainty associated with it, known as its certainty factor CF. Certainty factors serve a similar role to the *affirms* and *denies* weightings in Bayesian systems:

```
IF <evidence> THEN <hypothesis> WITH certainty factor CF
```

Part of the simplicity of certainty theory stems from the fact that identical measures of certainty are attached to rules and hypotheses. The certainty factor of a rule is modified to reflect the level of certainty of the evidence, such that the modified certainty factor CF' is given by:

$$CF' = CF \times C(E) \tag{3.27}$$

If the evidence is known to be present, i.e., $C(E) = 1$, then Equation 3.27 yields $CF' = CF$.

The technique for updating the certainty of hypothesis H, in the light of evidence E, involves the application of the following composite function:

if $C(H) \geq 0$ and $CF' \geq 0$:

$$C(H|E) = C(H) + [CF' \times (1 - C(H))] \tag{3.28}$$

if $C(H) \leq 0$ and $CF' \leq 0$:

$$C(H|E) = C(H) + [CF' \times (1 + C(H))] \tag{3.29}$$

if $C(H)$ and CF' have opposite signs:

$$C(H|E) = \frac{C(H) + CF'}{1 - \min(|C(H)|, |CF'|)} \tag{3.30}$$

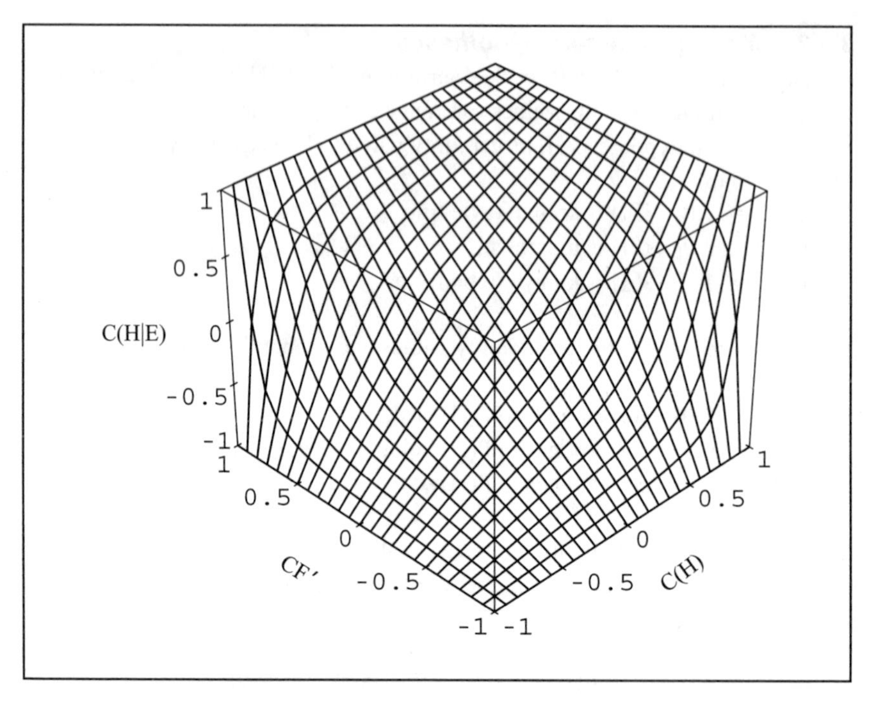

Figure 3.5 Equations 3.28–3.30 for updating certainties

where:

C(H|E) is the certainty of H updated in the light of evidence E;

C(H) is the initial certainty of H, i.e., 0 unless it has been updated by the previous application of a rule;

$|x|$ = the magnitude of x, ignoring its sign.

It can be seen from the above equations that the updating procedure consists of *adding* a positive or negative value to the current certainty of a hypothesis. This contrasts with Bayesian updating, where the odds of a hypothesis are *multiplied* by the appropriate likelihood ratio. The composite function represented by Equations 3.28 to 3.30 is plotted in Figure 3.5, and can be seen to have a broadly similar shape to the Bayesian updating equation (plotted in Figure 3.1).

In the standard version of certainty theory, a rule can only be applied if the certainty of the evidence C(E) is greater than 0, i.e., if the evidence is more likely to be present than not. EMYCIN restricts rule firing further by requiring that C(E) > 0.2 for a rule to be considered applicable. The justification for this heuristic is that it saves computational power and makes explanations clearer, as marginally effective rules are suppressed. In fact it is possible to allow rules to fire regardless of the value of C(E). The absence of supporting evidence, indicated by C(E) < 0, would then be taken into account since CF' would have the opposite sign to CF.

Although there is no theoretical justification for the function for updating certainty values, it does have a number of desirable properties:

(i) the function is continuous and has no singularities or steps;
(ii) the updated certainty C(H|E) always lies within the bounds –1 and +1;
(iii) if either C(H) or CF' is +1 (i.e., definitely true) then C(H|E) is also +1;
(iv) if either C(H) or CF' is –1 (i.e., definitely false) then C(H|E) is also –1;
(v) when contradictory conclusions are combined, they tend to cancel each other out, i.e., if $C(H) = -CF'$ then $C(H|E) = 0$;
(vi) several pieces of independent evidence can be combined by repeated application of the function, and the outcome is independent of the order in which the pieces of evidence are applied;
(vii) if C(H) = 0, i.e., the certainty of H is at its *a priori* value, then $C(H|E) = CF'$;
(viii) if the evidence is certain (i.e., C(E) = 1) then $CF' = CF$.
(ix) although not part of the standard implementation, the absence of evidence can be taken into account by allowing rules to fire when C(E) < 0.

3.3.3 Logical combinations of evidence

In Bayesian updating systems, each piece of evidence that contributes toward a hypothesis is assumed to be independent and is given its own *affirms* and *denies* weights. In systems based upon certainty theory, the certainty factor is associated with the rule as a whole, rather than with individual pieces of evidence. For this reason, certainty theory provides a simple algorithm for determining the value of the certainty factor that should be applied when more than one item of evidence is included in a single rule. The relationship between pieces of evidence is made explicit by the use of AND and OR. If separate pieces of evidence are intended to contribute toward a single hypothesis independently of each other, they must be placed in separate rules. The algorithm for combining items of evidence in a single rule is borrowed from Zadeh's possibility theory (Section 3.4). The algorithm covers the cases where evidence is conjoined (i.e., joined by AND), disjoined (i.e., joined by OR), and negated (using NOT).

Conjunction

Consider a rule of the form:

```
IF <evidence E1> AND <evidence E2> THEN <hypothesis>
   WITH certainty factor CF
```

The certainty of the combined evidence is given by $C(E_1 \text{ AND } E_2)$, where:

$$C(E_1 \text{ AND } E_2) = \min[C(E_1), C(E_2)] \tag{3.31}$$

Disjunction

Consider a rule of the form:

```
IF <evidence E1> OR <evidence E2> THEN <hypothesis>
   WITH certainty factor CF
```

The certainty of the combined evidence is given by $C(E_1 \text{ OR } E_2)$, where:

$$C(E_1 \text{ OR } E_2) = \max[C(E_1), C(E_2)] \tag{3.32}$$

Negation

Consider a rule of the form:

```
IF NOT <evidence E> THEN <hypothesis> WITH certainty factor CF
```

The certainty of the negated evidence, C(E), is given by C(~E) where:

$$C(\sim E) = -C(E) \tag{3.33}$$

3.3.4 A worked example of certainty theory

In order to illustrate the application of certainty theory, we can rework the example that was used to illustrate Bayesian updating. Four rules were used, which together could determine whether the release valve of a power station boiler needs cleaning (see Section 3.2.8). Each of the four rules can be rewritten with an associated certainty factor, which is estimated by the rule-writer:

```
/* Rule 3.1c */
IF release valve is stuck THEN release valve needs cleaning
WITH CERTAINTY FACTOR 1

/* Rule 3.2c */
IF warning light is on THEN release valve is stuck
WITH CERTAINTY FACTOR 0.2

/* Rule 3.3c */
IF pressure is high THEN release valve is stuck
WITH CERTAINTY FACTOR 0.9
```

```
/* Rule 3.4c */
IF temperature is high AND NOT(water level is low)
THEN pressure is high
WITH CERTAINTY FACTOR 0.5
```

Although the process of providing certainty factors might appear *ad hoc* compared with Bayesian updating, it may be no less reliable than estimating the probabilities upon which Bayesian updating relies. In the Bayesian example, the production Rule 3.1b had to be treated as a special case. In a system based upon uncertainty theory, Rule 3.1c can be made to behave as a production rule simply by giving it a certainty factor of 1.

As before, the following set of input data will be considered:

- `NOT(water level is low)`
- `warning light is on`
- `temperature is high`

We will assume that the rules fire in the order:

Rule 3.4c → Rule 3.3c → Rule 3.2c → Rule 3.1c

The resultant rule trace might then appear as follows:

```
Rule 3.4c                                     CF = 0.5
H = pressure is high;                         C(H) = 0
E1 = temperature is high;                     C(E1) = 1
E2 = water level is low; C(E2) = -1,  C(~E2) = 1
C(E1&~E2) = min[C(E1),C(~E2)] = 1
CF' = CF × C(E1&~E2) = CF
C(H|(E1&~E2)) = CF' = 0.5
/* Updated certainty of pressure is high is 0.5 */

Rule 3.3c                                     CF = 0.9
H = release valve is stuck;                   C(H) = 0
E = pressure is high;                         C(E) = 0.5
CF' = CF × C(E) = 0.45
C(H|(E)) = CF' = 0.45
/* Updated certainty of release valve is stuck is 0.45 */

Rule 3.2c                                     CF = 0.2
H = release valve is stuck;                   C(H) = 0.45
E = warning light is on;                      C(E) = 1
CF' = CF × C(E) = CF
C(H|(E)) = C(H) + [CF' × (1-C(H))] = 0.56
/* Updated certainty of release valve is stuck is 0.56 */
```

```
Rule 3.1c                                    CF = 1
H = release valve needs cleaning          C(H) = 0
E = release valve is stuck;               C(E) = 0.56
CF' = CF × C(E) = 0.56
C(H|(E)) = CF' = 0.56
/* Updated certainty of release valve needs cleaning is 0.56 */
```

3.3.5 Discussion of the worked example

Given the certainty factors shown, the example yielded the result `release valve needs cleaning` with a similar level of confidence to the Bayesian updating example.

Under Bayesian updating, Rules 3.2b and 3.3b could be combined into a single rule without changing their effect:

```
/* Rule 3.5b */
IF warning light is on          (AFFIRMS 2.20; DENIES 0.20)
AND pressure is high            (AFFIRMS 85.0; DENIES 0.15)
THEN release valve is stuck
```

With certainty theory, the weightings apply not to the individual pieces of evidence (as with Bayesian updating) but to the rule itself. If Rules 3.2c and 3.3c were combined in one rule, a single certainty factor would need to be chosen to replace the two used previously. Thus a combined rule might look like:

```
/* Rule 3.5c */
IF warning light is on AND pressure is high
THEN release valve stuck WITH CERTAINTY FACTOR 0.95
```

In the combined rule, the two items of evidence are no longer treated independently and the certainty factor is the adjudged weighting if *both* items of evidence are present. If our worked example had contained this combined rule instead of Rules 3.2c and 3.3c, then the rule trace would contain the following:

```
Rule 3.5c                                CF = 0.95
H = release valve is stuck;            C(H) = 0
E1 = warning light is on;             C(E1) = 1
E2 = pressure is high;                C(E2) = 0.5
C(E1 & E2) = min[C(E1),C(E2)] = 0.5
CF' = CF × C(E1 & E2) = 0.48
C(H|(E1 & E2)) = CF' = 0.48
/* Updated certainty of release valve is stuck is 0.48 */
```

With the certainty factors used in the example, the combined rule yields a lower confidence in the hypothesis `release valve stuck` than Rules 3.2c and 3.3c used separately. As a knock-on result, Rule 3.1c would yield the conclusion `release valve needs cleaning` with a diminished certainty of 0.48.

3.3.6 Relating certainty factors to probabilities

It has already been noted that there is a similarity between the certainty factors that are attached to hypotheses and the probabilities of those hypotheses, such that:

C(H) = 1.0 corresponds to P(H) = 1.0;
C(H) = 0.0 corresponds to P(H) being at its *a priori* value;
C(H) = –1.0 corresponds to P(H) = 0.0.

Additionally, a formal relationship exists between the certainty factor associated with a rule and the conditional probability P(H|E) of a hypothesis H given some evidence E. This is only of passing interest as certainty factors are not normally calculated in this way, but instead are simply estimated or chosen so as to give the right sort of results. The formal relationships are as follows.

If evidence E supports hypothesis H, i.e., P(H|E) is greater than P(H), then:

$$\left.\begin{array}{ll} CF = \dfrac{P(H|E) - P(H)}{1 - P(H)} & \text{if } P(H) \neq 1 \\ CF = 1 & \text{if } P(H) = 1 \end{array}\right\} \tag{3.34}$$

If evidence E opposes hypothesis H, i.e., P(H|E) is less than P(H), then:

$$\left.\begin{array}{ll} CF = \dfrac{P(H|E) - P(H)}{P(H)} & \text{if } P(H) \neq 0 \\ CF = -1 & \text{if } P(H) = 0 \end{array}\right\} \tag{3.35}$$

The shape of Equations 3.34 and 3.35 is shown in Figure 3.6.

3.4 Possibility theory: fuzzy sets and fuzzy logic

Bayesian updating and certainty theory are techniques for handling the uncertainty that arises, or is assumed to arise, from statistical variations or randomness. Possibility theory addresses a different source of uncertainty, namely vagueness in the use of language. Possibility theory, or fuzzy logic,

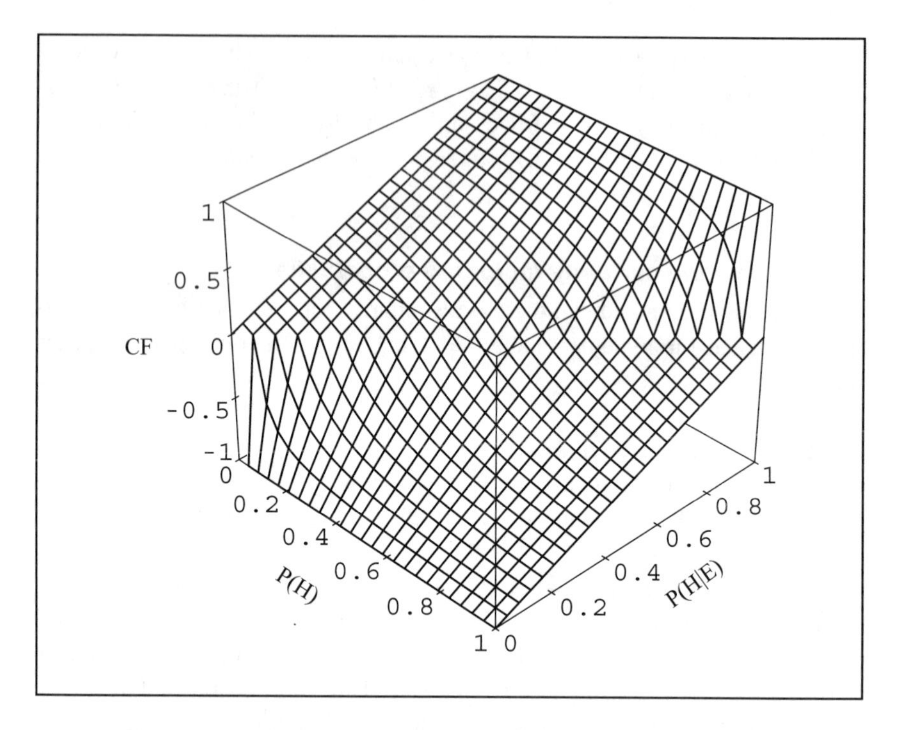

Figure 3.6 The relationship between certainty factors and probability

was developed by Zadeh [7, 8, 9] and builds upon his theory of fuzzy sets [10]. Zadeh asserts that while probability theory may be appropriate for measuring the likelihood of a hypothesis, it says nothing about the *meaning* of the hypothesis.

3.4.1 Crisp sets and fuzzy sets

The rules shown in this chapter and in Chapter 2 contain a number of examples of vague language where fuzzy sets might be applied, such as the following phrases:

- water level is low;
- temperature is high;
- pressure is high.

In conventional set theory, the sets high, medium and low — applied to a variable such as temperature — would be mutually exclusive. If a given temperature (say, 400°C) is high, then it is neither medium nor low. Such sets are said to be crisp or non-fuzzy (Figure 3.7). If the boundary between medium

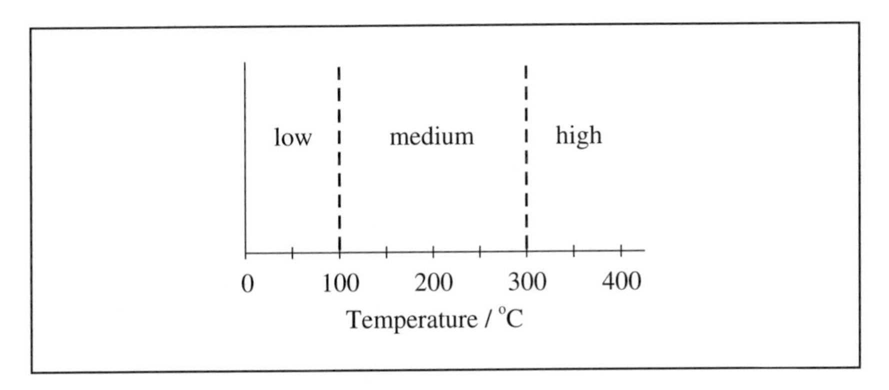

Figure 3.7 Conventional crisp sets applied to temperature.

and high is set at 300°C, then a temperature of 301°C is considered high, while 299°C is considered medium. This distinction is rather artificial, and means that a tiny difference in temperature can completely change the rule-firing, while a rise in temperature from 301°C to 1000°C has no effect at all.

Fuzzy sets are a means of smoothing out the boundaries. The theory of fuzzy sets expresses imprecision quantitatively by introducing characteristic membership functions that can assume values between 0 and 1 corresponding to degrees of membership from "not a member" through to "a full member." If F is a fuzzy set, then the membership function $\mu_F(x)$ measures the degree to which an absolute value x belongs to F. This degree of membership is sometimes called the *possibility* that x is described by F. The process of deriving these possibility values for a given value of x is called *fuzzification*.

Conversely, consider that we are given the imprecise statement `temperature is low`. If LT is the fuzzy set of low temperatures, then we might define the membership function μ_{LT} such that:

$$
\begin{aligned}
\mu_{LT}(250°C) &= 0.0 \\
\mu_{LT}(200°C) &= 0.0 \\
\mu_{LT}(150°C) &= 0.25 \\
\mu_{LT}(100°C) &= 0.5 \\
\mu_{LT}(50°C) &= 0.75 \\
\mu_{LT}(0°C) &= 1.0 \\
\mu_{LT}(-50°C) &= 1.0
\end{aligned}
$$

These values correspond with the linear membership function shown in Figure 3.8(a). Although linear membership functions like those in Figures 3.8(a) and (b) are convenient in many applications, the most suitable shape of the

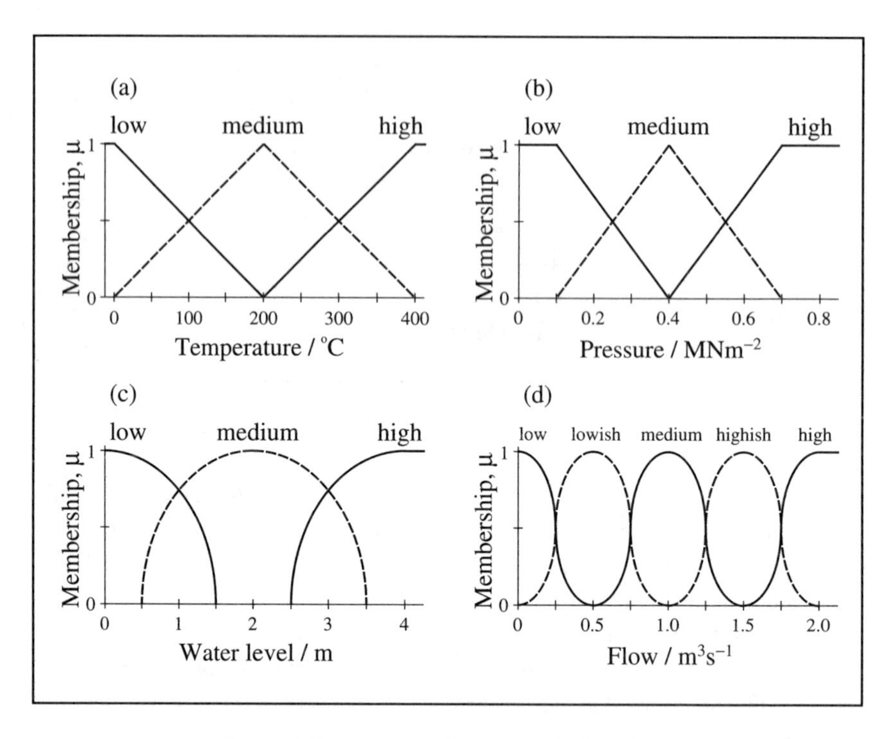

Figure 3.8 A variety of membership functions.

membership functions and the number of fuzzy sets depends on the particular application. Figures 3.8(c) and (d) show some nonlinear alternatives.

The key differences between fuzzy and crisp sets are that:

- an element has a degree of membership (0–1) of a fuzzy set;
- membership of one fuzzy set does not preclude membership of another.

Thus the temperature 350°C may have some (non-zero) degree of membership to both fuzzy sets `high` and `medium`. This is represented in Figure 3.8 by the overlap between the fuzzy sets. The sum of the membership functions for a given value can be arranged to equal 1, as shown for temperature and pressure in Figure 3.8, but this is not a necessary requirement.

Some of the terminology of fuzzy sets may require clarification. The statement `temperature is low` is an example of a *fuzzy statement* involving a *fuzzy set* (low temperature) and a *fuzzy variable* (temperature). A fuzzy variable is one that can take any value from a global set (e.g., the set of all temperatures), where each value can have a degree of membership of a fuzzy set (e.g., low temperature) associated with it.

Although the discussion so far has concentrated on continuous variables such as temperature and pressure, the same ideas can also be applied to discrete variables, such as the number of signals detected in a given time span.

3.4.2 Fuzzy rules

If a variable is set to a value by crisp rules, its value will change in steps as different rules fire. The only way to smooth those steps would be to have a large number of rules. However, only a small number of fuzzy rules is required to produce smooth changes in the outputs as the input values alter. The number of fuzzy rules required is dependent on the number of variables, the number of fuzzy sets, and the ways in which the variables are combined in the fuzzy rule conditions. Numerical information is explicit in crisp rules, e.g., `IF temperature > 300°C THEN ...` but in fuzzy rules it becomes implicit in the chosen shape of the fuzzy membership functions.

Consider a rule base that contains the following fuzzy rules:

```
/* Rule 3.6f */
IF temperature is high THEN pressure is high

/* Rule 3.7f */
IF temperature is medium THEN pressure is medium

/* Rule 3.8f */
IF temperature is low THEN pressure is low
```

Suppose the measured temperature is 350°C. As this is a member of both fuzzy sets `high` and `medium`, Rules 3.6f and 3.7f will both fire. The pressure, we conclude, will be somewhat high and somewhat medium. Suppose that the membership functions for temperature are as shown in Figure 3.8(a). The possibility that the temperature is high, μ_{HT}, is 0.75 and the possibility that the temperature is medium, μ_{MT}, is 0.25. As a result of firing the rules, the possibilities that the pressure is high and medium, μ_{HP} and μ_{MP}, are set as follows:

$$\mu_{HP} = \max[\mu_{HT}, \mu_{HP}]$$
$$\mu_{MP} = \max[\mu_{MT}, \mu_{MP}]$$

The initial possibility values for pressure are assumed to be zero if these are the first rules to fire, and thus μ_{HP} and μ_{MP} become 0.75 and 0.25, respectively. These values can be passed on to other rules that might have `pressure is high` or `pressure is medium` in their condition clauses.

The Rules 3.6f, 3.7f and 3.8f contain only simple conditions. Possibility theory provides a recipe for computing the possibilities of compound conditions. The formulas for conjunction, disjunction, and negation are similar to those used in certainty theory (Section 3.3.3):

$$\left.\begin{aligned} \mu_{X \text{ AND } Y}(x) &= \min[\mu_X(x), \mu_Y(x)] \\ \mu_{X \text{ OR } Y}(x) &= \max[\mu_X(x), \mu_Y(x)] \\ \mu_{\text{NOT } X}(x) &= 1 - \mu_X(x) \end{aligned}\right\} \tag{3.36}$$

To illustrate the use of these formulas, suppose that water level has the fuzzy membership functions shown in Figure 3.8c and that Rule 3.6f is redefined as follows:

```
/* Rule 3.9f */
IF temperature is high AND water level is NOT low
THEN pressure is high
```

For a water level of 1.2m, the possibility of the water level being low, μ_{LW}(1.2m), is 0.6. The possibility of the water level not being low is therefore 0.4. As this is less than 0.75, the combined possibility for the temperature being high and the water level not being low is 0.4. Thus the possibility that the pressure is high, μ_{HP}, becomes 0.4 if it has not already been set to a higher value.

If several rules affect the same fuzzy set of the same variable, they are equivalent to a single rule whose conditions are joined by the disjunction OR. For example, these two rules:

```
/* Rule 3.6f */
IF temperature is high THEN pressure is high

/* Rule 3.10f */
IF water level is high THEN pressure is high
```

are equivalent to this single rule:

```
/* Rule 3.11f */
IF temperature is high OR water level is high
THEN pressure is high
```

Aoki and Sasaki [11] have argued for treating OR differently when it involves two fuzzy sets of the same fuzzy variable, for example, high and medium temperature. In such cases, the memberships are clearly dependent on

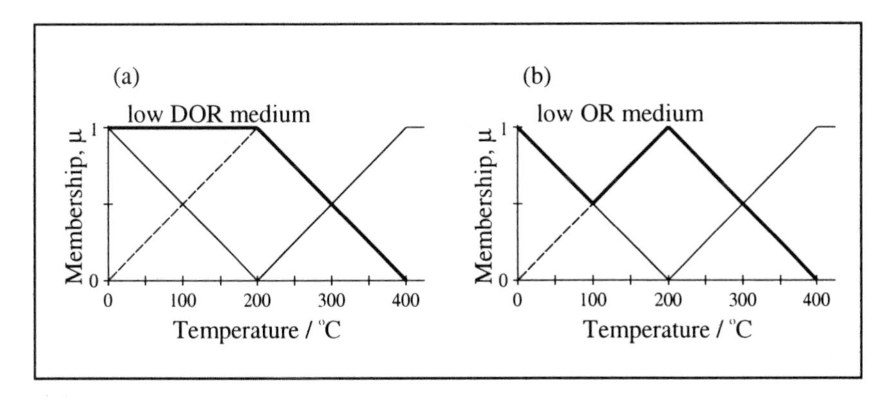

Figure 3.9 (a) Dependent OR; (b) standard OR

each other. Therefore, we can introduce a new operator `DOR` for `dependent OR`. For example, given the rule:

```
/* Rule 3.12f */
IF temperature is low DOR temperature is medium
THEN pressure is lowish
```

the combined possibility for the condition becomes:

$$ {}_{LT\ \mathrm{DOR}\ MT}(x) = \min[1, \quad {}_{LT}(x) \qquad {}_{MT}(x)] \tag{3.37}$$

Given the fuzzy sets for temperature shown in Figure 3.8(a), the combined possibility would be the same for any temperature below 200°C, as shown in Figure 3.9(a). This is consistent with the intended meaning of fuzzy Rule 3.12f. If the OR operator had been used, the membership would dip between 0°C and 200°C, with a minimum at 100°C, as shown in Figure 3.9(b).

3.4.3 Defuzzification

In the above example, at a temperature of 350°C the possibilities for the pressure being high and medium, μ_{HP} and μ_{MP}, are set to 0.75 and 0.25, respectively, by the fuzzy rules 3.6f and 3.7f. It is assumed that the possibility for the pressure being low, μ_{LP}, remains at 0. These values can be passed on to other rules that might have `pressure is high` or `pressure is medium` in their condition clauses without any further manipulation. However, if we want to interpret these membership values in terms of a numerical value of pressure, they would need to be *defuzzified*. Defuzzification is particularly important when the fuzzy variable is a control action such as "set current," where a specific setting is required. The use of fuzzy logic in control systems is

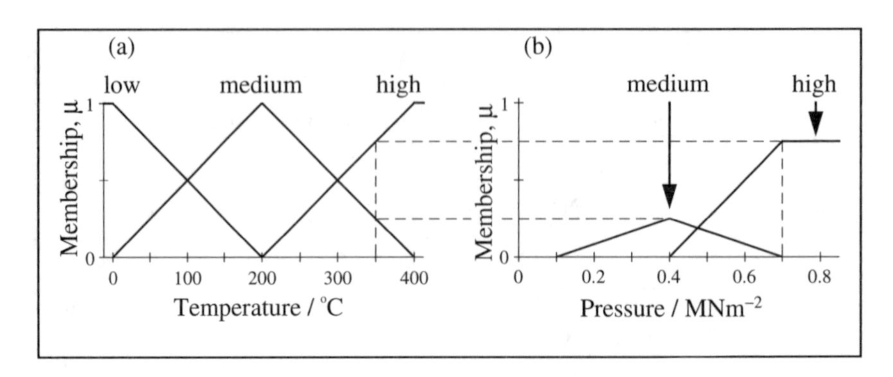

Figure 3.10 Larsen's product operation rule for calculating membership functions from fuzzy rules. Membership functions for pressure are shown, derived from Rules 3.6f and 3.7f, for a temperature of 350°C

discussed further in Chapter 14. Defuzzification takes place in two stages, described below.

Stage 1: scaling the membership functions

The first step in defuzzification is to adjust the fuzzy sets in accordance with the calculated possibilities. A commonly used method is Larsen's product operation rule [12, 13], in which the membership functions are multiplied by their respective possibility values. The effect is to compress the fuzzy sets so that the peaks equal the calculated possibility values, as shown in Figure 3.10. Some authors [14] adopt an alternative approach in which the fuzzy sets are truncated, as shown in Figure 3.11. For most shapes of fuzzy set, the difference between the two approaches is small, but Larsen's product operation rule has the advantages of simplifying the calculations and allowing fuzzification

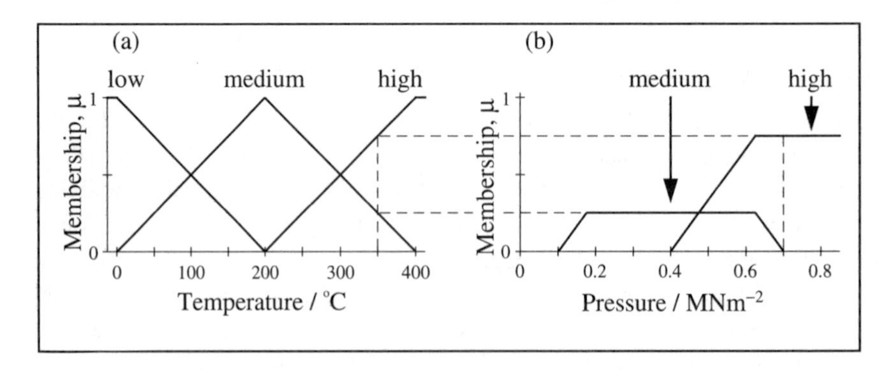

Figure 3.11 Truncation method for calculating membership functions from fuzzy rules. Membership functions for pressure are shown, derived from Rules 3.6f and 3.7f, for a temperature of 350°C

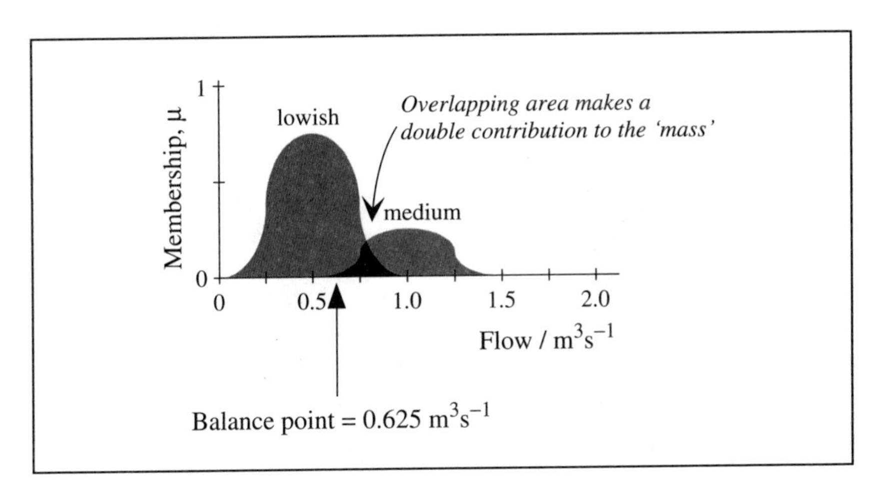

Figure 3.12 Defuzzification by the centroid method

followed by defuzzification to return the initial value, except as described in *A defuzzification anomaly* below.

Stage 2: finding the centroid

The most commonly used method of defuzzification is the *centroid* method, sometimes called the center of gravity, center of mass, or center of area method. The defuzzified value is taken as the point along the fuzzy variable axis that is the centroid, or balance point, of all the scaled membership functions taken together for that variable (Figure 3.12). One way to visualize this is to imagine the membership functions cut out from stiff card and pasted together where (and if) they overlap. The defuzzified value is the balance point along the fuzzy variable axis of this composite shape. When two membership functions overlap, both overlapping regions contribute to the mass of the composite shape. Figure 3.12 shows a simple case, involving neither the `low` nor `high` fuzzy sets. The example that we have been following concerning boiler pressure is more complex and is described in *Defuzzifying at the extremes* below.

If there are N membership functions with centroids c_i and areas a_i then the combined centroid C, i.e., the defuzzified value, is:

$$C = \frac{\sum_{i=1}^{N} a_i c_i}{\sum_{i=1}^{N} a_i} \tag{3.38}$$

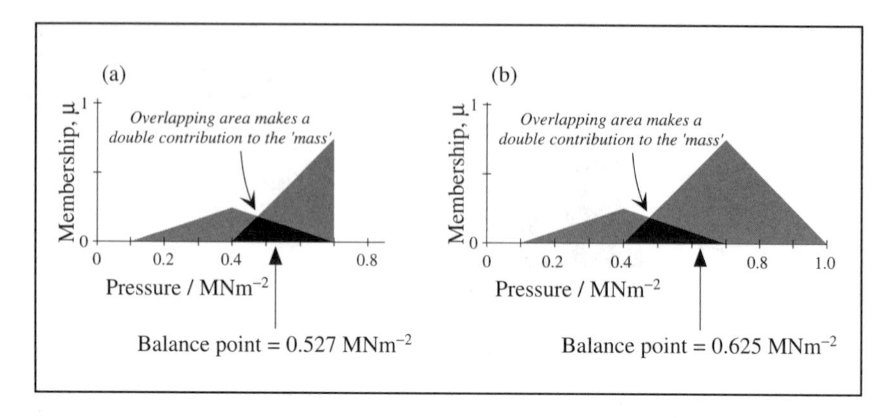

Figure 3.13 Defuzzification at the extremes: (a) bounded range, (b) mirror rule

When the fuzzy sets are compressed using Larsen's product operation rule, the values of c_i are unchanged from the centroids of the uncompressed shapes, C_i, and a_i is simply $\mu_i A_i$ where A_i is the area of the membership function prior to compression. (This is not the case with the truncation method shown in Figure 3.11, which causes the centroid of asymmetrical membership functions to shift along the fuzzy variable axis.) The use of triangular membership functions or other simple geometries simplifies the calculations further. For triangular membership functions, A_i is one half of the base length multiplied by the height. For isosceles triangles C_i is the midpoint along the base, and for right-angle triangles C_i is approx. 29% of the base length from the upright.

Defuzzifying at the extremes

There is a complication in defuzzifying whenever the two extreme membership functions are involved, i.e., those labeled `high` and `low` here. Given the fuzzy sets shown in Figure 3.8b, any pressure above 0.7MNm^{-2} has a membership of `high` of 1. Thus the membership function continues indefinitely toward the right and we cannot find a balance point using the centroid method. Similarly, any pressure below 0.1MNm^{-2} has a membership of `low` of 1, although in this case the membership function is bounded because the pressure cannot go below 0.

One solution to these problems might be to specify a range for the fuzzy variable, *MIN–MAX*, or 0.1–0.7MNm^{-2} in this example. During fuzzification, a value outside this range can be accepted and given a membership of 1 for the fuzzy sets `low` or `high`. However, during defuzzification, the `low` and `high` fuzzy sets can be considered bounded at *MIN* and *MAX* and defuzzification by the centroid method can proceed. This method is shown in Figure 3.13(a) using the values 0.75 and 0.25 for μ_{HP} and μ_{MP}, respectively, as calculated in Section 3.4.2, yielding a defuzzified pressure of 0.527MNm^{-2}. A drawback of this solution is that the defuzzified value can never reach the extremes of the range. For example, if we know that a fuzzy variable has a membership of 1 for the fuzzy set `high` and 0 for the other fuzzy sets, then its actual value could be any value greater than or equal to *MAX*. However, its defuzzified value using this scheme would be the centroid of the `high` fuzzy set, in this case 0.612MNm^{-2}, which is considerably below *MAX*.

An alternative solution is the *mirror rule*. During defuzzification only, the `low` and `high` membership functions are treated as symmetrical shapes centered on *MIN* and *MAX* respectively. This is achieved by reflecting the `low` and `high` fuzzy sets in imaginary mirrors. This method has been used in Figure 3.13(b), yielding a significantly different result, i.e., 0.625MNm^{-2}, for the same possibility values. The method uses the full range *MIN–MAX* of the fuzzy variable during defuzzification, so that a fuzzy variable with a membership of 1 for the fuzzy set `high` and 0 for the other fuzzy sets would be defuzzified to *MAX*. In the example shown in Figure 3.13(b), all values of A_i became identical as a result of adding the mirrored versions of the `low` and `high` fuzzy sets. Because of this, and given that the fuzzy sets have been compressed using Larsen's product operation rule, the equation for defuzzification (3.38) can be simplified to:

$$C = \frac{\sum_{i=1}^{N} \mu_i C_i}{\sum_{i=1}^{N} \mu_i} \tag{3.39}$$

A defuzzification anomaly

It is interesting to investigate whether defuzzification can be regarded as the inverse of fuzzification. In the example considered above, a pressure of 0.625MNm^{-2} would fuzzify to a membership of 0.25 for `medium` and 0.75 for `high`. When defuzzified by the method shown in Figure 3.13(b), the original value of 0.625MNm^{-2} is returned. This observation provides strong support for defuzzification based upon Larsen's product operation rule combined with the mirror rule for dealing with the fuzzy sets at the extremes (Figure 3.13(b)). No such simple relationship exists if the membership functions are truncated (Figure 3.11) or if the extremes are handled by imposing a range (Figure 3.13(a)).

However, even the use of Larsen's product operation rule and the mirror rule cannot always guarantee that fuzzification and defuzzification will be straightforward inverses of each other. For example, as a result of firing a set

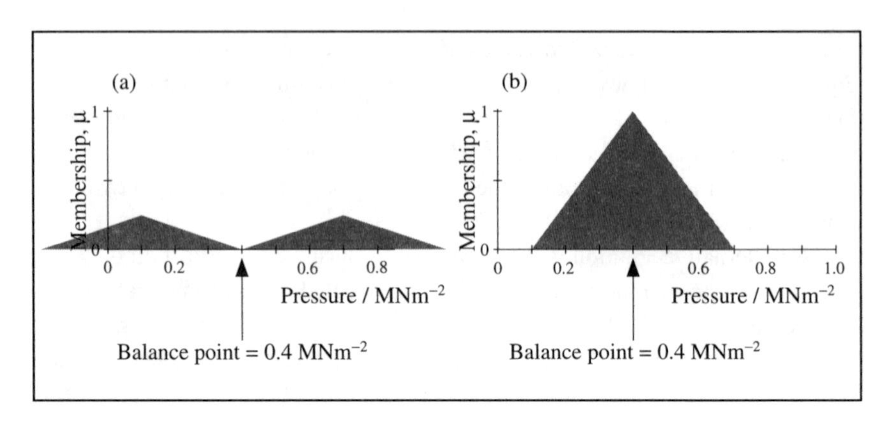

Figure 3.14 Different combinations of memberships can defuzzify to the same value.

of fuzzy rules, we might end up with the following memberships for the fuzzy variable pressure:

Low membership = 0.25
Medium membership = 0.0
High membership = 0.25

Defuzzification of these membership values would yield an absolute value of $0.4 MNm^{-2}$ for the pressure (Figure 3.14(a)). If we were now to look up the fuzzy memberships for an absolute value of $0.4 MNm^{-2}$, i.e., to fuzzify the value, we would obtain:

Low membership = 0.0
Medium membership = 1.0
High membership = 0.0

The resulting memberships values are clearly different from the ones we started with, although they still defuzzify to $0.4 MNm^{-2}$, as shown in Figure 3.14(b). The reason for this anomaly is that, under defuzzification, there are many different combinations of membership values that can yield an absolute value such as $0.4 MNm^{-2}$. The above sets of membership values are just two examples. However, under fuzzification, there is only one absolute value, namely $0.4 MNm^{-2}$, that can yield fuzzy membership values for `low`, `medium`, and `high` of 0.0, 1.0, and 0.0, respectively. Thus, defuzzification is said to be a "many-to-one" relationship, whereas fuzzification is a "one-to-one" relationship.

This observation poses a dilemma for implementers of a fuzzy system. If `pressure` appears in the condition part of further fuzzy rules, different membership values could be used depending on whether or not it is defuzzified and refuzzified before being passed on to those rules.

A secondary aspect of the anomaly is the observation that in the above example we began with possibility values of 0.25 and, therefore, apparently rather weak evidence about the pressure. However, as a result of defuzzification followed by fuzzification, these values are transformed into evidence that appears much stronger. Johnson and Picton [14] have labeled this "Hopgood's defuzzification paradox." The paradox arises because, unlike probabilities or certainty factors, possibility values need to be interpreted relative to each other rather than in absolute terms.

3.5 Other techniques

Possibility theory occupies a distinct position among the many strategies for handling uncertainty, as it is the only established one that is concerned specifically with uncertainty arising from imprecise use of language. Techniques have been developed for dealing with other specific sources of uncertainty. For example, plausibility theory [15] addresses the problems arising from unreliable or contradictory sources of information. Other techniques have been developed in order to overcome some of the perceived shortcomings of Bayesian updating and certainty theory. Notable among these are the Dempster–Shafer theory of evidence and Quinlan's Inferno, both of which are briefly reviewed here.

None of the more sophisticated techniques for handling uncertainty overcomes the most difficult problem, namely, obtaining accurate estimates of the likelihood of events and combinations of events. For this reason, their use is rarely justified in practical knowledge-based systems.

3.5.1 Dempster–Shafer theory of evidence

The theory of evidence [16] is a generalization of probability theory that was created by Dempster and developed by Shafer [17]. It addresses two specific deficiencies of probability theory that have already been highlighted, namely:

- the single probability value for the truth of a hypothesis tells us nothing about its precision;
- because evidence for and against a hypothesis are lumped together, we have no record of how much there is of each.

Rather than representing the probability of a hypothesis H by a single value P(H), Dempster and Shafer's technique binds the probability to a subinterval L(H)–U(H) of the range 0–1. Although the exact probability P(H) may not be known, L(H) and U(H) represent lower and upper bounds on the probability, such that:

$$L(H) \leq P(H) \leq U(H) \tag{3.40}$$

The precision of our knowledge about H is characterized by the difference U(H)–L(H). If this is small, our knowledge about H is fairly precise, but if it is large, we know relatively little about H. A clear distinction is therefore made between uncertainty and ignorance, where uncertainty is expressed by the limits on the value of P(H), and ignorance is represented by the size of the interval defined by those limits. According to Buchanan and Duda [4], Dempster and Shafer have pointed out that the Bayesian agony of assigning prior probabilities to hypotheses is often due to ignorance of the correct values, and this ignorance can make any particular choice arbitrary and unjustifiable.

The above ordering (3.40) can be interpreted as two assertions:

- the probability of H is at least L(H);
- the probability of ~H is at least 1.0 – U(H).

Thus a separate record is kept of degree of belief and disbelief in H. Like Bayesian updating, the theory of evidence benefits from the solid basis of probability theory for the interpretation of L(H) and U(H). When L(H) = U(H), the theory of evidence reduces to the Bayesian updating method. It is, therefore, not surprising that the theory of evidence also suffers from many of the same difficulties.

3.5.2 Inferno

The conclusions that can be reached by the Dempster–Shafer theory of evidence are of necessity weaker than those that can be arrived at by Bayesian updating. If the available knowledge does not justify stronger solutions, then drawing weaker solutions is desirable. This theme is developed further in Inferno [18], a technique that its creator, Quinlan, has subtitled: "a cautious approach to uncertain inference." Although Inferno is based upon probability theory, it avoids assumptions about the dependence or independence of pieces of evidence and hypotheses. As a result, the correctness of any inferences can be guaranteed, given the available knowledge. Thus, Inferno deliberately errs on the side of caution.

The three key motivations for the development of Inferno were as follows:

(i) Other systems often make unjustified assumptions about the dependence or independence of pieces of evidence or hypotheses. Inferno allows users to state any such relationships when they are known, but it makes no assumptions.

(ii) Other systems take a measure of belief (e.g., probability or certainty) in a piece of evidence, and calculate from it a measure of belief in a hypothesis or conclusion. In terms of an inference network (Figures 3.3 and 3.4), probabilities or certainty values are always propagated in one direction, namely, from the bottom (evidence) to the top (conclusions). Inferno allows users to enter values for any node on the network and to observe the effects on values at all other nodes.

(iii) Inferno informs the user of inconsistencies that might be present in the information presented to it and can make suggestions of ways to restore consistency.

Quinlan [18] gives a detailed account of how these aims are achieved and provides a comprehensive set of expressions for propagating probabilities throughout the nodes of an inference network.

3.6 Summary

A number of different schemes exists for assigning numerical values to assertions in order to represent levels of confidence in them, and for updating the confidence levels in the light of supporting or opposing evidence. The greatest difficulty lies in obtaining accurate values of likelihood, whether measured as a probability or by some other means. The certainty factors that are associated with rules in certainty theory, and the *affirms* and *denies* weightings in Bayesian updating, can be derived from probability estimates. However, a more pragmatic approach is frequently adopted, namely, to choose values that produce the right sort of results, even though the values cannot be theoretically justified. As the more sophisticated techniques (e.g., Dempster–Shafer theory of evidence and Inferno) also depend upon probability estimates that are often dubious, their use is rarely justified.

Bayesian updating is soundly based on probability theory, whereas many of the alternative techniques are *ad hoc*. In practice, Bayesian updating is also an *ad hoc* technique because:

- linear interpolation of the *affirms* and *denies* weighting is frequently used as a convenient means of compensating for uncertainty in the evidence;
- the likelihood ratios (or the probabilities from which they are derived) and prior probabilities are often based on estimates rather than statistical analysis;
- separate items of evidence that support a single assertion are assumed to be statistically independent, although this may not be the case in reality.

Neural networks (see Chapter 8) represent an alternative approach that avoids the difficulties in obtaining reliable probability estimates. Neural networks can be used to train a computer system using many examples, so that it can draw conclusions weighted according to the evidence supplied. Of course, given a large enough set of examples, it would also be possible to calculate accurately the prior probabilities and weightings needed in order to make Bayesian updating or one of its derivatives work effectively.

Fuzzy logic is also closely associated with neural networks, as will be discussed in Chapter 9. Fuzzy logic provides a precise way of handling vague terms such as low and high. As a result, a small set of rules can produce output values that change smoothly as the input values change.

References

1. Hart, P. E., Duda, R. O., and Einaudi, M. T., "PROSPECTOR; a computer-based consultation system for mineral exploration," *Math Geology*, vol. 10, pp. 589–610, 1978.
2. Duda, R., Gashnig, J., and Hart, P., "Model design in the PROSPECTOR consultant system for mineral exploration," in *Expert Systems in the Micro-electronic Age*, Michie, D. (Ed.), Edinburgh University Press, 1979.
3. Duda, R. O., Hart, P. E., and Nilsson, N. J., "Subjective Bayesian methods for rule-based inference systems," National Computer Conference, vol. 45, pp. 1075–1082, AFIPS, 1976.
4. Buchanan, B. G. and Duda, R. O., "Principles of rule-based expert systems," in *Advances in Computers*, vol. 22, Yovits, M. C. (Ed.), Academic Press, 1983.
5. Shortliffe, E. H. and Buchanan, B. G., "A model of inexact reasoning in medicine," *Mathematical Biosciences*, vol. 23, pp. 351–379, 1975.
6. Shortliffe, E. H., *Computer-Based Medical Consultations: MYCIN*, Elsevier, 1976.

7. Zadeh, L. A., "Fuzzy logic and approximate reasoning," *Synthese*, vol. 30, pp. 407–428, 1975.
8. Zadeh, L. A., "Commonsense knowledge representation based on fuzzy logic," *IEEE Computer*, vol. 16, issue 10, pp. 61–65, 1983.
9. Zadeh, L. A., "The role of fuzzy logic in the management of uncertainty in expert systems," *Fuzzy Sets and Systems*, vol. 11, pp. 199–227, 1983.
10. Zadeh, L. A., "Fuzzy sets," *Information and Control*, vol. 8, pp. 338–353, 1965.
11. Aoki, H. and Sasaki, K., "Group supervisory control system assisted by artificial intelligence," *Elevator World*, pp. 70–91, February 1990.
12. Lee, C. C., "Fuzzy logic in control systems: fuzzy logic controller – part I," *IEEE Transactions on Systems, Man and Cybernetics*, vol. 20, pp. 404–418, 1990.
13. Lee, C. C., "Fuzzy logic in control systems: fuzzy logic controller – part II," *IEEE Transactions on Systems, Man and Cybernetics*, vol. 20, pp. 419–435, 1990.
14. Johnson, J. H. and Picton, P. D., *Concepts in Artificial Intelligence*, Butterworth–Heinemann, 1995.
15. Rescher, N., *Plausible Reasoning*, Van Gorcum, 1976.
16. Barnett, J. A., "Computational methods for a mathematical theory of evidence," 7th International Joint Conference on Artificial Intelligence (IJCAI'81), Vancouver, pp. 868–875, 1981.
17. Schafer, G., *A Mathematical Theory of Evidence*, Princeton University Press, 1976.
18. Quinlan, J. R., "Inferno: a cautious approach to uncertain inference," *The Computer Journal*, vol. 26, pp. 255–269, 1983.

Further reading

- Bacchus, F., *Representing and Reasoning with Probabilistic Knowledge*, MIT Press, 1991.
- Buchanan, B. G. and Shortliffe, E. H. (Eds.), *Rule-Based Expert Systems: the MYCIN experiments of the Stanford Heuristic Programming Project*, Addison-Wesley, 1984.
- Hajek, P., Havranek, T., and Jirousek, R., *Uncertain Information Processing in Expert Systems*, CRC Press, 1992.
- Kandel, A., *Fuzzy Expert Systems*, CRC Press, 1991.

- Klir, G. J. and Wierman, M. J., *Uncertainty-based Information*, Physica–Verlag, 2000.
- Li, H. X. and Yen, V. C., *Fuzzy Sets and Fuzzy Decision-making*, CRC Press, 1995.
- Polson, N. G. and Tiao, G. C. (Eds.), *Bayesian Inference*, Edward Elgar, 1995.
- Ralescu, A. L. and Shanahan, G. J. (Eds.), *Fuzzy Logic in Artificial Intelligence*, Springer–Verlag, 1999.

Chapter four

Object-oriented systems

4.1 Objects and frames

System design usually involves breaking down complex problems into simpler constituents. Objects and frames are two closely related ways of achieving this while maintaining the overall integrity of the system. Frame-based programming is usually associated with the construction and organization of knowledge-based systems. Object-oriented programming (OOP) is used in a wide variety of software systems including, but by no means limited to, intelligent systems. OOP has been developed as "a better way to program," while frames were conceived as a versatile and expressive way of representing and organizing information.

Both techniques assist in the design of software and make the resultant software more maintainable, adaptable, and reusable. They provide a structure for breaking down a world representation into manageable components such as *house*, *owner*, and *dog*. In a frame-based system, each of these components could be represented by a frame, containing information about itself. Frames are passive in the sense that, like entries in a database, they don't perform any tasks themselves. Objects are similar, but as well as storing information about themselves they also have the ability to perform certain tasks. When they receive an instruction, they will perform an action as instructed, provided it is within their capability. They are therefore like obedient servants.

Objects and frames share many similarities. In this chapter, OOP is described in some detail, then a review of frames focuses on their differences from objects. Section 4.2 introduces an example of a problem that has been successfully tackled using OOP. This example will be used extensively during this chapter in order to illustrate the features and benefits of OOP.

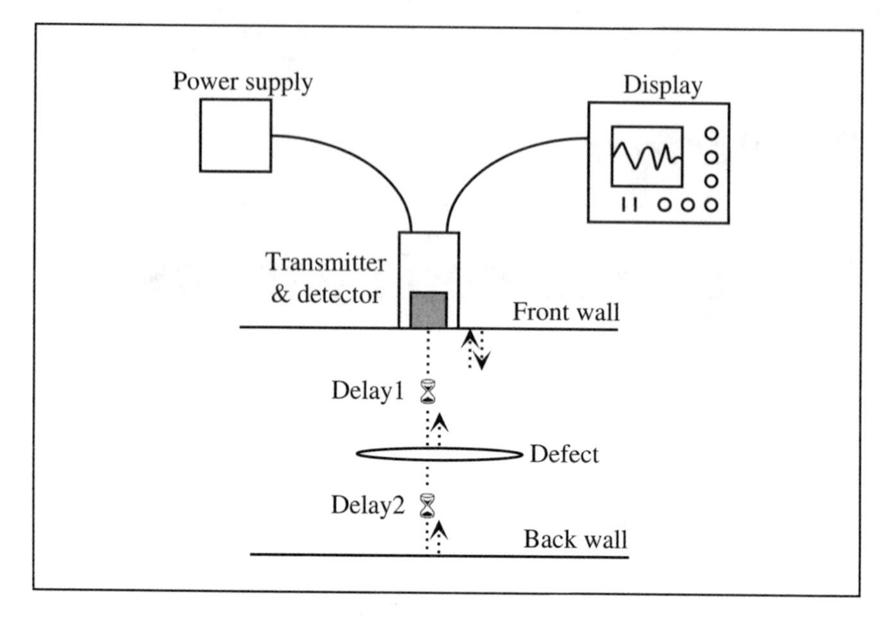

Figure 4.1 Detection of defects using ultrasonics; delays simulate the depth of the defect and the back wall

4.2 An illustrative example

We will illustrate the features of object-oriented programming by considering a small simulation of ultrasonic imaging. The physical arrangement to be simulated is shown in Figure 4.1. It comprises an ultrasonic probe, containing a detector and a transmitter, which sits on the surface of a component under test. The probe emits a single pulse of ultrasound, i.e., sound at a higher frequency than the human audible range. The pulse is strongly reflected by the front and back walls of the component and is weakly reflected by defects within the component. The intensity and time of arrival of reflected pulses reaching the detector are plotted on an oscilloscope. There are two possible types of ultrasonic pulses — longitudinal and shear — that travel differently through a solid. The atoms of the material travel to-and-fro, parallel to the direction of a longitudinal pulse, but they move perpendicular to the direction of a shear pulse (Figure 4.2). The two types of pulses travel at different speeds in a given material and behave differently at obstructions or edges. The emitted pulse is always longitudinal, but reflections may generate both a longitudinal and a shear pulse (Figure 4.3).

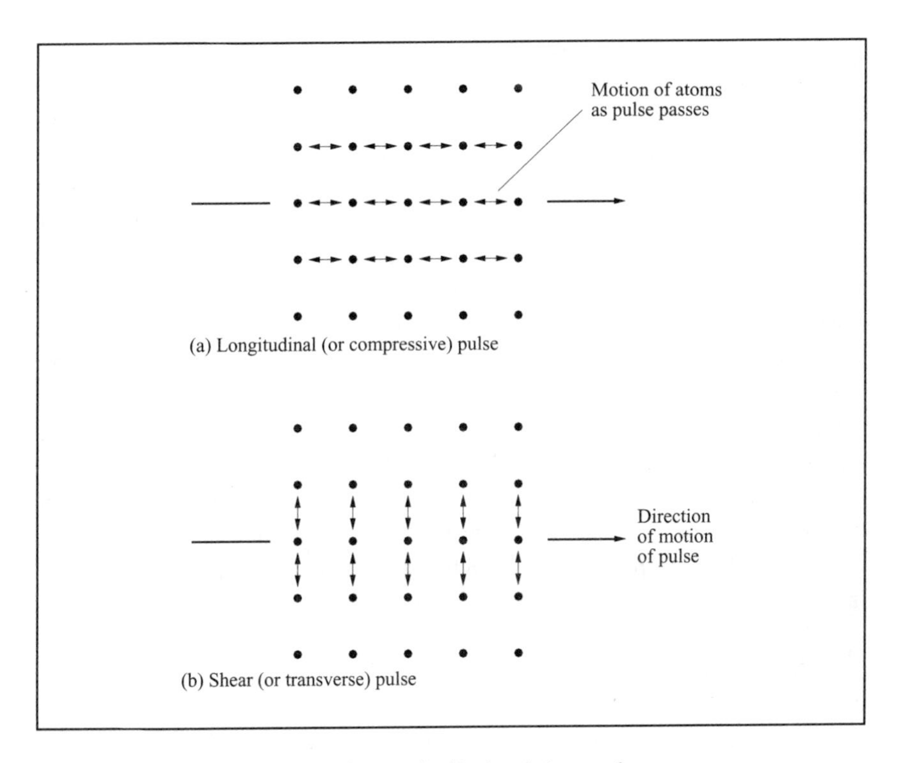

Figure 4.2 Longitudinal and shear pulses

This system, like so many others, lends itself to modeling with objects. Each of the components of the system can be treated as an independent object, containing data and code that define its own particular behavior. Thus, there will be objects representing each pulse, the transmitter, the detector, the component, the front wall, the back wall, the defects, and the oscilloscope. These objects stand in well-defined relationships with one another, e.g., a transmitter object can generate a pulse object.

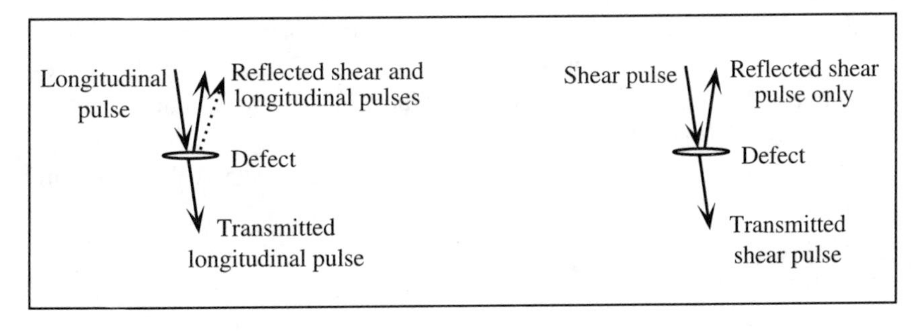

Figure 4.3 Reflection of longitudinal and shear pulses at a defect

4.3 Introducing OOP

As well as the practical advantages of maintainability, adaptability, and reusability, OOP is also a natural way of representing many real-world problems within the confines of a computer. Programs of interest to engineers and scientists normally perform calculations or make decisions about physical entities. Such programs must contain a model of those entities, and OOP is a convenient way to carry out this modeling. Every entity can be represented by a self-contained "object," where an object contains data and the code that can be performed on those data. Thus, a system for simulating ultrasonic imaging might include an ultrasonic pulse that is reflected between features in a steel component. The pulse, the features, and the component itself can all be represented by objects. Systems theory leads directly to such an approach, since a system can be defined as "an assembly of components, separated from their environment by a boundary, but related to each other and organized to achieve a clear purpose." Thus, OOP can be thought of as a computer implementation of systems theory, where each component is itself a subsystem that can be represented by an object.

Modeling with objects is not restricted to physical things, but extends to include entities within the programming environment such as windows, icons, menus, clocks, and just about anything else that you can think of. In fact, the development of graphical user interfaces has been one of the main driving forces for OOP. This demonstrates that OOP is a powerful technique in its own right and is by no means restricted to knowledge-based systems. However, OOP does have a critical role in many knowledge-based systems, as it offers a way of representing the things that are being reasoned about, their properties and behaviors, and the relationships among them.

There are many languages and programming environments that offer object-orientation, some of which are supplied as extensions to existing languages. Four widely used OOP languages are C++, Smalltalk, Java, and CLOS (Common Lisp Object System). Smalltalk is a programming environment rather than just a language. It includes a number of menus, windows, browsers (see Section 4.5.4) and built-in classes (see Section 4.4.1). C++ and CLOS are extended versions of the C and Lisp languages respectively, but many modern implementations also include similar support tools to those of Smalltalk.

According to Pascoe [1], an OOP language offers at least the following facilities:

- data abstraction
- inheritance

- encapsulation (or information hiding)
- dynamic (late) binding

The following sections describe how these facilities relate to the ultrasonics model and the advantages that they confer.

4.4 Data abstraction

4.4.1 Classes

The ultrasonic example involves several different types of object. Shear pulses are objects of the same type as each other, but of a different type from the detector. Data abstraction allows us to define these types and the functions that go with them. These object types are called *classes*, and they form templates for the objects themselves. Objects which represent the same idea or concept are grouped together as a class. The words *type* and *class* are generally equivalent and are used interchangeably here.

Most computer languages include some simple built-in data types, such as integer or real. An object-oriented language allows us to define additional data types (i.e., classes) that are treated identically, or nearly identically, to the built-in types. The user-defined classes are sometimes called *abstract data types*. These may be classes which have a specific role within a particular domain, such as `Shear_pulse`, `Circle`, or `Polymer`, or they may describe a specialized version of a standard data type, such as `Big_integer`.

The definition of a class contains the class name, its attributes (Section 4.4.3), its operations (Section 4.4.4) and its relationships to other classes (Sections 4.5 and 4.7). C++ and some other OOP languages require that the attribute types and the visibility from other classes also be specified.

4.4.2 Instances

It was stated in Section 4.4.1 that classes form templates for "the objects themselves," meaning object *instances*. Once a class has been defined, instances of the class can be created that have the properties defined for the class. For example, `pulse#259` might be an ultrasonic pulse whose location at a given moment is (x = 112mm, y = 294mm, z = 3.5mm). We would represent `pulse#259` as an instance of the class `Longitudinal_pulse`. As a more tangible example, my car is an instance of the class of car. The class specifies the characteristics and behavior of its instances. A class can, therefore, be thought of as the blueprint from which instances are built. The terms *object* and *instance* are often used interchangeably, although it is helpful to use the latter to stress the distinction from a class.

This is just an extension of the concepts of data types and variables that exist in other programming languages. For instance, most languages include a built-in definition of the type integer. However, we can only draw upon the properties of an integer by creating an instance, i.e., an integer variable. In C, this would look like this:

```
int x;       /* create an instance, x, of type int */
x = 3;       /* manipulate x */
```

Similarly, once we have defined a class in an object-oriented language, its properties can only be used by creating one or more instances. Consider the class `Longitudinal_pulse`. One instance of this class is generated by the transmitter object (which is itself an instance of the class `Transmitter`). This instance represents one specific pulse, which has position, amplitude, phase, speed, and direction. When this pulse interacts with a defect (another instance), a new instance of `Longitudinal_pulse` must be created, since there will be both a transmitted and a reflected pulse (see Figure 4.3). The new pulse will have the same attributes as the old one because they are both derived from the same class, but some of these attributes will have different values associated with them (e.g., the amplitude and direction will be different).

4.4.3 Attributes (or data members)

A class definition includes its attributes and operations. The attributes are particular quantities that describe instances of that class. They are sometimes described as *slots* into which values are inserted. Thus the class `Shear_pulse` might have attributes such as amplitude, position, speed, and direction. Only the names and, depending on the language, the types of the attributes need to be declared within the class, although values can optionally be supplied too. The class acts as a template, so that when a new shear pulse is created, it will contain the names of these attributes. The attribute values can be supplied or calculated for each instance, or a value provided within the class can be used as a default. The attributes can be of any type, including abstract data types, i.e., classes. Some languages, such as C++, require that the class definition defines the attribute types in advance. Amplitude and speed might be of type float, whereas position and direction would be of type vector.

If a value for an attribute is supplied in the class definition, any new instances would carry those values by default unless specifically overwritten. Most OOP languages distinguish between *class attributes* (or class variables) and *instance attributes* (or instance variables). The value of a class attribute remains the same for all instances of that class. In contrast, instances contain their own copies of each instance attribute. If a default has been specified, each instance of given class has the same initial value for an instance attribute. However, the values of the instance attribute may subsequently vary from one instance to another. Thus, in the above example, `speed` might be a class attribute for `Shear_pulse`, since the speed will be the same for all shear pulses in a given material. On the other hand, `amplitude` and `position` are properties of each individual pulse and are represented as instance attributes.

In C++, attributes are called *data members*. All data members are assumed to be instance attributes unless they are declared static. Static data members are stored at the same memory location for every instance of a class, so there is only one copy, regardless of how many instances there might be. Static data members are, therefore, equivalent to class attributes.

4.4.4 Operations (or methods or member functions)

Each object has a set of operations or functions that it can perform. For example, a `Shear_pulse` object may contain the function `move`. This function might take as its parameters the amount and direction of movement, and return a new value for its position attribute. In some OOP languages, operations belonging to objects are called *methods*. In C++, they are called *member functions*. Operations are defined for a class, and can then be used by all instances of that class. It may also be possible for instances of other classes to access these operations (see Sections 4.5 and 4.6).

4.4.5 Creation and deletion of instances

Creating a new instance requires the computer to set aside some of its memory for storing the instance. Given a class definition, `Myclass`, the creation of new instances is similar in Smalltalk and in C++. In Smalltalk we might write:

```
Myinstance := Myclass new.       "Myinstance is a global variable"
```

The words in quotes are a comment and are ignored by the compiler. The equivalent in C++ would be:

```
Myclass* myinstance;
        // declare a pointer to objects of type Myclass
myinstance = new Myclass;
        // pointer now points to a new instance
```

Here the // symbol indicates a comment. The C++ version can be abbreviated to:

```
Myclass* myinstance = new Myclass;
       // new pointer points to a new instance.
```

Large numbers of instances may be created while running a typical object-oriented system. An important consideration, therefore, is the release of memory that is no longer required. In the example of the ultrasonic simulation, new pulse instances are generated through reflections, but they must be removed when they reach the detector. The memory that is occupied by these unwanted instances must be released again if we are to avoid building a system with an insatiable appetite for computer memory. In the Smalltalk example above, `Myinstance` is a global variable, indicated by the capitalization of the first letter of the name. Global variables are accessible from any part of the program and are retained in the programming environment until explicitly deleted. It is more common to attach instances to temporary variables, shown here in Smalltalk:

```
|myinstance|      "myinstance is declared as a temporary variable"
myinstance := Myclass new.
```

Temporary variables exist only within the method in which they are declared, and their lifetime is that of the method activation. When execution of the method has finished, these variables (and deleted global variables) leave behind an area of "unowned" memory. The Smalltalk system automatically reclaims this memory — a process known as *garbage collection*. Depending on the particular implementation, garbage collection may cause the program to momentarily "freeze" while the system carries out its memory management. Garbage collection is a feature of Smalltalk, CLOS, and other OOP languages. Some implementations allow the programmer to influence the timing of garbage collection, while in other implementations garbage collection is an unnoticed background process.

In C++, the responsibility for memory management rests with the programmer. Objects which are created as described above must be explicitly destroyed when they are no longer needed:

```
delete myinstance;
```

This operation releases the corresponding memory. C++ also allows an alternative method of object creation and deletion which relies on the *scope* of objects. A new instance, `myinstance`, may be created within a block of code in the following way:

```
Myclass myinstance;
```

The instance exists only within that particular block of code, which is its scope. When the flow of execution enters the block of code, the object is automatically created. Similarly, it is deleted as soon as the flow of execution leaves the block.

C++ allows the programmer to define special functions to be performed when a new instance is created or deleted. These are known respectively as the *constructor* and the *destructor*. The constructor is a member function whose name is identical to the name of the class, and is typically used to set the initial values of some attributes. The destructor is defined in a similar way to the constructor, and its name is that of the class preceded by a tilde (~). It is mostly used to release memory when an instance is deleted. As an example, consider the definition for `Sonic_pulse` in C++:

```
// class definition:
class Sonic_pulse
{
  protected:
    float amplitude;
  public:
    Sonic_pulse(float initial_amplitude);          // constructor
    ~Sonic_pulse();                                 // destructor
};

// constructor definition:
Sonic_pulse::Sonic_pulse(float initial_amplitude)
{
  amplitude=initial_amplitude;              // set up initial value
}

// destructor definition:
Sonic_pulse::~Sonic_pulse()
{
      // perform any tidying up that may be necessary before
      // deleting the object
}
```

The class has a single attribute, `amplitude`, and this is set to an initial value by the constructor. The constructor is automatically called immediately after a new instance of `Sonic_pulse` is created. To create a new sonic pulse whose initial amplitude is 131.4 units, we would write in C++ either:

```
Sonic_pulse* myinstance = new Sonic_pulse(131.4);    //technique 1
```

or:

```
Sonic_pulse myinstance(131.4);                        //technique 2
```

If a destructor has been defined for a class, it is automatically called whenever an instance is deleted. If the instance was created by technique 1, it must be explicitly deleted by the programmer. If it was created by technique 2, it is automatically deleted when it becomes out of scope from the program's thread of control.

Note that data members in C++ are normally given the prefix `m_`, so that `amplitude`, above, would become `m_amplitude`. However, this convention has not been adopted here as it would confuse the comparisons between segments of C++ code and Smalltalk code.

4.5 Inheritance

4.5.1 Single inheritance

Returning again to our example of the ultrasonic simulation, note that there are two classes of sonic pulses, namely `Longitudinal_pulse` and `Shear_pulse`. While there are some differences between the two classes, there are also many similarities. It would be most unwieldy if all of this common information had to be specified twice. This problem can be avoided by the use of *inheritance*, sometimes called *derivation*. A class `Sonic_pulse` is defined that encompasses both types of pulses. All of the attributes and methods that are common can be defined here. The classes `Longitudinal_pulse` and `Shear_pulse` are then designated as subclasses or specializations of `Sonic_pulse`. Conversely, `Sonic_pulse` is said to be the superclass or generalization of the other two classes. The sub/super class relationship can be thought of as *is-a-kind-of*, i.e., a shear pulse is-a-kind-of sonic pulse. The following expressions, commonly used to describe the is-a-kind-of relationship, are equivalent:

- a *subclass* is-a-kind-of *superclass*;
- an *offspring* is-a-kind-of *parent*;
- a *derived* class is-a-kind-of *base* class;
- a *specialized* class is-a-kind-of *generalized* class;
- a specialized class *inherits from* a generalized class.

In defining a derived class, it is necessary to specify only the name of the base class and the *differences* from the base class. Attributes and their initial values (where supplied) and methods are inherited by the derived class. (In C++ there are a few exceptions that are not inherited, notably the constructor, destructor, and an overloaded = operator, described in Section 4.11.3.) Any attributes or methods that are redefined in the derived class are said to be

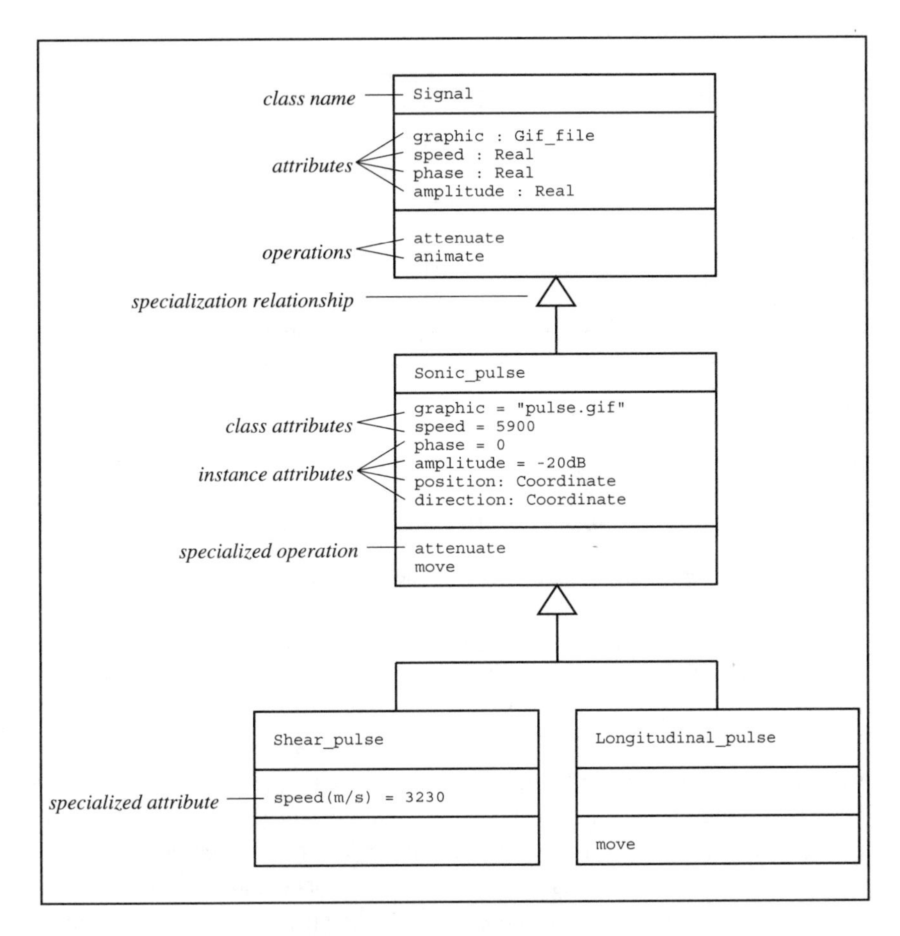

Figure 4.4 An example of inheritance

specialized, just as the derived class is said to be a specialization of the base class. Another form of specialization is the introduction of extra attributes and methods in the derived class. C++ gives the user some control over which methods and attributes of the base class are accessible from the derived class and which are not (Section 4.6). Many other OOP languages assume that *all* attributes and methods of the base class are accessible from the derived class, apart from those that are explicitly specialized.

Figure 4.4 shows the class definitions for `Sonic_pulse`, its superclasss `Signal`, and its subclasses `Shear_pulse` and `Longitudinal_pulse`. The attributes `graphic`, `speed`, `phase`, and `amplitude` are all declared at the highest level class and inherited by the other classes. The class variable `graphic`, which determines the screen representation of an instance, is assigned a value at the `Sonic_pulse` level. This value, as well as the

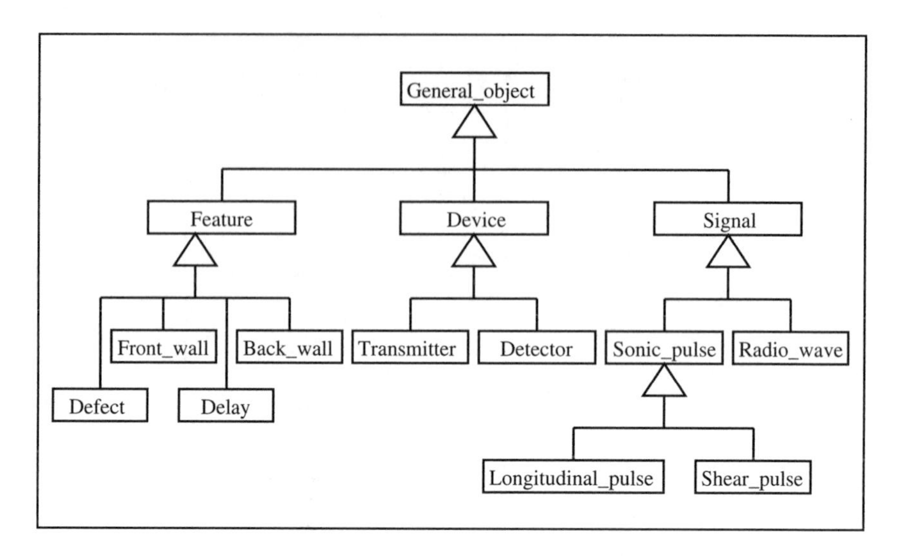

Figure 4.5 Single inheritance defines a class hierarchy

declaration of the attribute, is inherited by the subclasses. Values for the instance variables `amplitude` and `phase` are inherited in the same way. Inherited values may be overridden either when an instance is created or subsequently. The class variable `speed` for `Shear_pulse` is assigned a different value from the one that would otherwise be inherited.

Operations as well as attributes are inherited. Therefore, instances of `Shear_pulse` and `Longitudinal_pulse` have access to the operations `animate` and `attenuate`. The former is inherited across two generations, while the latter is specialized at the `Sonic_pulse` class level. The operation `move` is defined at the `Sonic_pulse` class level and inherited by `Shear_pulse`, but redefined for `Longitudinal_pulse`.

Figure 4.5 shows the inheritance between other objects in the ultrasonic simulation. Note that the is-a-kind-of relationship is *transitive*, i.e.:

if x is-a-kind-of y and y is-a-kind-of z, then x is-a-kind-of z.

Therefore, the class `Defect`, for example, inherits information that is defined at the `Feature` level and at the `General_object` level.

4.5.2 Multiple and repeated inheritance

In the example shown in Figure 4.5, each offspring has only one parent. The specialization relationships therefore form a hierarchy. Some OOP languages insist upon hierarchical inheritance. However, others allow an offspring to inherit from more than one parent. This is known as *multiple inheritance.*

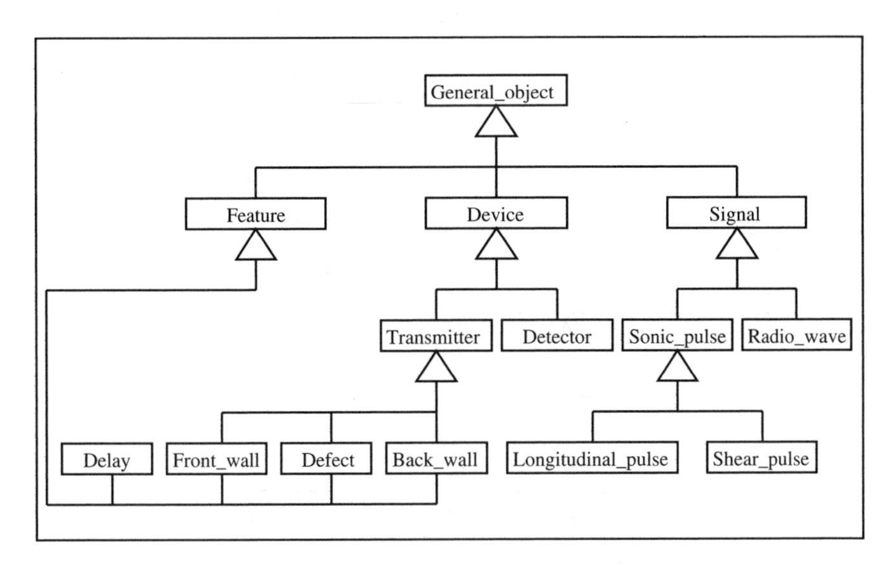

Figure 4.6 Multiple inheritance defines a network of classes

Where multiple inheritance occurs, the specialization relationships between classes define a network rather than a hierarchy.

Figure 4.6 shows how multiple inheritance might be applied to the ultrasonic simulation. The parts of the component that interact with sonic pulses all inherit from the class `Feature`. Three of the classes derived from `Feature` are required to simulate partial reflection of pulses. This is done by generating one or more reflected pulses, while the existing pulse is propagated in the forward direction with diminished amplitude. Code for generating pulses is contained within the class definition for `Transmitter`. Multiple inheritance allows those classes that need to generate pulses (`Front_wall`, `Back_wall`, and `Defect`) to inherit this capability from `Transmitter`, while inheriting other functions and attributes from `Feature`.

While multiple inheritance is useful, it can cause ambiguities. This is illustrated in Figure 4.7, where `Defect` inherits the graphic `feature.gif` from `Feature` and at the same time inherits the graphic `transmitter.gif` from `Transmitter`. This raises two questions. Do the two attributes with the same name refer to the same attribute? If so, the class `Defect` can have only one value corresponding to the attribute `graphic`, so which value should be selected? Similarly, `Defect` inherits conflicting definitions for the operation `send_pulse`.

The most reliable way to resolve such conflicts is to have them detected by the language compiler, so that the programmer can then state explicitly the intended meaning. Some OOP environments allow the user to set a default strategy for resolving conflicts. An example might be to give preference to the

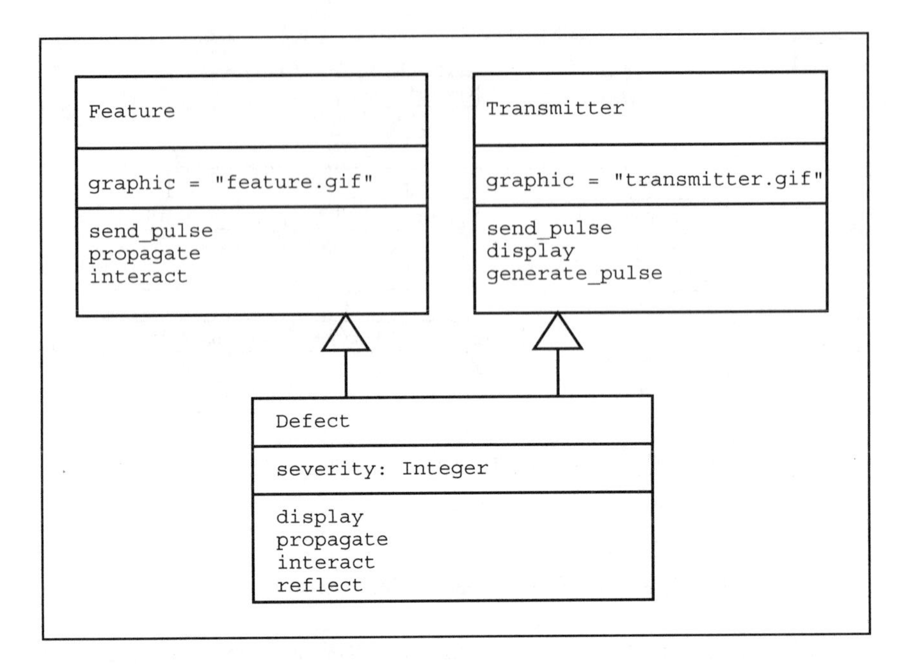

Figure 4.7 Conflicts arising from multiple inheritance: each parent has a different value for `graphic` and a different definition for `send_pulse`

parent class that is either closest to, or furthest from, the root of the inheritance tree. This would not help with the example in Figure 4.7 as the two parents are equidistant from the root. Alternatively the class names may be deemed significant in some way, or preference may be given to the most recently defined specialization.

A further problem is that one class may find itself indirectly inheriting from another via more than one route, as shown in Figure 4.8. This is known as *repeated inheritance*. Although the meaning may be clear to the programmer, an OOP language must have some strategy for recognizing and dealing with repeated inheritance, if it allows it at all. C++ offers the programmer a choice of two strategies. Class D can have two copies of Class A, one for each inheritance route. Alternatively it can have just a single copy, if both Class B and Class C are declared to have Class A as a *virtual* base class.

Multiple inheritance gives rise to the idea of *mixins*, which are classes designed solely to help organize the inheritance structure. Instances of mixins cannot be created. Consider, for example, a class hierarchy for engineering materials. The materials polyethylene, Bakelite, gold, steel, and silicon nitride can be classified as polymers, metals, or ceramics. Single inheritance would allow us to construct these hierarchical relationships. Under multiple inheritance we can categorize the materials in a variety of ways at the same

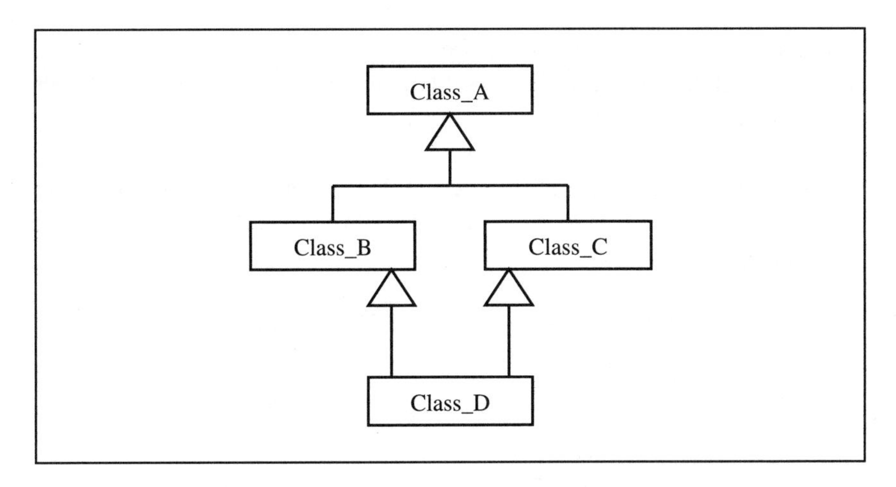

Figure 4.8 Repeated inheritance

time. Figure 4.9 shows the use of the mixins `Cheap` and `Brittle` for this purpose.

4.5.3 Specialization of methods

We have already seen that specialization can involve the introduction of new methods or attributes, overriding default assignments to attributes, or

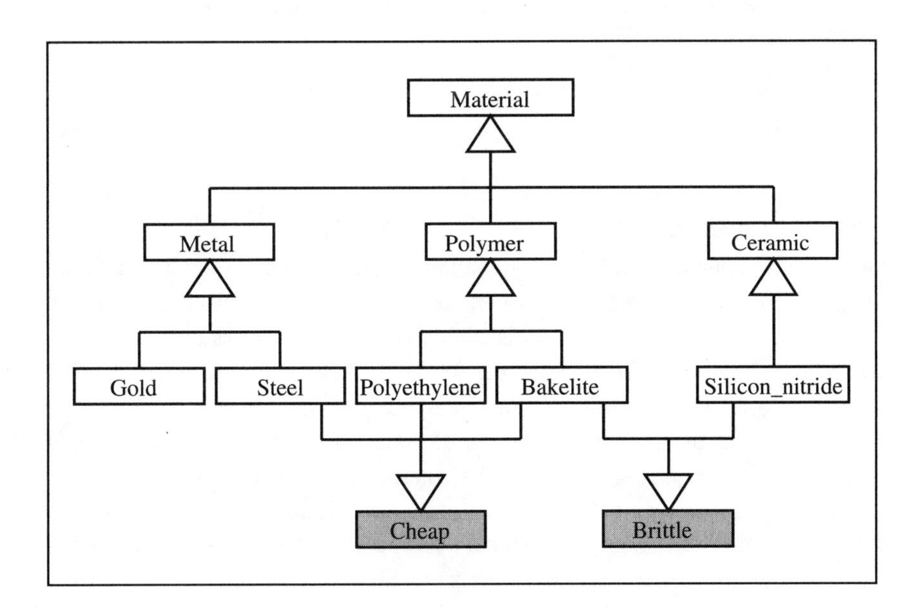

Figure 4.9 An example of the use of mixins

redefinition of inherited operations. If an operation needs to be redefined, i.e., specialized, it is not always necessary to rewrite it from scratch. In our example, a definition of the method `propagate` is inherited by the class `Defect` from the class `Feature`. The specialized version of `propagate` is the same, except that the sonic pulse must be attenuated. This can be achieved by calling the inherited definition from within the specialized one, and then adding the attenuation instruction, shown here in C++:

```
void Defect::propagate(Sonic_pulse* inst1)
{
  Feature::propagate(inst1);
    // propagate a pulse, inst1, using inherited version of
    // 'propagate'
  inst1->attenuate(0.5);                  // Attenuate the pulse.
    // The member function 'attenuate' must be defined for the
    // class 'Sonic_pulse'.
}
```

We might wish to call the specialized method `propagate` from within the definition of `interact`, a function which handles the overall interaction of a pulse with a defect:

```
void Defect::interact(Sonic_pulse* inst1)
{
  propagate(inst1);
    // propagate a pulse, inst1, using the locally defined
    // member function
  reflect(inst1);
    // generate a new pulse and send it in the opposite
    // direction
}
```

4.5.4 Browsers

Many OOP systems provide not only a language, but an environment in which to program. The environment may include tools that make OOP easier, and one of the most important of these tools is a class-browser. A class-browser shows inheritance relationships, often in the form of a tree such as those in Figures 4.5 and 4.6. If the programmer wants to alter a class or to specialize it to form a new one, he or she simply chooses the class from the browser and then selects the type of change from a menu. Incidentally, browsers themselves are invariably built using OOP. Thus, the class-browser may be an instance of a class called `Class_browser`, which may be a specialization of `Browser`, itself a specialization of `Window`. The class `Class_browser` could be further specialized to provide different display or editing facilities.

4.6 Encapsulation

Encapsulation, or information-hiding, is a term used to express the notion that the instance attributes and the methods that define an object belong to that object and to no other. The methods and attributes are therefore private and are said to be encapsulated within the object. The interface to each object reveals as little as possible of the inner workings of the object. The object has control over its own data, and those data cannot be directly altered by other objects. Class attributes are an exception, as these are not encapsulated within any one instance but are shared between all instances of the same class.

In general, Smalltalk adheres to the principle of encapsulation, whereas C++ adopts a more liberal approach. In Smalltalk, object A can only influence object B by sending B a message telling it to call one of its methods (see Section 4.9). Apart from this mechanism, secrecy is maintained between objects. The method on B can access and change its own data, but it cannot access the data of any other objects.

C++ offers flexibility in its enforcement of encapsulation through access controls. All data members and member functions (collectively known as *members*) are allocated one of four access levels:

- private: access from member functions of this class only (the default);
- protected: access from member functions of this class and of derived classes;
- public: access from any part of the program;
- friend: access from member functions of nominated classes.

Access controls are illustrated in Figure 4.10. C++ allows two types of derivation (i.e., inheritance), public and private. In private derivation, protected and public members in the base class become private members of the derived class. In public derivation, the access level of members in the derived class is unchanged from the base class. In neither case does the derived class have access to private members of the base class. The two types of derivation are shown in Figure 4.11.

4.7 Unified Modeling Language (UML)

The Unified Modeling Language (UML) [2] provides a means of specifying the relationships between classes and instances, and representing them diagrammatically. We have already come across two such relationships:

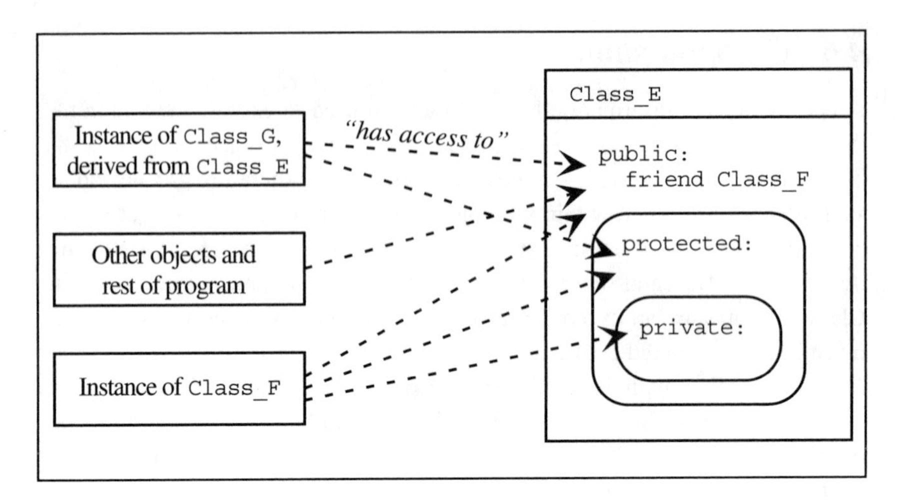

Figure 4.10 Access control in C++

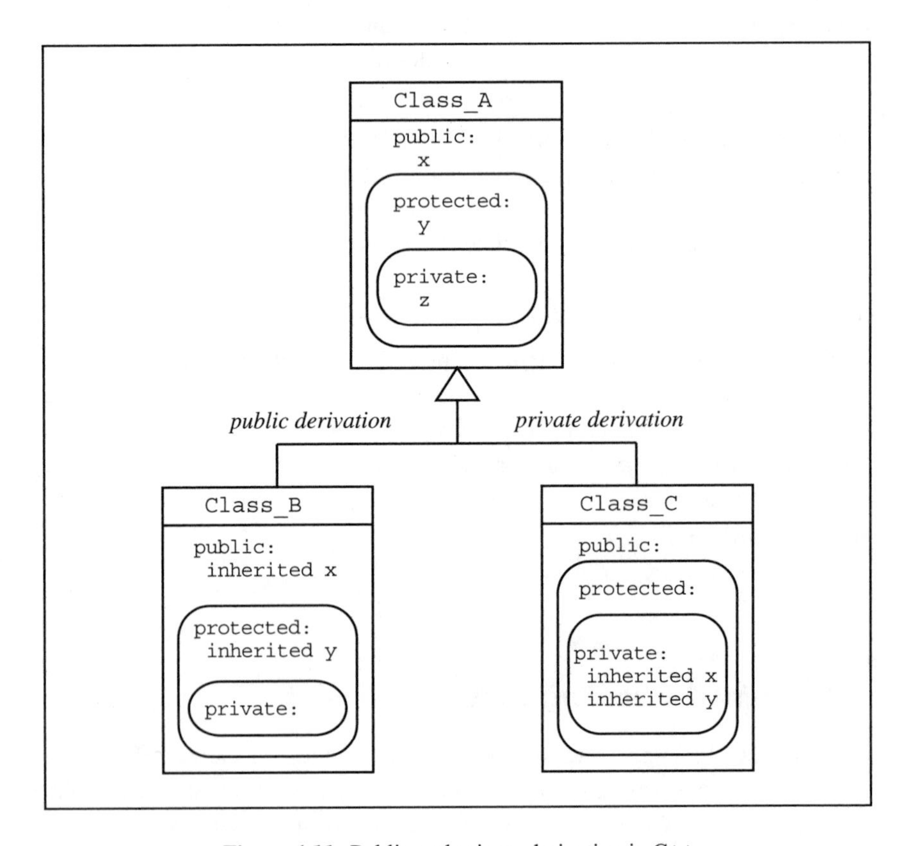

Figure 4.11 Public and private derivation in C++

- specialization/generalization;
- instantiation.

Specialization describes the is-a-kind-of relationship between a subclass and superclass, and involves the inheritance of common information. Instantiation describes the relation between a class and an instance of that class. An instance of a class and a subclass of a class are both said to be *clients* of that class, since they both derive information from it, but in different ways.

We will now consider three more types of relationships:

- aggregation;
- composition;
- association.

An *aggregation* relationship exists when an object can be viewed as comprising several subobjects. The world of software itself provides a good example, as a software library can be seen as an aggregation of many component modules. The degree of ownership of the modules by the software library is rather weak, in the sense that the same module could also belong to another software library. The *composition* relationship is a special case in which there is a strong degree of ownership of the component parts. For instance, a car comprises an engine, chassis, doors, seats, wheels, etc. This can be recognized as a composition relation since duplication or deletion of a car would require duplication or deletion of these component objects. Returning to our ultrasonic example, we can regard the imaging equipment as a composition of a transmitter, detector, display, and image processing software. The image processing software is an aggregation of various modules. Figure 4.12 shows that assembly and composition relationships allow problems to be viewed at different levels of abstraction.

Associations are loose relationships between objects. For instance, they might be used to represent the spatial layout of objects, possibly by naming the related instance as an instance attribute. Another form of association arises when one object makes use of another by sending messages to it (or, in the case of C++, calling its member functions or accessing its data members) — see Section 4.9. For example, pulses may send messages to the display object so that they can be redisplayed. The senders of messages are termed *actors*, and the recipients are *servers*. Objects that both send and receive messages are sometimes termed *agents*, although this is a confusing use of the word, given that agents have a different meaning described in Chapter 5.

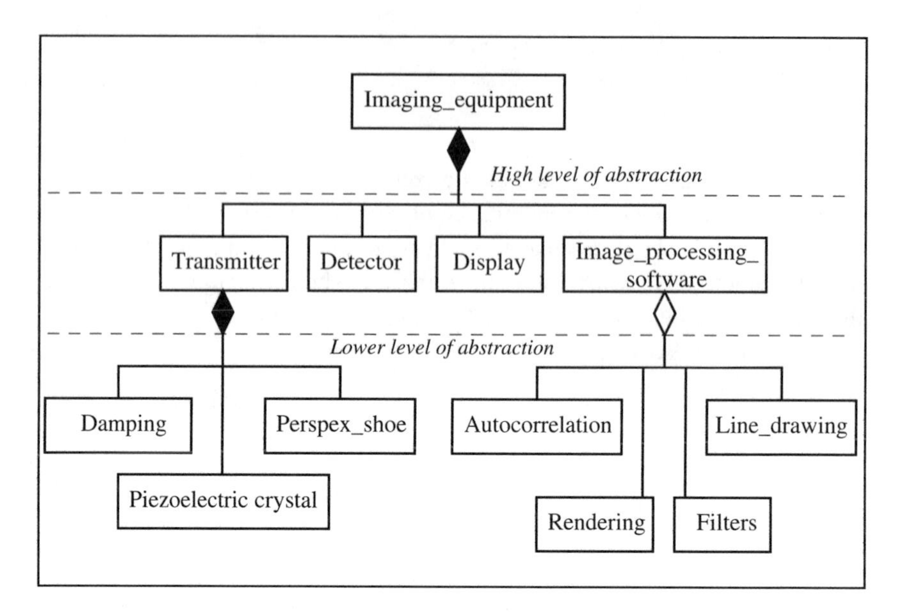

Figure 4.12 The aggregation and composition relationships allow problems to be viewed on different levels of abstraction. Filled diamonds indicate composition; unfilled diamonds indicate aggregation

The Unified Modeling Language (UML) [2] includes a diagrammatic representation for the relationships between classes, summarized in Figure 4.13. Wherever appropriate, this notation has been used throughout this chapter.

4.8 Dynamic (or late) binding

Three necessary features of an OOP language, as defined in Section 4.3, are data abstraction, inheritance, and encapsulation. The fourth and final necessary feature is *dynamic binding* (or *late binding*). Although the parent classes of objects may be known at compilation time, the actual (derived) classes may not be. The actual class is not bound to an object name at compilation time, but instead the binding is postponed until run-time. This is known as dynamic binding and its significance is best shown by example.

Suppose that we have a method, `propagate`, defined in C++ for the class, `Feature`:

```
void Feature::propagate(Sonic_pulse* p)
{
  float x, y;
  x = getx();
```

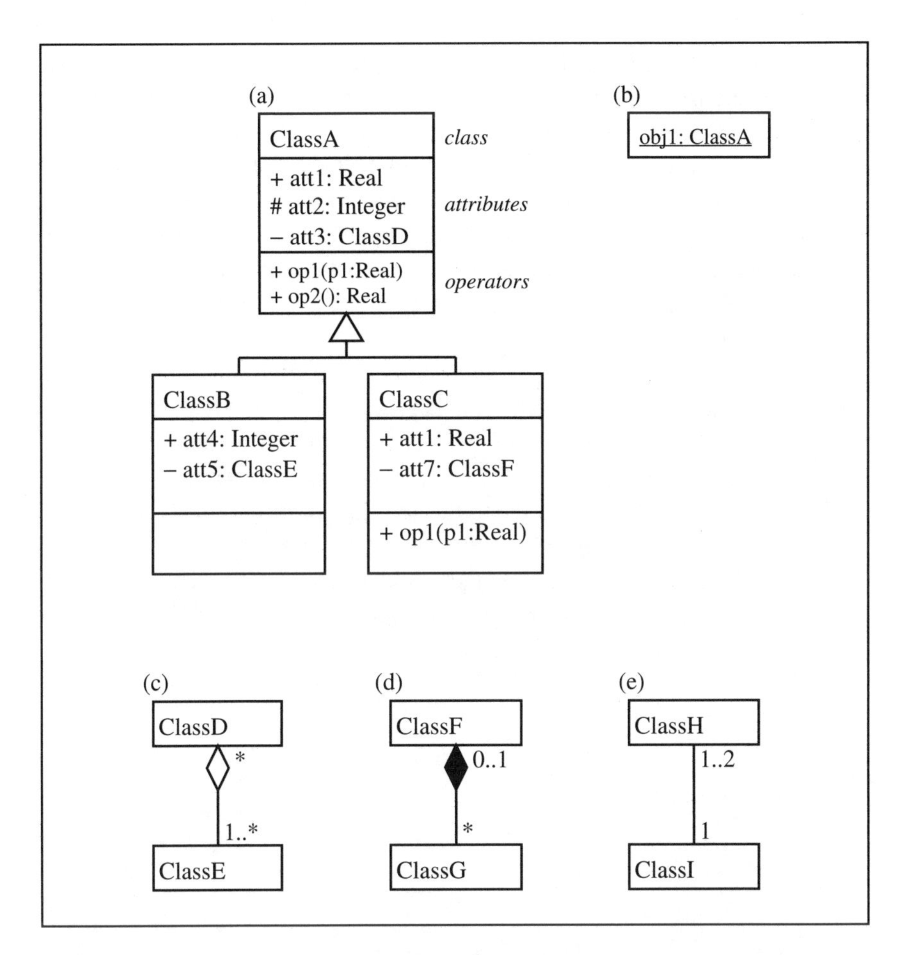

Figure 4.13 Class relationships in Unified Modeling Language (UML).
(a) ClassA is a generalization of ClassB and ClassC; (b) obj1 is an instance of ClassA;
(c) aggregation; (d) composition; (e) association.
Numbers represent permissible number of objects; * = any number
Access controls: + indicates public, # indicates protected, and – indicates private

```
  y = gety();              // calculate new position for pulse, p
  p->move(x,y);                // move pulse p to its new position
}
```

Suppose also that `shear1` and `long1` are instances of `Shear_pulse` and `Longitudinal_pulse`, respectively. When the program is run, transmission of `shear1` and `long1` by a feature is achieved by calls to `propagate`, with pointers to each pulse passed as parameters. The parameter types are correct, since `shear1` is an instance of `Shear_pulse`, which is derived from `Sonic_pulse`. Similarly `long1` is an instance of `Longitudinal_pulse`, which is also derived from `Sonic_pulse`. The method `propagate` calls the method `move`, but this may be specialized differently for `shear1` and for `long1`. Nevertheless, the correct definition will be chosen in each case. This is an example of late binding, since the actual method `move` that will be used is determined each time that `propagate` is called, rather than when the program is compiled.

The combined effect of inheritance and dynamic binding is that the same function call (`move` in the above example) can have more than one meaning, and the actual meaning is not interpreted until run-time. This effect is known as *polymorphism.**

To see why polymorphism is so important, we should consider how we would tackle the above problem in a language where binding is static (or early). In such languages, the exact meaning of each function call is determined at compilation time, and is known as *monomorphism*. In the method `propagate`, we would have to test the class of its argument and invoke a behavior accordingly. Depending on the language, we might need to include a class variable on `Sonic_pulse` to tag each instance with the name of its class (which may be `Sonic_pulse` or a class derived from it). We might then use a `case` statement, shown here in C++:

```
void Feature::propagate(Sonic_pulse* p)
{
    float x, y;
    char class_label;
    x = getx(); y = gety(); // calculate new position for pulse p
    class_label = p->tag; // identify the class of p from its tag
    switch (class_label) {
    case 's':
        ....;                            // code to move a Shear_pulse
        break;
    case 'l':
        ....;                     // code to move a Longitudinal_pulse
        break;
    default:
        ....;                                 // print an error message
    }
}
```

---

* C++ distinguishes between those member functions (i.e., methods) that can be redefined polymorphically in a derived class and those that cannot. Functions that can be redefined polymorphically are called *virtual functions*. In this example, `move` must be a virtual function if it is to be polymorphic. A pure virtual function is one that is declared in the base class, but no definition is supplied there. (The declaration merely states the existence of a function; the definition is the chunk of code that makes up the function.) A definition must therefore appear in the derived classes. Any class containing one or more pure virtual functions is termed an *abstract base class*. Instances of an abstract base class are not allowed, since an instance (if it were allowed) would "know" that it had access to a virtual function but would not have a definition for it.

Two drawbacks are immediately apparent. First, the code that uses monomorphism is much longer and is likely to include considerable duplication, as the code for moving a `Shear_pulse` will be similar to the code for moving a `Longitudinal_pulse`. In contrast, polymorphism avoids duplication and allows the commonality between classes to be made explicit. Second, the code is more difficult to maintain than the polymorphic code, because subsequent addition of a new class of pulse would require changes to the method `propagate` in the class `feature`. This runs against the philosophy of encapsulation, because the addition of a class should not require changes to existing classes.

The effect of polymorphism is that the language will always select the sensible meaning for a function call. Thus, once its importance has been understood, polymorphism need not vex the programmer.

4.9 Message passing and function calls

In C++, member functions (equivalent to methods) are accessed in a similar way to any other function, except that we need to distinguish the object to which the member function belongs. In the following C++ example, defects `d1` and `d2` are created using the two techniques introduced in Section 4.4.5, and the member function `propagate` is then accessed. A call to a conventional function is also included for comparison:

```
result = somefunc(parameter);
    // call a conventional function, defined elsewhere.

Defect* d1 = new Defect();    // make a new instance d1 of Defect.
Sonic_pulse* p1 = new Sonic_pulse();
Shear_pulse* p2 = new Shear_pulse();
                        // make two new instances of pulses, p1 and p2.
d1->propagate(p1);
    // call member function, 'propagate', with parameter p1.

Defect d2; // make a new instance d2 of Defect.
d2.propagate(p2);
    // call member function, 'propagate', with parameter p2.
```

In Smalltalk and other OOP languages, methods are invoked by passing *messages*. This terminology emphasizes the concept of encapsulation. Each object is independent and is in charge of its own methods. Nonetheless, objects can interact. One object can stimulate another to fire up a method by sending a message to it. An example of message passing in Smalltalk is:

```
d1 propagate: p1.
```

This is interpreted as:

send to the instance `d1` *the message* `propagate:`, *with parameter* `p1`.

The Smalltalk syntax becomes a little more confusing when there are many parameters:

```
p1 move: x and: y.
```

The arguments are interspersed with the method name, making the message read rather like English. This example means:

send to the instance `p1` *the message* `move:and:` *with parameters* `x` *and* `y`.

Upon receiving a message, an object calls up the corresponding method using the supplied parameters.

4.9.1 Pseudovariables

Smalltalk and some other OOP languages include reserved words which provide a shorthand form for "this object" and "the parent class of this object," thereby avoiding the need to name classes explicitly. In Smalltalk, the reserved words are `self` and `super`, respectively. They are known as *pseudovariables*; they differ from normal variables because they cannot be assigned a value directly by the programmer. C++ includes the pseudovariable `this`, which is equivalent to Smalltalk's `self`. In order to illustrate the use of `super`, consider the example of specializing the method `propagate`, which was shown in Section 4.5.3 using C++. Here is a Smalltalk equivalent, defined for the class `Defect`:

```
propagate: inst1
    super propagate: inst1.
"Propagate a pulse, inst1, using inherited version of propagate"

    inst1 attenuate: 0.5.     "Attenuate the pulse"
"The method 'attenuate' must be defined for the class of inst1"
```

The version of `propagate` defined for the parent class `Feature` is first called. Then the `inst1` is sent the message `attenuate:`, which instructs it to apply the method `attenuate:`, thereby reducing its amplitude. It is assumed that `inst1` is a sonic pulse.

Similarly, the method `interact` that is described for the class `Defect` describes the overall interaction between a pulse and the defect. This method includes calls to the locally defined versions of `propagate` and `reflect`, specified in Smalltalk by the use of `self`:

```
interact: inst1
    self propagate: inst1.
"Propagate a pulse, inst1, using the locally defined method"

    self reflect: inst1.
"Generate a new pulse and send it in the opposite direction"
```

4.9.2 Metaclasses

It was emphasized in Section 4.4 that a class defines the characteristics (methods and attributes) that are available to instances. Methods can therefore be used only by instances and not by the class itself. This is a sensible constraint, as it is clearly appropriate that methods such as `move`, `propagate`, and `reflect` should be performed by instances in the ultrasonics example. However, a problem arises with the creation of new instances. In Smalltalk, new instances are created by sending the message `new`. This message cannot be sent to an instance, as we have not yet created one. Instead it is sent to the class. This apparent paradox is overcome by imagining that each class is an instance of a metaclass, which is a class of classes (Figure 4.14). Thus, the method `new` is defined within one or more metaclasses and made available to each class. The notion of metaclasses helps to address a philosophical problem, but for most practical purposes it is sufficient to remember that messages are always sent to instances, except for the message `new`, which is sent to a class.

Metaclasses also provide a means by which class variables can be implemented. In some OOP languages, although not in Smalltalk, class variables are simply instance variables that are declared at the metaclass level. However, these implementation details are rarely of concern to the programmer. Although C++ does not explicitly include metaclasses, it supports class variables in the form of static data members. C++ also provides static member functions. Unlike ordinary member functions, which are encapsulated within each instance, there is only one copy of a static member function for a class. The single copy is shared by all instances of the class and is not associated with any particular instance. A static member function is, therefore, equivalent to a method (such as `new`) that is defined within a metaclass.

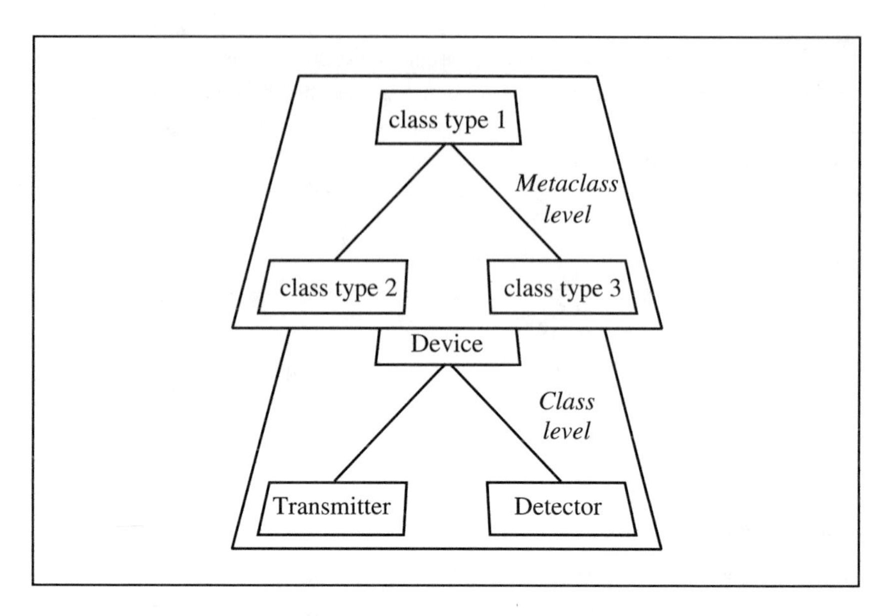

Figure 4.14 Each class can be treated as an instance of a metaclass

4.10 Type checking

Smalltalk and C++ both offer the key features of OOP, namely, data abstraction, inheritance, encapsulation, and dynamic binding. There are, however, some significant differences between the languages, as well as their obvious differences in syntax. An important difference is that C++ uses static type checking whereas Smalltalk uses dynamic type checking. Thus a C++ programmer must explicitly state the type (or class) of all variables, and the compiler checks for consistency. The Smalltalk programmer does not need to specify the classes of variables, and no checks are made during compilation. Static type checking is sometimes called *strong* type checking, and the variables are said to be *manifestly* typed. Conversely, dynamic type checking is sometimes called *weak* type checking, and the variables are said to be *latently* typed. The words *strong* and *weak* are perhaps misleading, since they might alternatively indicate the strictness or level of detail of static type checking.

Static type checking describes the ability of the compiler to check that objects of the right class are supplied to all functions and operators. In C++, the programmer must explicitly state the class of all objects (including built-in types such as `int` or `float`). Whenever assignments are made, the compiler checks that types are compatible. Assignments can be made directly by a statement such as:

```
    x = "a string";
```

or indirectly by passing an object as a parameter to a function. Consider again our definition of the method `propagate` in C++:

```
void Feature::propagate(Sonic_pulse* p)
{
  float x, y;
  x = getx();
  y = gety();          // calculate new position for Sonic_pulse p
  p->move(x,y);          // move Sonic_pulse p to its new position
}
```

Let us now create some object instances:

```
Sonic_pulse* p1 = new Sonic_pulse();
Shear_pulse* p2 = new Shear_pulse();
Defect* d1 = new Defect();
Defect* d2 = new Defect();
```

If we pass as a parameter to `propagate` any object whose type is neither a pointer to `Sonic_pulse` nor a pointer to a class publicly derived from `Sonic_pulse`, we will get an error at compilation time:

```
d1->propagate(p1);        // OK because p1 is of type Sonic_pulse.
d1->propagate(p2);            // OK because p2's type is publicly
                                   // derived from Sonic_pulse.
d1->propagate(d2);             // ERROR: d2 is not a Sonic_pulse.
```

Contrast this with the Smalltalk equivalent, where the method `propagate` may be defined as follows for the class `Feature`:

```
propagate: aSonicpulse
| x y |
x:= self getx.
y:= self gety.
aSonicpulse move: x and: y.
```

Now we create some instances as before and try passing them as parameters to `propagate`. The following would be typed into a temporary workspace, selected, and Smalltalk's "Do It" command invoked:

```
|p1 p2 d1 d2|
p1 := Sonicpulse new.
p2 := Shearpulse new.
d1 := Defect new.
d2 := Defect new.
```

```
d1 propagate: p1.      "OK because method move is defined for p1"
d1 propagate: p2.      "OK because method move is defined for p2"
d1 propagate: d2.       "Compiles OK but causes a run-time error"
```

All of this code will compile, but the last line will cause an error at run-time, producing a message similar to:

```
"Message not understood - move:and:"
```

The type of the argument to `propagate` is neither specified nor checked, and therefore the compiler allows us to supply anything. In our definition of `propagate`, the argument was given the name `aSonicpulse`. This helps us to recognize the appropriate argument type, but the name has no significance to the compiler. When we passed `d2` (an instance of `Defect`) as the parameter, a run-time error came about because Smalltalk tried to find the method `move:and:` for `d2`, when this method was only defined for the class `Sonicpulse` and its derivatives.

The difference between static and dynamic typing represents a difference in programming philosophy. C++ insists that all types be stated, and checks them all at compilation time to ensure that there are no incompatibilities. This means extra work for the programmer, especially if he or she is only performing a quick experiment. On the other hand, it leads to clearly defined interactions between objects, thereby helping to document the program and making error detection easier. Static typing tends to be preferred for building large software systems that involve more than one programmer. Dynamically typed languages such as Smalltalk (and also Lisp and Prolog, introduced in Chapter 10) are ideal for trying out ideas and building prototypes, but they are more likely to contain latent bugs that may show up unpredictably at run-time. CLOS offers a compromise between static and dynamic typing by allowing the programmer to make type declarations and giving the option to enforce or ignore them.

4.11 Further aspects of OOP

This section examines some further issues that may arise in object-oriented programming. The features described here may be available in some OOP languages, but they are not essential according to our definition of an OOP language (Section 4.3).

4.11.1 Persistence

We have already discussed the creation and deletion of objects. The lifetime of an object is the time between its creation and deletion. So far our discussion has implicitly assumed that instances are created and destroyed in a time frame corresponding to one of the following:

- the evaluation of a single expression;
- running a method or other block of code;
- specific creation and deletion events during the running of a program;
- the duration of the program.

However, object instances can outlive the run of the program in which they were created. A program can create an object and then store it. That object is said to be persistent across time (because it is stored) and space (because it can be moved elsewhere). Persistent objects are particularly important in database applications. For example, a payroll system might store instances of the class `Employee`. Such applications require careful design, since the stored instances are required not only between different executions of the program, but also between different versions of the program.

4.11.2 Concurrency

Each object acts independently, on the basis of messages that it receives. Objects can, therefore, perform their own individual tasks concurrently. This makes them strong candidates for implementation on parallel processing computers. Returning once more to our object-oriented ultrasonic simulation, a pulse could arrive at one feature (say the front wall) at the same time as another arrives at a different feature (say a defect), and both would require processing by the respective features. Although this could be achieved on a parallel machine, the actual implementation was on a serial computer, where concurrency was simulated. A clock (implemented as an object) marked simulated time. Thus if two pulses arrived at different features at the same time t, it would not matter which was processed first, as according to the simulation clock, they would both be processed at time t.

4.11.3 Overloading

Most high-level languages provide "in-line" operators such as +, –, *, / that are placed between the arguments to which they refer, e.g., in C++:

```
a=b+c;
```

This is roughly equivalent to calling a function named `plus` with the arguments `b` and `c` and assigning the result to `a`:

```
a=plus(b,c);
```

It is only roughly equivalent since the use of an operator may be more efficient than a function call. From the perspective of the compiler, the operator + has to fulfill a different task depending on whether its arguments are integers or floats. However, it is convenient from the programmer's perspective for the operator to be called + irrespective of whether integer or float addition is required. The compiler "knows" which meaning of + is intended by examining the type of the arguments. This is termed operator *overloading* — the same operator has a different meaning depending on the type and number of its arguments. This is similar to polymorphism (Section 4.8), except that the meaning of an overloaded operator is determined at compile-time rather than at run-time. Some languages, such as C++, allow operator overloading to be extended to classes as well as to built in data types. A common example of this is to define the class `Complex`, describing complex numbers. The class definition might include a definition for + such that it would carry out complex number addition.

Consider now the expression:

```
pulse1 + pulse2
```

where `pulse1` and `pulse2` are instances of `Shear_pulse`. This could have a sensible meaning if we defined + within the definition of class `Shear_pulse` so that it performed constructive interference where the two pulses were in phase and destructive interference where they were out of phase. This is another example of operator overloading.

Overloading is not just restricted to operators, but can apply to function names as well (depending on the language). In C++, functions can be overloaded independent of whether they are member functions (functions defined as part of a class definition) or other functions. For instance, we might have separate functions called `print` for printing an integer, a character, or a string. In C++ this would look like this:

```
void print(int x)
{
  /* code for printing an integer */
}

void print(char x)
{
```

```
  /* code for printing a character */
}

void print(char* x)
{
  /* code for printing a string */
}

print(74); // uses 1st definition of print
print('A');// uses 2nd definition of print
print("have a nice day");        // uses 3rd definition of print
```

Overloading is a convenient feature of some object-oriented and conventional languages. It is not, however, essential that an OOP language should have this facility.

4.11.4 Active values and daemons

So far we have considered programs in which methods are explicitly called and where these methods may involve reading or altering data attached to object attributes. Control is achieved through function (method) calls and data are accessed as a side effect. Active values and daemons allow us to achieve the opposite effect, namely, function calls are made as a side effect of accessing data. Attributes that can trigger function calls in this way are said to be *active*, and their data are *active values*. The functions that are attached to the data are called *daemons*. A daemon (sometimes spelled "demon") can be thought of as a piece of code that lies dormant, watching an active value, and which springs to life as soon as the active value is accessed.

Daemons can pose some problems for the flow of control in a program. Suppose that a method `M` accesses an active value that is monitored by daemon `D`. If `D` were to fire immediately, before `M` has finished, then `D` might disrupt some control variables upon which `M` relies. The safer solution is to treat the methods and daemons as indivisible "atomic" actions, so that `M` has to run to completion before `D` fires.

Active values need not be confined to attributes (data members) but may also include methods (member functions). Thus, a daemon may be set to fire when one or more of the following occur:

- an attribute is read;
- an attribute is set; or
- a method is called.

An example of the use of daemons is in the construction of gauges in graphical user interfaces (Figure 4.15). A gauge object may be created to

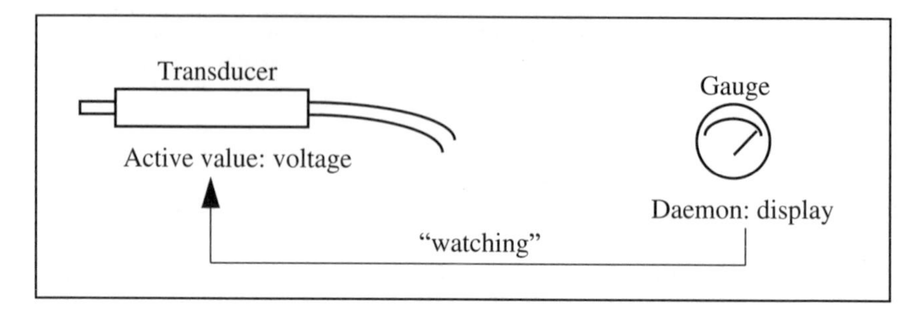

Figure 4.15 Using a daemon to monitor an active value

monitor some attribute of an object in a physical model, such as the voltage across a transducer. As soon as the voltage value is altered, the gauge should update its display, regardless of what caused the change in voltage. The `voltage` attribute on the object `transducer` is active, and the `display` method on the object `gauge` is a daemon.

4.12 Frame-based systems

Frame-based systems are closely allied to object-oriented systems. Frame-based systems evolved from artificial intelligence research, whereas OOP has its origins in more general software engineering. Frames provide a means of representing and organizing knowledge. They use some of the key features of object-orientation for organizing information, but they do not have the behaviors or methods of objects. As relations can be created between frames, they can be used to model physical or abstract connections. Just as object instances are derived from object classes, so frame instances can be derived from frame classes, taking advantage of inheritance.

Section 4.3 quoted Pascoe's [1] definition of OOP. Frame-based systems are less rigidly defined, so a corresponding definition might be:

> *Frame-based systems offer data abstraction and inheritance. In general, they do not offer either encapsulation or dynamic (late) binding.*

The absence of encapsulation is the key feature that distinguishes frames from objects. Since frames do not have ownership of their own functions, they are passive structures. Without encapsulation, the concept of dynamic binding becomes meaningless. This does not, however, detract from their usefulness. In fact they provide an extremely useful way of organizing data and knowledge in a knowledge-based system.

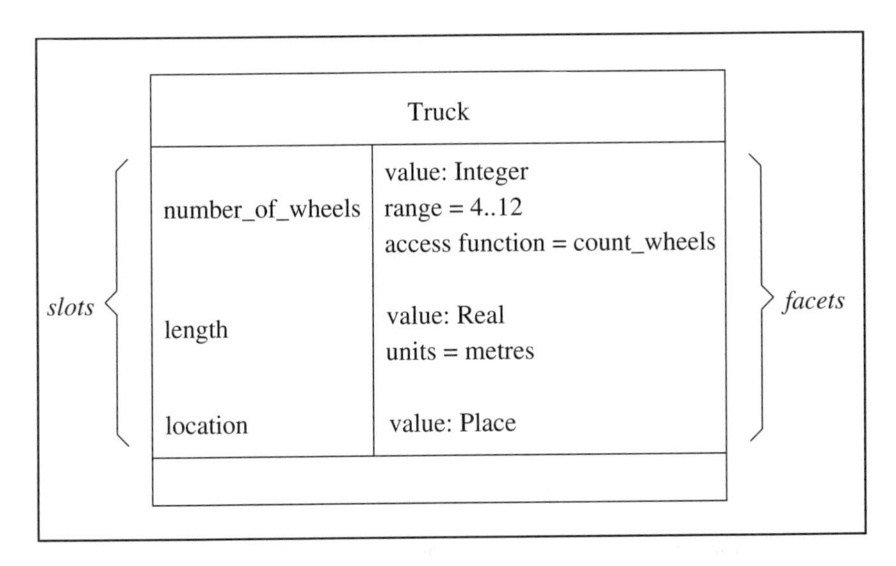

Figure 4.16 An example of a frame-based representation

The frame-based view is that an item such as my truck can be represented by a data structure, i.e., a frame instance, which we can call `my_truck`. This instance can inherit from a class of frame such as `Truck`, which can itself inherit from a parent class such as `Vehicle`. Thus, frame-based systems support data abstraction by allowing the definition of new classes. They also support inheritance between classes and from a class to its instances.

We can hang extra information onto a frame, such as the number of wheels on my truck. Thus `number_of_wheels` could be a slot (see Section 4.4.3) associated with the frame instance `my_truck`. This slot could use the default value of 4 inherited from `Vehicle` or it may be a locally defined value that overrides the default. The value associated with a slot can be a number, a description, a number range, a procedure, another frame, or anything allowed by the particular implementation. Some frame-based systems allow us to place multiple values in a slot. In such systems, the different pieces of information that we might want to associate with a slot are known as its *facets* . Each facet can have a value associated with it, as shown in Figure 4.16. For instance, we may wish to specify limits on the number of wheels, provide a default, or calculate a value using a function known as *an access function*. In this example, an access function `count_wheels` could calculate the number of wheels when a value is not previously known.

It was pointed out above that frame-based systems do not, in general, provide encapsulation. Consider the following example, written using the syntax of Flex, a knowledge-based and frame-based toolkit [3]:

```
/* define a frame */
frame vehicle;
  default location is garage and
  default number_of_wheels is 4
  and default mass_in_tonnes is 1 .

/* define another frame */
frame truck is a kind of vehicle;
  default mass_in_tonnes is 10 .

/* create an instance of frame truck */
instance my_truck is a truck .

/* create another instance of frame truck */
instance your_truck is a truck;
  number_of_wheels is my_truck`s number_of_wheels + 2 .
```

In this example, the default number of wheels is inherited by `my_truck`. The number of wheels on `your_truck` is derived by an access function defined within the frame and which does not need to be explicitly named. The access function in `your_truck` makes use of a value in `my_truck`, though this would be forbidden under the principle of encapsulation. As the access function is not calculated until run-time, the example also demonstrates that Flex offers dynamic binding.

Frames can be applied to physical things such as trucks or abstract things such as plans or designs. A special kind of frame, known as a *script*, can be used to describe actions or sequences of actions. For instance, in a system for fixing plumbing faults we might construct scripts called `changing_a_washer` or `unblocking_a_pipe`.

The term *frame* has been used so far here to imply a framework onto which information can be hung. However, frames are also analogous to the frames of a movie film or video tape. In associating values with the slots of a frame we are taking a snapshot of the world at a given instant. At one moment the slot `location` on `my_truck` might contain the value `smallville`, while sometime later it might contain the value `largetown`.

4.13 Summary

We began with the premise that object-oriented programming requires the following language capabilities:

- *Data Abstraction*: New types (classes) can be defined and a full set of operations provided for each, so that the new classes behave like built-in ones. Operator overloading enhances this perception, since it allows the new data types to be used as though they were built-in types such as `int` and `float`.

- *Inheritance*: Class definitions can be treated as specializations of other (parent) classes. This ability maximizes code reuse and minimizes duplication. OOP therefore helps us to build a world model that maps directly onto the real world. This model is expressed in terms of the classes of objects that the world manipulates, and OOP helps us to successively refine our understanding of these classes.

- *Encapsulation*: Data and code are bundled together into objects and are hidden from other objects. This simplifies program maintenance, helps to ensure that interactions between objects are clearly defined and predictable, and provides a realistic world model.

- *Dynamic (late) binding*: The object (and, hence, its class) that is associated with a given variable is not determined until run-time. Although the parent class may be specified and checked at compile-time (depending on the language), the instance used at run-time could legitimately belong to a subclass of that parent. Therefore, it cannot be known at compile-time whether the object will use methods defined within the parent class or specialized versions of them. Different objects can, therefore, respond to a single command in a way that is appropriate for those objects, and which may be different from the response of other objects. This property is known as *polymorphism*.

Stroustrup [4] has pointed out that although it is *possible* to build these capabilities in many computer languages, a language can only be described as object-oriented if it *supports* these features, i.e., it makes these features convenient for the programmer.

Class definitions act as templates from which multiple instances (i.e., objects) can be created. Classes and instances are easily inspected, modified, and reused. All of these capabilities assist in the construction of large, complex software systems by breaking the system down into smaller, independent, manageable chunks. The problem representation is also more natural, allowing more time to be spent designing a system and less on coding, testing, and integration.

We saw that frame-based systems support data abstraction and inheritance, but that they do not necessarily offer encapsulation or dynamic binding. As a result, frames are passive in the sense that, like entries in a database, they do not perform any tasks themselves. Their ability to calculate values for slots through access functions might be regarded as an exception to this generality. Although frames are more limited than a fully fledged object-oriented system, they nonetheless provide a useful way of organizing and managing knowledge in a knowledge-based system.

References

1. Pascoe, G. A., "Elements of object-oriented programming," *Byte*, pp. 139–144, August 1986.
2. Booch, G., Rumbaugh, J., and Jacobson, I., *The Unified Modeling Language User Guide*, Addison-Wesley, 1999.
3. Vasey, P., Westwood, D., and Johns, N., *Flex Reference Manual*, Logic Programming Associates Ltd., 1996.
4. Stroustrup, B., "What is object-oriented programming?" *IEEE Software*, pp. 10–20, May 1988.

Further reading

- Booch, G., *Object-Oriented Analysis and Design with Applications*, 2nd ed., Addison-Wesley, 1994.
- Budd, T., *An Introduction to Object-Oriented Programming*, 2nd ed., Addison-Wesley, 1996.
- Cox, B. and Novobilski, A., *Object-Oriented Programming: an evolutionary approach*, 2nd ed., Addison-Wesley, 1991.
- Goldberg, A. and Robson, D., *Smalltalk-80: the language*, Addison-Wesley, 1989.
- Lalonde, W. R., *Discovering Smalltalk*, Addison-Wesley, 1994.
- Meyer, B., *Object-Oriented Software Construction*, 2nd ed., Prentice-Hall, 1997.
- Mullin, M., *Object-Oriented Program Design, with Examples in C++*, Addison-Wesley, 1989.
- Stevens, P. and Pooley, R., *Using UML: software engineering with objects and components*, Addison-Wesley, 2000.
- Stroustrop, B., *The C++ Programming Language*, 3rd ed., Addison-Wesley, 1997.

Touretzky, D. S., *The Mathematics of Inheritance Systems*, Pitman/Morgan Kaufmann, 1986.

Chapter five

Intelligent agents

5.1 Characteristics of an intelligent agent

As the information world expands, people are becoming less and less able to act upon the escalating quantities of information presented to them. A way around this problem is to build *intelligent agents* — software assistants that take care of specific tasks for us. For instance, if you want to search the World Wide Web for a specific piece of information, you might use an intelligent agent to consult a selection of search engines and filter the web pages for you. In this way you are presented with only two or three pages that precisely match your needs. An intelligent agent of this type, that personalizes itself to your individual requirements by learning your habits and preferences, is called a *user agent.*

Similarly, much of the trading on the world's stock exchanges is performed by intelligent agents. Reaping the benefits of some types of share dealing relies on reacting rapidly to minor price fluctuations. By the time a human trader has assimilated the data and made a decision, the opportunity would be lost.

Just as intelligent agents provide a way of alleviating complexity in the real world, they also fulfill a similar role within computer systems. As software systems become larger and more complex, it becomes progressively less feasible to maintain them as centralized systems that are designed and tested against every eventuality. An alternative approach is to take the idea of modular software toward its apotheosis, namely, to turn the modules into autonomous agents that can make their own intelligent decisions in a wide range of circumstances. Such agents allow the system to be largely self-managing, as they can be provided with knowledge of how to cope in particular situations, rather than being explicitly programmed to handle every foreseeable eventuality.

While noting that not all agents are intelligent, Wooldridge [1] gives the following definition for an agent:

An agent is an encapsulated computer system that is situated in some environment, and that is capable of flexible, autonomous action in that environment in order to meet its design objectives.

From this definition we can see that the three key characteristics of an agent are autonomy, persistence, and the ability to interact with its environment. *Autonomy* refers to an agent's ability to make its own decisions based on its own experience and circumstances, and to control its own internal state and behavior. The definition implies that an agent functions continuously within its environment, i.e., it is *persistent* over time. Agents are also said to be *situated*, i.e., they are responsive to the demands of their environment and are capable of acting upon it. Interaction with a physical environment requires perception through sensors, and action through actuators or effectors. Interaction with a purely software environment is more straightforward, requiring only access to and manipulation of data and programs.

Intelligence can be added to a greater or lesser degree, but we might reasonably expect an *intelligent* agent to be all of the following:

- reactive,
- goal-directed,
- adaptable,
- socially capable.

Social capability refers to the ability to cooperate and negotiate with other agents (or humans), which forms the basis of Section 5.4 below. It is quite easy to envisage an agent that is purely reactive, e.g., one whose only role is to place a warning on your computer screen when the printer has run out of paper. This behavior is akin to a daemon (see Chapter 4). Likewise, modules of conventional computer code can be thought of as goal-directed in the limited sense that they have been programmed to perform a specific task regardless of their environment. Since it is autonomous, an intelligent agent can decide its own goals and choose its own actions in pursuit of those goals. At the same time, it must also be able to respond to unexpected changes in its environment. It, therefore, has to balance reactive and goal-directed behavior, typically through a mixture of problem solving, planning, searching, decision making, and learning through experience.

There is no reason why an agent should remain permanently on a single computer. If a computer is connected to a network, a *mobile agent* can travel to remote computers to carry out its designated task before returning back home with the task completed. A typical task for a mobile agent might be to determine a person's travel plan. This will require information about train and airline timetables, together with hotel availability. Instead of transferring large quantities of data across the network from the train companies, airlines, and hotels, it is more efficient for the agent to transfer itself to these remote sites, find the information it needs, and return. Clearly, there is potential for malicious use of mobile agents, so security is a prime consideration for sites that accept them.

5.2 Agents and objects

We saw in Chapter 4 that objects and frames allow complex problems to be broken down into simpler constituents while maintaining the integrity of the overall system. Although the history of agent-based programming can be traced back to the 1970s, it is now seen by many as the next logical development from object-oriented programming (OOP). If objects are viewed as obedient servants, then intelligent agents can be seen as independent beings. Indeed they are often referred to as *autonomous* agents. When an agent receives a request to perform an action, it will make its own decision, based on its beliefs and in pursuit of its goals. Thus, it behaves more like an individual with his or her own personality. In consequence, agent-based systems are analogous to human societies or organizations.

As an encapsulated software entity, an intelligent agent bears some resemblance to an object. However, it is different in three key ways:

- *Autonomy*: Once an object has declared a method as public, as it must for the method to be useful, it loses its autonomy. Other objects can then invoke that method by sending a message. In contrast, agents cannot invoke the actions of another agent, they can only make *requests*. The decision over what action to take rests with the receiver of the message, not the sender. Autonomy is not required in an object-oriented system because each object is designed to perform a task in pursuit of the developer's overall goal. However, it cannot necessarily be assumed that agents will share a common goal.

- *Intelligence*: Although intelligent behavior can be built into an object, it is not a requirement of the OOP model.

- *Persistence*: It was stated in Chapter 4 that objects could be made to persist from one run of a program to another by storing them. In contrast, agents persist in the sense that they are constantly "switched on" and thus they operate concurrently. They are said to have their own thread of control, which can be thought of as another facet of autonomy since an agent decides for itself when it will do something. In contrast, a standard object-oriented system has a single thread of control, with objects performing actions sequentially. One exception to this is the idea of an active object [2] that has its own thread of control and is, therefore, more akin to an agent in this respect.

While OOP languages such as C++ and Smalltalk are quite closely defined, there are no standard approaches to the implementation of agent-based systems. Indeed, an object-oriented approach might typically be used in the implementation of an agent-based system, while the reverse is unlikely.

5.3 Agent architectures

Any of the techniques met so far in this book could be used to provide the internal representation and reasoning capabilities of an agent. Nevertheless, several different types of approaches can be identified. There are at least four different schools of thought about how to achieve an appropriate balance between reactive and goal-directed behavior. These are reviewed below.

5.3.1 Logic-based architectures

At one extreme, the purists favor logical deduction based on a symbolic representation of the environment [3, 4]. This approach is elegant and rigorous, but it relies on the environment's remaining unchanged during the reasoning process. It also presents particular difficulties in symbolically representing the environment and reasoning about it.

5.3.2 Emergent behavior architectures

In contrast, other researchers propose that logical deduction about the environment is inappropriately detailed and time-consuming. For instance, if a heavy object is falling toward you, the priority should be to move out of the way rather than to analyze and prove the observation. These researchers suggest that agents need only a set of reactive responses to circumstances, and that intelligent behavior will emerge from the combination of such responses. This kind of architecture is based on *reactive agents*, i.e., agents that include neither a symbolic world model nor the ability to perform complex symbolic reasoning [5]. A well-known example of this approach is Brooks' *subsumption architecture* [6], containing behavior modules that link actions to observed situations without any reasoning at all. The behaviors are arranged into a

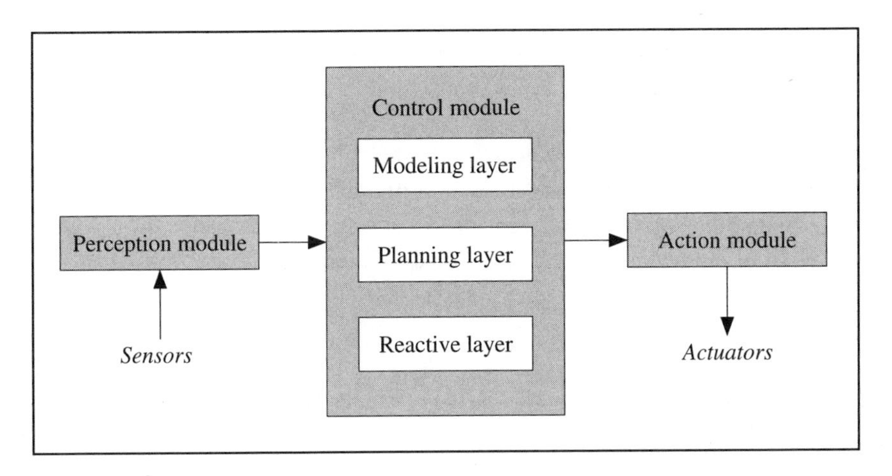

Figure 5.3 Touring Machine architecture

intelligent agents working together. In Chapter 1, it was stated that the study of artificial intelligence has been inspired by attempts to mimic the logical reasoning of a human. In the same way, a branch of AI known as *distributed artificial intelligence* (DAI) has been inspired by attempts to mimic a society of humans working together. *Multiagent systems* (MASs), sometimes called *agent-oriented* or *agent-based systems*, are one of the most important approaches to DAI; blackboard systems (see Chapter 9) are another.

A multiagent system can be defined as a system in which several interacting, intelligent agents pursue a set of individually held goals or perform a set of individual tasks. Though this definition is a useful starting point, it raises a number of key questions that will be addressed in the remainder of this chapter, notably:

- What are the benefits?
- How do intelligent agents interact?
- How do they pursue goals and perform tasks?

5.4.1 Benefits of a multiagent system

There are some problems for which multiagent systems offer the only practicable approach:

- *Inherently complex problems*: Such problems are simply too large to be solved by a single hardware or software system. As the agents are provided with the intelligence to handle a variety of circumstances, there is some uncertainty as to exactly how an agent system will perform in a specific situation. Nevertheless, well-designed agents will ensure that every circumstance is handled in an appropriate manner even though it may not have been explicitly anticipated.

- *Inherently distributed problems*: Here the data and information may exist in different physical locations, or at different times, or may be clustered into groups requiring different processing methods or semantics. These types of problems require a distributed solution, which can be provided by agents running concurrently, each with its own thread of control (see Section 5.2 above).

Furthermore, a multiagent system (MAS) offers the following general benefits:

- A more *natural view of intelligence.*

- *Speed and efficiency gains*, brought about because agents can function concurrently and communicate asynchronously.

- *Robustness and reliability*: No agent is vital provided there are others that can take over its role in the event of its failure. Thus, the performance of an MAS will degrade gracefully if individual agents fail.

- *Scalability*: DAI systems can generally be scaled-up simply by adding additional components. In the case of an MAS, additional agents can be added without adversely affecting those already present.

- *Granularity*: Agents can be designed to operate at an appropriate level of detail. Many "fine-grained" agents may be required to work on the minutiae of a problem, where each agent deals with a separate detail. At the same time, a few "coarse-grained," more sophisticated agents can concentrate on higher-level strategy.

- *Ease of development*: As with OOP, encapsulation enables individual agents to be developed separately and to be re-used wherever applicable.

- *Cost*: A system comprising many small processing agents is likely to be cheaper than a large centralized system.

Thus Jennings has argued that MASs, on the one hand, are suited to the design and construction of complex, distributed software systems and, on the other, are appropriate as a mainstream software engineering paradigm [12].

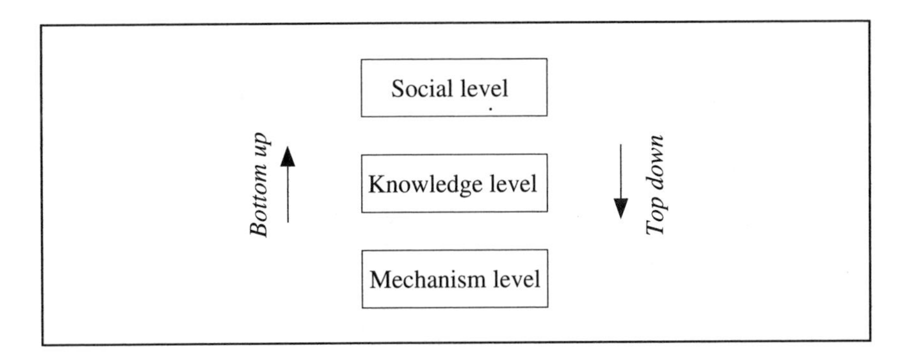

Figure 5.4 Agent levels of abstraction

5.4.2 Building a multiagent system

A multiagent system is dependent on interactions between intelligent agents. There are, therefore, some key design decisions to be made, e.g., when, how, and with whom should agents interact? In *cooperative* models, several agents try to combine their efforts to accomplish as a group what the individuals cannot. In *competitive* models, each agent tries to get what only some of them can have. In either type of model, agents are generally assumed to be honest.

In order to achieve coherency, multiagent systems can be designed bottom-up or top-down. In a bottom-up approach, agents are endowed with sufficient capabilities, including communication protocols, to enable them to interact effectively. The overall system performance then emerges from these interactions. In a top-down approach, conventions — sometimes called *societal norms* — are applied at the group level in order to define how agents should interact. An example might be the principle of democracy, achieved by giving agents the right to vote. If we view an agent as having a knowledge level abstracted above its inner mechanisms, then these conventions can be seen as residing at a still higher level of abstraction, namely the *social level* (Figure 5.4).

Multiagent systems are often designed as computer models of human functional roles. For example, we may have a hierarchical control structure in which one agent is the superior of other subordinate agents. Peer group relations, such as may exist in a team-based organization, are also possible. This section will address three models for managing agent interaction, known as *contract nets* [13], *cooperative problem solving* (CPS) [14, 15] and *shifting matrix management* (SMM) [16]. After considering each of these models in turn, the semantics of communication between agents will be addressed.

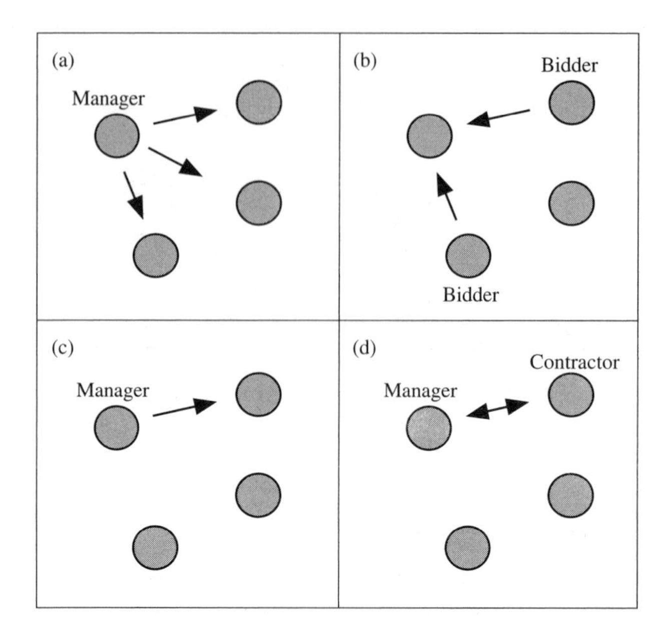

Figure 5.5 Contract nets [13]:
(a) Manager advertises a task; (b) potential contractors bid for the task;
(c) manager awards the contract; (d) manager and contractor communicate privately

Contract nets

Imagine that you have decided to build your own house. You are unlikely to undertake all the work yourself. You will probably employ specialists to draw up the architectural plans, obtain statutory planning permission, lay the foundations, build the walls, install the floors, build the roof, and connect the various utilities. Each of these specialists may in turn use a subcontractor for some aspect of the work. This arrangement is akin to the contract net framework [13] for agent cooperation (Figure 5.5). Here, a manager agent generates tasks and is responsible for monitoring their execution. The manager enters into explicit agreements with contractor agents willing to execute the tasks. Individual agents are not designated *a priori* as manager or contractor. These are only roles, and any agent can take on either role dynamically during problem solving.

To establish a contract, the manager agent advertises the existence of the tasks to other agents. Agents that are potential contractors evaluate the task announcements and submit bids for those to which they are suited. The manager evaluates the bids and awards contracts for execution of the task to the agents it determines to be the most appropriate. The manager and contractor are thus linked by a contract and communicate privately while the

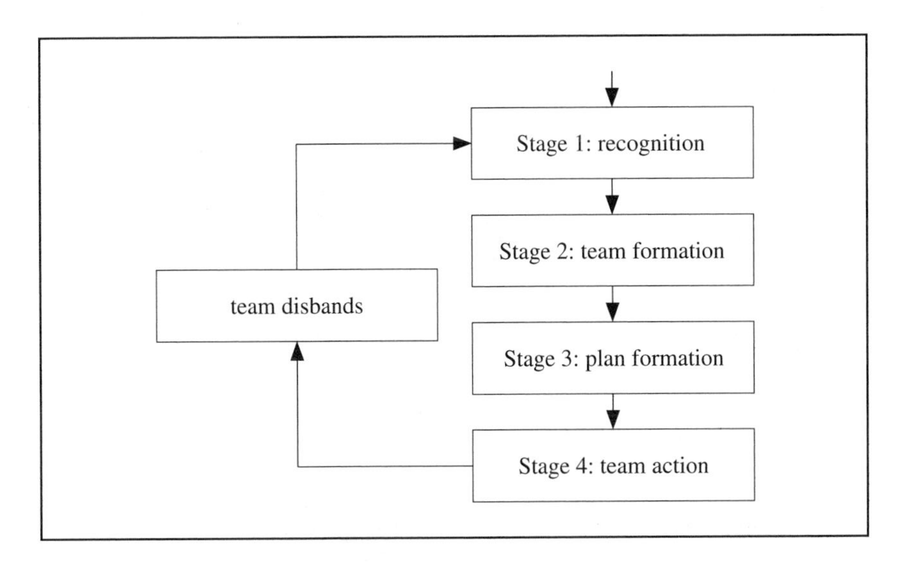

Figure 5.6 CPS framework

contract is being executed. The managers supply mostly task information and the contractor reports progress and the eventual result of the task. The negotiation process may recur if a contractor subdivides its task and awards contracts to other agents, for which it is the manager.

CPS framework

The cooperative problem-solving (CPS) framework is a top-down model for agent cooperation. As in the BDI model, an agent's intentions play a key role. They determine the agent's personal behavior at any instant, while joint intentions control its social behavior [17]. An agent's intentions are shaped by its *commitment*, and its joint intentions by its social *convention*. The framework comprises the following four stages, also shown in Figure 5.6:

Stage 1: recognition. Some agents recognize the potential for cooperation with an agent that is seeking assistance, possibly because it has a goal it cannot achieve in isolation.
Stage 2: team formation. An agent that recognized the potential for cooperative action at Stage 1 solicits further assistance. If successful, this stage ends with a group having a joint commitment to collective action.
Stage 3: plan formation. The agents attempt to negotiate a joint plan that they believe will achieve the desired goal.
Stage 4: team action. The newly agreed plan of joint action is executed. By adhering to an agreed social convention, the agents maintain a close-knit relationship throughout.

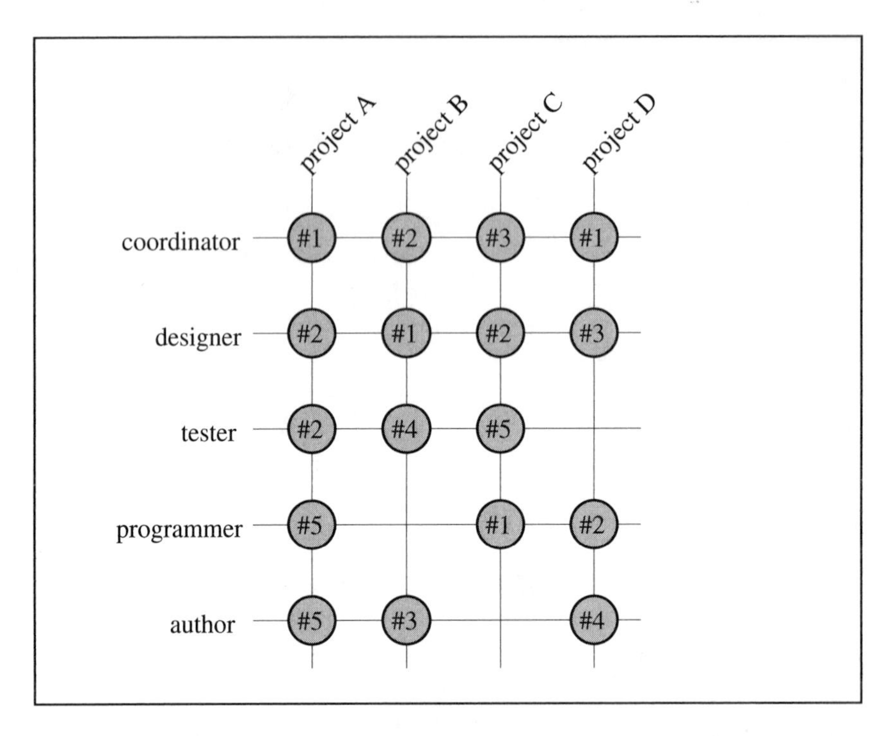

Figure 5.7 Shifting Matrix Management (SMM): the nodes represent people [18]

Shifting Matrix Management (SMM)

SMM [16] is a model of agent coordination that has been inspired by Mintzberg's Shifting Matrix Management model of organizational structures [18], as illustrated in Figure 5.7. Unlike the traditional management hierarchy, matrix management allows multiple lines of authority, reflecting the multiple functions expected of a flexible workforce. *Shifting* matrix management takes this idea a stage further by regarding the lines of authority as temporary, typically changing as different projects start and finish. For example, in Figure 5.7, individual #1 is the coordinator for project A and the designer for project B. In order to apply these ideas to agent cooperation, a six-stage framework has been devised (Figure 5.8) and outlined below. The agents are distinguished by their different motives, functionality, and knowledge. These differences define the agents' variety of mental states with respect to goals, beliefs, and intentions.

Stage 1: goal selection. Agents select the tasks they want to perform, based on their initial mental states.

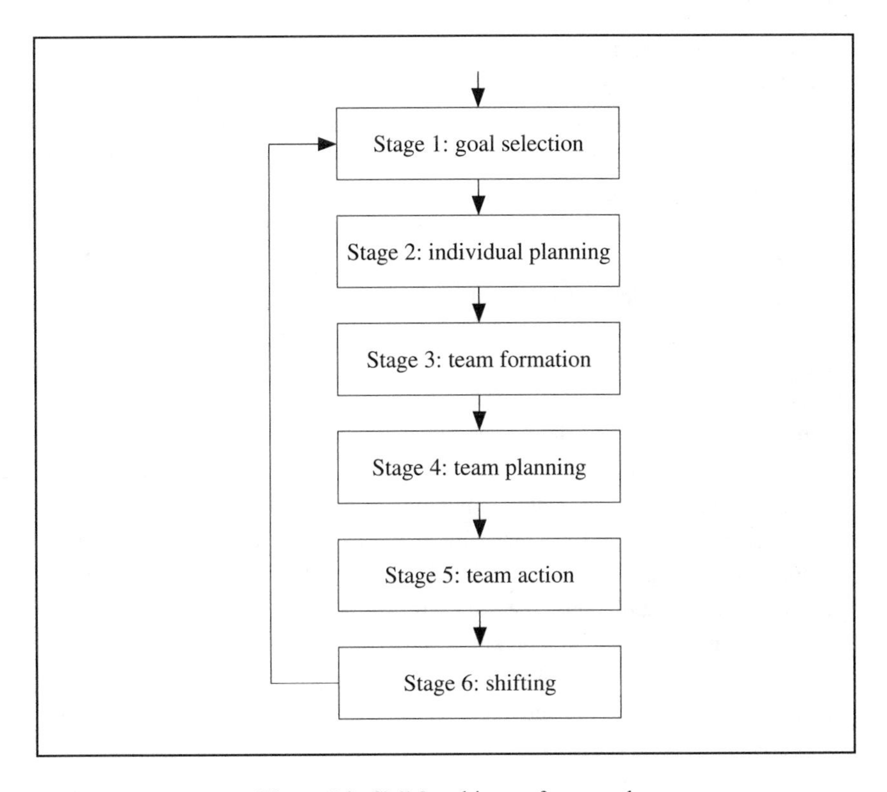

Figure 5.8 SMM multiagent framework

Stage 2: individual planning. Agents select a way to achieve their goals. In particular, an agent that recognizes its intended goal is common to other agents would have to decide whether to pursue the goal in isolation or in collaboration with other agents.

Stage 3: team formation. Agents that are seeking cooperation attempt to organize themselves into a team. The establishment of a team requires an agreed code of conduct, a basis for sharing resources, and a common measure of performance.

Stage 4: team planning. The workload is distributed among team members.

Stage 5: team action. The team plan is executed by the members under the team's code of conduct.

Stage 6: shifting. The last stage of the cooperation process, which marks the disbanding of the team, involves shifting agents' goals, positions, and roles. Each agent updates its probability of team-working with other agents, depending on whether or not the completed team-working experience with that agent was successful. This updated knowledge is important, as iteration through the six stages takes place until all the tasks are accomplished.

5.4.3 Communication between agents

So far we have seen how intelligent agents can be designed and implemented, and the ways in which they can cooperate and negotiate in a society. In this final section on multiagent systems, we will examine how agents communicate with each other, both synchronously and asynchronously. Synchronous communication is rather like a conversation — after sending a message, the sending agent awaits a reply from the recipient. Asynchronous communication is more akin to sending an email or a letter — although you might expect a reply at some future time, you do not expect the recipient to read or act upon the message immediately.

Agents may be implemented by different people at different times on different computers, yet still be expected to communicate with each other. Consequently, there has been a drive to standardize the structure of messages between agents, regardless of the domain in which they are operating.

A generally accepted premise is that the *form* of the message should be understandable by all agents regardless of their domain, even if they do not understand its content. Thus, the structure needs to be standardized in such a way that the domain-specific content is self-contained within it. Only specialist agents need to understand the content, but all agents need to be able to understand the form of the message. Structures for achieving this are called *agent communication languages* (ACLs), which include *Knowledge Query and Manipulation Language* (KQML) [19].

In KQML, the structure contains at least the following components:

- A *performative*. This is a single word that describes the purpose of the message, e.g., tell, cancel, evaluate, advertise, ask-one, register, reply.

- The identity of the agent that is the *sender*.

- The identity of the agent that is the *receiver*.

- The *language* used in the content of the message. Although KQML defines the overall form of the message, any programming language can be used for the domain-specific content.

- The *ontology*, or vocabulary, of the message. This provides the context within which the message content is to be interpreted. For example, the problem of selecting a polymer to meet an engineering design requirement is used in Chapter 10 to demonstrate the Lisp and Prolog languages. At the programming language level, a program to tackle this problem is merely a collection of words and symbols organized as statements. It would, for instance, remain syntactically correct if each polymer name were replaced by the name of a separate type of fruit. The statements only become meaningful once they are interpreted in the vocabulary of engineering polymers.

- The message *content*.

In the polymer selection world mentioned above, `agent1` might wish to tell `agent2` about the properties of polystyrene, encoded in Prolog. Using KQML, it could do so with the following message.

```
(tell
  :sender    agent1
  :receiver  agent2
  :language  prolog
  :ontology  polymer-world
  :content
    "materials_database(polystyrene, thermoplastic,
      [[impact_resistance, 0.02],
       [flexural_modulus, 3.0],
       [maximum_temperature, 50]]).")
```

5.5 Summary

Intelligent agents extend the ideas of objects by giving them autonomy, intelligence, and persistence. Intelligent agents act in pursuit of their personal goals, which may or may not be the same as those of other agents. Intelligent agents can be made mobile, so they can travel across a network — possibly the Internet — to perform a set of tasks. This usually involves the transfer of much less data than if the whole task had to be performed on the originating computer.

Multiagent systems contain several interacting intelligent agents, often designed as computer models of human functional roles. The overall effect is to mimic human teams or societies. There are a variety of ways for organizing a multiagent system, and the agents within it may work cooperatively or competitively. There are also a variety of ways of achieving inter-agent communication, including the Knowledge Query and Manipulation Language (KQML).

References

1. Wooldridge, M. J., "Agent-based software engineering," *IEE Proc. Software Engineering*, vol. 144, pp. 26–37, 1997.
2. Booch, G., *Object-oriented Analysis and Design with Applications*, 2nd ed., Addison-Wesley, 1994.
3. Ulrich, I., Mondada, F., and Nicoud, J. D., "Autonomous vacuum cleaner," *Robotics and Autonomous Systems*, vol. 19, pp. 233–245, 1997.
4. Russell, S. and Norvig, P., *Artificial Intelligence: a modern approach*, Prentice-Hall, 1995.
5. Wooldridge, M. J. and Jennings, N. R., "Intelligent agents: theory and practice," *Knowledge Engineering Review*, vol. 10, pp. 115–152, 1995.
6. Brooks, R. A., "Intelligence without reason," 12th International Joint Conference on Artificial Intelligence (IJCAI'91), Sydney, pp. 569–595, 1991.
7. Newell, A., "The knowledge level," *Artificial Intelligence*, vol. 18, pp. 87–127, 1982.
8. Bratman, M. E., Israel, D. J., and Pollack, M. E., "Plans and resource-bounded practical reasoning," *Computational Intelligence*, vol. 4, pp. 349–355, 1988.
9. Wooldridge, M. J., "Intelligent agents," in *Multiagent Systems: a modern approach to distributed artificial intelligence*, Weiss, G. (Ed.), pp. 27–77, MIT Press, 1999.
10. Kinny, D. and Georgeff, M., "Commitment and effectiveness in situated agents," 12th International Joint Conference on Artificial Intelligence (IJCAI'91), Sydney, pp. 82–88, 1991.
11. Ferguson, I. A., "Integrated control and coordinated behaviour: a case for agent models," in *Intelligent Agents: Theories, Architectures and Languages*, vol. 890, *Lecture Notes in AI*, Wooldridge, M. and Jennings, N. R. (Eds.), pp. 203–218, Springer-Verlag, 1995.
12. Jennings, N. R., "On agent-based software engineering," *Artificial Intelligence*, vol. 117, pp. 277–296, 2000.
13. Smith, R. G. and Davis, R., "Frameworks for cooperation in distributed problem solving," *IEEE Transactions on Systems, Man, and Cybernetics*, vol. 11, pp. 61–70, 1981.
14. Wooldridge, M. J. and Jennings, N. R., "Formalising the cooperative problem solving process," 13th International Workshop on Distributed

Artificial Intelligence (IWDAI'94), Lake Quinalt, WA, pp. 403–417, 1994.

15. Wooldridge, M. J. and Jennings, N. R., "Towards a theory of cooperative problem solving," 6th European Conference on Modelling Autonomous Agents in a Multi-Agent World (MAAMAW'94), Odense, Denmark, pp. 15–26, 1994.
16. Li, G., Weller, M. J., and Hopgood, A. A., "Shifting Matrix Management — a framework for multi-agent cooperation," 2nd International Conference on Practical Applications of Intelligent Agents and Multi-Agents (PAAM'97), London, 1997.
17. Bratman, M. E., *Intentions, Plans, and Practical Reason*, Harvard University Press, 1987.
18. Mintzberg, H., *The Structuring of Organizations*, Prentice-Hall, 1979.
19. Finin, T., Labrou, Y., and Mayfield, J., "KQML as an agent communication language," in *Software Agents*, Bradshaw, J. M. (Ed.), pp. 291–316, MIT Press, 1997.

Further reading

- Ferber, J., *Multi-Agent Systems: an introduction to distributed artificial intelligence*, Addison Wesley, 1999.
- Jennings, N. and Lesperance, Y. (Eds.), *Intelligent Agents VI: agent theories, architectures, and languages*, Springer-Verlag, 2000.
- Weiss, G. (Ed.), *Multiagent Systems: a modern approach to distributed artificial intelligence*, MIT Press, 1999.

Chapter six

Symbolic learning

6.1 Introduction

The preceding chapters have discussed ways of representing knowledge and drawing inferences. It was assumed that the knowledge itself was readily available and could be expressed explicitly. However, there are many circumstances where this is not the case, such as those listed below.

- The software engineer may need to obtain the knowledge from a domain expert. This task of knowledge acquisition is extensively discussed in the literature, mainly as an exercise in psychology.
- The rules that describe a particular domain may not be known.
- The problem may not be expressible explicitly in terms of rules, facts or relationships. This category includes *skills*, such as welding or painting.

One way around these difficulties is to have the system learn for itself from a set of example solutions. Two approaches can be broadly recognized — *symbolic learning* and *numerical learning*. Symbolic learning describes systems that formulate and modify rules, facts, and relationships, explicitly expressed in words and symbols. In other words, they create and modify their own knowledge base. Numerical learning refers to systems that use numerical models; learning in this context refers to techniques for optimizing the numerical parameters. Numerical learning includes artificial neural networks (Chapter 8) and a variety of optimization algorithms such as genetic algorithms and simulated annealing (Chapter 7).

A learning system is normally given some feedback on its performance. The source of this feedback is called the *teacher* or the *oracle*. Often the teacher role is fulfilled by the environment within which the knowledge-based system is working, i.e., the reaction of the environment to a decision is sufficient to indicate whether the decision was right or wrong. Learning with a teacher is sometimes called *supervised* learning. Learning can be classified as follows, where each category involves a different level of supervision:

(i) *Rote learning*
The system receives confirmation of correct decisions. When it produces an incorrect decision it is "spoon-fed" with the correct rule or relationship that it should have used.

(ii) *Learning from advice*
Rather than being given a specific rule that should apply in a given circumstance, the system is given a piece of general advice, such as "gas is more likely to escape from a valve than from a pipe." The system must sort out for itself how to move from this high-level abstract advice to an immediately usable rule.

(iii) *Learning by induction*
The system is presented with sets of example data and is told the correct conclusions that it should draw from each. The system continually refines its rules and relations so as to correctly handle each new example.

(iv) *Learning by analogy*
The system is told the correct response to a similar, but not identical, task. The system must adapt the previous response to generate a new rule applicable to the new circumstances.

(v) Explanation-based learning (EBL)
The system analyzes a set of example solutions and their outcomes to determine *why* each one was successful or otherwise. Explanations are generated, which are used to guide future problem solving. EBL is incorporated into PRODIGY, a general-purpose problem-solver [1].

(vi) *Case-based reasoning (CBR)*
Any case about which the system has reasoned is filed away, together with the outcome, whether it be successful or otherwise. Whenever a new case is encountered, the system adapts its stored behavior to fit the new circumstances. Case-based reasoning is discussed in further detail in Section 6.3 below.

(vii) *Explorative or unsupervised learning*
Rather than having an explicit goal, an explorative system continuously searches for patterns and relationships in the input data, perhaps marking

some patterns as interesting and warranting further investigation. Examples of the use of unsupervised learning include:

- *data mining*, where patterns are sought among large or complex data sets;
- identifying *clusters*, possibly for compressing the data;
- learning to *recognize* fundamental features, such as edges, from pixel images;
- *designing* products, where innovation is a desirable characteristic.

In rote learning and learning from advice, the sophistication lies in the ability of the teacher rather than the learning system. If the teacher is a human expert, these two techniques can provide an interactive means of eliciting the expert's knowledge in a suitable form for addition to the knowledge base. However, most of the interest in symbolic learning has focussed on learning by induction and case-based reasoning, discussed in Sections 6.2 and 6.3 below. Reasoning by analogy is similar to case-based reasoning, while many of the problems and solutions associated with learning by induction also apply to the other categories of symbolic learning.

6.2 Learning by induction

6.2.1 Overview

Rule induction involves generating from specific examples a general rule of the type:

```
IF <general circumstance> THEN <general conclusion>
```

Since it is based on trial-and-error, induction can be said to be an empirical approach. We can never be certain of the accuracy of an induced rule, since it may be shown to be invalid by an example that we have not yet encountered. The aim of induction is to build rules that are successful as often as possible, and to modify them quickly when they are found to be wrong. Whatever is being learned — typically rules and relationships — should match the positive examples but not the negative ones.

The first step is to generate an initial prototype rule that can subsequently be refined. The initial prototype may be a copy of a general-purpose template, or it can be generated by hypothesizing a causal link between a pair of observations. For instance, if a specific valve `valve_1` in a particular plant is open and the flow rate through it is $0.5m^3s^{-1}$, we can propose two initial prototype rules:

```
IF valve_1 is open
THEN flow rate through valve_1 is 0.5

IF flow rate through valve_1 is 0.5
THEN valve_1 is open
```

The prototype rules can then be modified in the light of additional example data or rejected. Rule modifications can be classified as either strengthening or weakening. The condition is made stronger (or *specialized*) by restricting the circumstances to which it applies, and conversely it is made weaker (or *generalized*) by increasing its applicability. The pair of examples above could be made more general by considering other valves or a less precise measure of flow. On the other hand, they could be made more specific by specifying the necessary state of other parts of the boiler. A rule needs to be generalized if it fails to fire for a given set of data, where we are told by the teacher that it should have fired. Conversely, a rule needs to be specialized if it fires when it should not.

Assume for the moment that we are dealing with a rule that needs to be generalized. The first task is to spot where the condition is deficient. This is easy when using pattern matching, provided that we have a suitable representation. Consider the following prototype rule:

```
IF ?x is open AND ?x is a Gas_valve
THEN flow rate through ?x is high
```

This rule would fire given the scenario:

```
valve_1 is open
valve_1 is a Gas_valve
```

The rule is able to fire because `valve_1 is open` matches `?x is open` and `valve_1 is a Gas_valve` matches `?x is a Gas_valve`. The conclusion `flow rate through valve_1 is high` would be drawn. Now consider the following scenario:

```
valve_2 is open
valve_2 is a Water_valve
```

The rule would not fire. However, the teacher may tell us that the conclusion `flow rate through valve_2 is high` should have been drawn. We now look for matches as before and find that `valve_1 is open` matches `?x is`

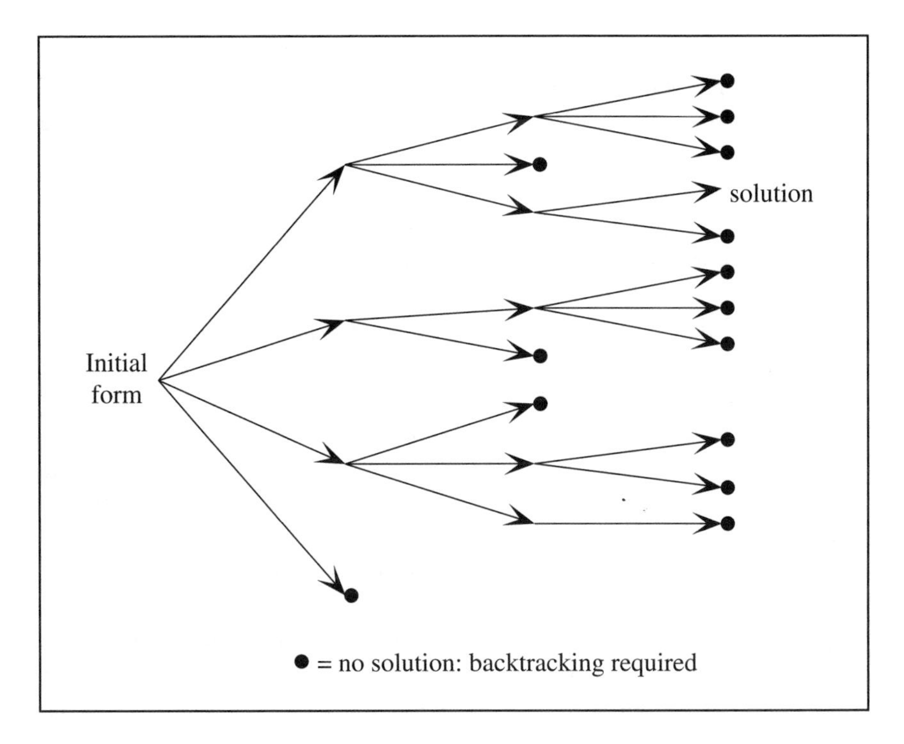

Figure 6.1 A search tree for rules

`open`. However, there is no match to `?x is a Gas_valve`, so this part of the condition needs to be generalized to embrace the circumstance `?x is a Water_valve`.

We can therefore recognize where a rule is deficient by pattern-matching between the condition part of the rule and the scenario description. This is analogous to *means–ends analysis*, which is used to determine a plan for changing the world from its current state to a goal state (see Sections 5.3 and 13.3). Means–ends analysis typically uses pattern-matching to determine which rules can lead to such a change.

6.2.2 Learning viewed as a search problem

The task of generalizing or specializing a rule is not straightforward, as there are many alternative ways in which a rule can be changed. Finding the correctly modified rule is a search problem, where the search field can be enormous. Figure 6.1 shows a possible form of the search tree, where each branch represents a generalization or specialization that would correctly handle the most recently encountered input. Subsequent inputs may reveal an incorrect choice (indicated by a dot at the end of a branch). The system must keep track of its current position in the search tree, as it must backtrack whenever an unsuitable choice is found to have been made.

Recording all of the valid options (or branches) that make up a search tree is likely to be unwieldy. The Lex system [2] offers a neat solution to this problem. Rather than keeping track of all the possible solutions, it simply records the most general and the most specific representations that fit the examples. These boundaries define the range of acceptable solutions. The boundaries move as further examples are encountered, converging on a smaller and smaller choice of rules, and ideally settling on a single rule.

There are other difficulties associated with the search problem. Suppose that a rule-based system is controlling a boiler. It makes a series of adjustments, but ultimately the boiler overheats, i.e., the control objective has not been achieved. In this case, the feedback of the temperature reading to the learning system serves as the teacher. The system is faced with the difficulty of knowing where it went wrong, i.e., which of its decisions were good and which ones were at fault. This is the *credit-assignment problem*. Credit assignment applies not only to negative examples such as this (which could be termed "blame assignment"), but also to cases where the overall series of decisions was successful. For example, a boiler controller which succeeds in keeping the temperature within the specified range might do so because several good decisions compensate for poorer ones.

The *frame problem* (or *situation-identification problem*), introduced in Chapter 1, affects many areas of artificial intelligence, particularly planning (Chapter 13). It is also pertinent here, where the problem is to determine which aspects of a given example situation are relevant to the new rule. A system for control of a boiler will have access to a wide range of information. Suppose that it comes across a set of circumstances where it is told by the teacher that it should shut off `valve_2`. The current world state perceived by the system is determined by stored data and sensor values, for example:

```
valve_1:                          shut
valve_2:                          open
gas_flow_rate:                    high
gas_temperature:                  300°C
water_temperature (pressurized):  150°C
fuel oil stored:                  100 gallons
preferred supplier of fuel oil:   ACME inc.
```

In order to derive a rule condition such as:

```
IF gas_flow_rate is high AND valve_2 is open THEN shut valve_2
```

the system must be capable of deducing which parts of its world model to ignore — such as the information about fuel oil — and which to include. The ability to distinguish the relevant information from the rest places a requirement that the system must already have some knowledge about the domain.

6.2.3 Techniques for generalization and specialization

We can identify at least five methods of generalizing (or specializing through the inverse operation):

(i) universalization;
(ii) replacing constants with variables;
(iii) using disjunctions (generalization) and conjunctions (specialization);
(iv) moving up a hierarchy (generalization) or down it (specialization);
(v) chunking.

Universalization

Universalization involves inferring a new general rule from a set of specific cases. Consider the following series of separate scenarios:

```
valve_1 is open AND flow rate for valve_1 is high
valve_2 is open AND flow rate for valve_2 is high
valve_3 is open AND flow rate for valve_3 is high
```

From these we might induce the following general rule:

```
/* Rule 6.1 */
IF ?x is open THEN flow rate for ?x is high
```

Thus we have generalized from a few specific cases to a general rule. This rule turns out to be *too* general as it does not specify that `x` must be a valve. However, this could be fixed through subsequent specialization based on some negative examples.

Replacing constants with variables

Universalization shows how we might induce a rule from a set of example scenarios. Similarly, general rules can be generated from more specific ones by replacing constants with variables (see Section 2.6). Consider, for example, the following specific rules:

```
IF gas_valve_1 is open THEN flow rate for gas_valve_1 is high
IF gas_valve_2 is open THEN flow rate for gas_valve_2 is high
```

```
IF gas_valve_3 is open THEN flow rate for gas_valve_3 is high
IF gas_valve_4 is open THEN flow rate for gas_valve_4 is high
IF gas_valve_5 is open THEN flow rate for gas_valve_5 is high
```

From these specific rules we might induce a more general rule:

```
/* Rule 6.1 */
IF ?x is open THEN flow rate for ?x is high
```

However, even this apparently simple change requires a certain amount of metaknowledge (knowledge about knowledge). The system must "know" to favor Rule 6.1 rather than, say:

```
/* Rule 6.2 */
IF ?x is open THEN flow rate for ?y is high
```

or:

```
/* Rule 6.3 */
IF ?x is open THEN ?y for ?x is high
```

Rule 6.2 implies that if any valve is open (we'll assume for now that we are only dealing with valves) then the flow rate through all valves is high. Rule 6.3 implies that if a valve is open then everything associated with that valve (e.g., cost and temperature) is high.

Using conjunctions and disjunctions

Rules can be made more specific by adding conjunctions to the condition and more general by adding disjunctions. Suppose that Rule 6.1 is applied when the world state includes the information `office_door is open`. We will draw the nonsensical conclusion that `flow rate through office_door is high`. The rule clearly needs to be modified by strengthening the condition. One way to achieve this is by use of a conjunction (AND):

```
/* Rule 6.4 */
IF ?x is open AND ?x is a Gas_valve THEN flow rate for ?x is
high
```

We are relying on the teacher to tell us that `flow rate through office_door is high` is not an accurate conclusion. We have already noted that there may be several alternative ways in which variables might be introduced. The number of alternatives greatly increases when compound conditions are used. Here are some examples:

```
IF valve_1 is open and valve_1 is a Gas_valve THEN ...
IF valve_1 is open and ?x is a Gas_valve THEN ...
IF ?x is open and valve_1 is a Gas_valve THEN ...
IF ?x is open AND ?x is a Gas_valve THEN ...
IF ?x is open AND ?y is a Gas_valve THEN ..
```

The existence of these alternatives is another illustration that learning by rule induction is a search process in which the system searches for the correct rule.

Suppose now that we wish to extend Rule 6.4 so that it includes water valves as well as gas valves. One way of doing this is to add a disjunction (`OR`) to the condition part of the rule:

```
/* Rule 6.5 */
IF ?x is open AND (?x is a Gas_valve OR ?x is a Water_valve)
THEN flow rate for ?x is high
```

This is an example of generalization by use of disjunctions. The use of disjunctions in this way is a "cautious generalization," as it caters for the latest example, but does not embrace any novel situations. The other techniques risk overgeneralization, but the risky approach is necessary if we want the system to learn to handle data that it may not have seen previously. Overgeneralization can be fixed by specialization at a later stage when negative examples have come to light. However, the use of disjunctions should not be avoided altogether, as there are cases where a disjunctive condition is correct. This dilemma between the cautious and risky approaches to generalization is called the *disjunctive-concept* problem. A reasonable approach is to look for an alternative form of generalization, only resorting to the use of disjunctions if no other form can be found that fits the examples.

Moving up or down a hierarchy

Rule 6.5 showed a cautious way of adapting Rule 6.4 to include both water valves and gas valves. Another approach would be to modify Rule 6.4 so that it dealt with valves of any description:

```
/* Rule 6.6 */
IF ?x is open AND ?x is a Valve THEN flow rate for ?x is high
```

Here we have made use of an is-a-kind-of relationship in order to generalize (see Section 4.5). In the class hierarchy for valves, `Water_valve` and `Gas_valve` are both specializations of the class `Valve`, as shown in Figure 6.2.

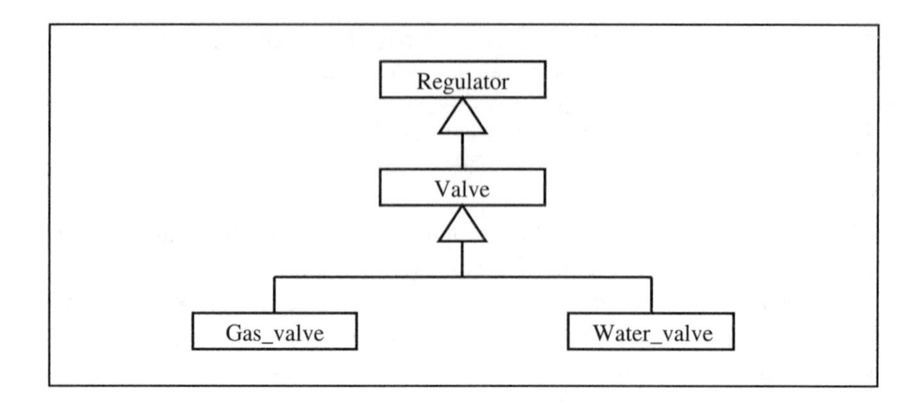

Figure 6.2 Class hierarchy for valves

Chunking

Chunking is a mechanism for automated learning that is used in SOAR [3, 4]. SOAR is a production system (i.e., one that uses production rules — see Chapter 2) that has been modeled on a theory of human cognition. It works on the premise that, given an overall goal, every problem encountered along the way can be regarded as a subgoal. Problems are, therefore, tackled hierarchically — if a goal cannot be met at one level, it is broken down into subgoals. These automatically generated subgoals may be satisfied through the application of production rules stored in a long-term memory from previous runs of the system. However, the application of such rules may be slow, as several of them may need to be fired successively to satisfy the subgoal. Once SOAR has recognized the series of rules required to satisfy a subgoal, it can collapse them down into a single production rule. This is the process of *chunking*. The new rule, or *chunk*, is then stored so that it can rapidly solve the same subgoal if it should arise again in the future.

SOAR also offers a novel method of conflict resolution (see Chapter 2). The usual role of a production rule is to change some aspect of the system's state, which models the problem being tackled. In SOAR, any rule that can fire does so, but the changes to the current state proposed by the rule are not applied immediately. Instead they are added to a list of suggested changes. A separate set of production rules is then applied to arrange conflicting suggestions in order of preference — a process known as *search control*. Once conflict has been resolved through search control, changes are made to the actual state elements.

6.3 Case-based reasoning (CBR)

The design of intelligent systems is often inspired by attempts to emulate the characteristics of human intelligence. One such characteristic is the ability to recall previous experience whenever a similar problem arises. This is the essence of *case-based reasoning* (CBR). Consider the example of diagnosing a fault in a refrigerator, an example that will be revisited in Chapter 11. If an expert system has made a successful diagnosis of the fault, given a set of symptoms, it can file away this information for future use. If the expert system is subsequently presented with details of another faulty refrigerator of exactly the same type, displaying exactly the same symptoms in exactly the same circumstances, then the diagnosis can be completed simply by recalling the previous solution. However, a full description of the symptoms and the environment would need to be very detailed, and the chances of it ever being exactly reproduced are remote. What we need is the ability to identify a previous case, the solution to which can be modified to reflect the slightly altered circumstances. Thus case-based reasoning involves two difficult problems:

- determining which cases constitute a similar problem to the current one;
- adapting a case to the current circumstances.

6.3.1 Storing cases

The stored cases form a *case base*. An effective way of representing the relevance of cases in the case base is by storing them as objects. Riesbeck and Schank [5] have defined a number of types of link between classes and instances in order to assist in locating relevant cases. These links are described below and examples of their use are shown in Figure 6.3.

Abstraction links and index links

The classes may form a structured hierarchy, in which the different levels correspond to the level of detail of the case descriptions. Riesbeck and Schank [5] distinguish two types of link between classes and their specializations: abstraction links and index links. A subclass connected by an abstraction link provides additional detail to its superclass, without overriding any information. An index link is a specialization in which the subclass has an attribute value that is different from the default defined in its superclass. For example, the class `Refrigerator_fault` might have an index link to the class `Electrical_fault`, whose defaults relate to faults in vacuum cleaners (Figure 6.3(a)).

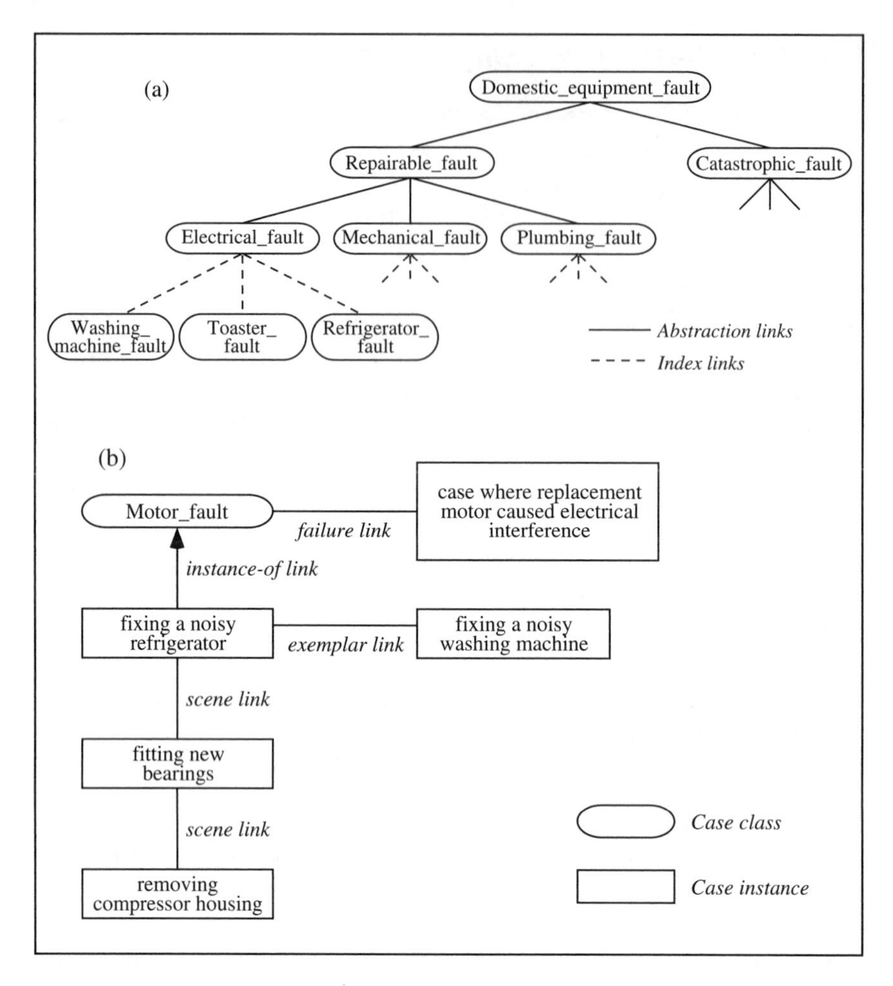

Figure 6.3 Classifying case histories:
(a) abstraction links and index links between classes;
(b) links between instances

Suppose that we wish to save an instance of fixing a refrigerator, where the refrigerator exhibited the following symptoms:

- failed to chill food;
- made no noise;
- the light would not come on.

This case might be stored as an instance of the class `Refrigerator fault`, which, as already noted, has an index link to the class `Electrical fault`.

Instance-of links

Each case is an instance of a specific class of cases (see Chapter 4 for a discussion of object classes and instances). Thus, each case has an *instance-of* relation with its parent class, as shown in Figure 6.3(b).

Scene links

These are used to link a historical event to its subevents. For example, removing the outer casing of a refrigerator's compressor is a subevent of the "changing compressor bearings" event. Both the event and subevent are stored as case instances, with a scene link between them (Figure 6.3(b)).

Exemplar links

These are used to link instances to other similar instances. Suppose that a refrigerator motor fault was diagnosed by referring to a previous case involving a failed washing machine motor. This new case could have an exemplar link to the case from which it was adapted (Figure 6.3(b)).

Failure links

A failure link is a special type of class–instance relation, where the instance represents a specific case where things did not turn out as expected. This is a convenient way of storing cases that form exceptions to their general category. For example, the class `Motor fault` might have a failure link to a case in which a replacement motor created a new problem, such as radio interference (Figure 6.3(b)).

6.3.2 Retrieving cases

The problem of retrieving cases that match a new case, termed the *probe* case, is eased if the case base is carefully indexed using links as described above. Such links are often used as a basis for the first stage of a two-stage interrogation of the case base. The second stage involves ranking the matched cases according to a measure of similarity to the probe case [6].

For a given probe case, a suitable similarity measure S_i for case i in the case bases could be:

$$S_i = \sum_{j=1}^{P} w_j m_{ij}$$

where P is the number of parameters considered, w_j is an importance weighting applied to parameter j, and m_{ij} is a degree of match between case i and the probe case for the parameter j. For some parameters, such as the presence or absence of a given observation, m_{ij} is binary i.e., it is either 0 or 1. Other parameters may concern continuous variables such as temperature, or nearly continuous variables such as cost. These can still yield binary values for m_{ij} if the measure is whether the parameter is within an acceptable range or tolerance. Alternatively, a sliding scale between 0 and 1 can be applied to m_{ij}. This is akin to using a fuzzy membership value rather than a crisp one (see Chapter 3).

6.3.3 Adapting case histories

There are two distinct categories of techniques for adapting a case history to suit a new situation, namely, structural and derivational adaptation. *Structural* adaptation describes techniques that use a previous solution as a guide and adapt it to the new circumstances. *Derivational* adaptation involves looking at the *way* in which the previous solution was derived, rather than the solution itself. The same reasoning processes are then applied to the set of data describing the new circumstances, i.e., the probe case. Four structural techniques and one derivational technique are outlined below.

Null adaptation

This is the simplest approach and, as the name implies, involves no adaptation at all of the previous solution. Instead, the previous solution is given as the solution to the new case. Suppose that the case history selector decides that a failed refrigerator is similar to the case of a failed washing machine. If the washing machine failure was found to be due to a severed power lead, then this is offered as the solution to the refrigerator problem.

Parameterization

This is a structural adaptation technique that is applicable when both the symptoms and the solution have an associated magnitude or extent. The previous solution can then be scaled up or down in accordance with the severity of the symptoms. Suppose, for example, that a case history was as follows:

symptom: fridge cabinet temperature is 15°C, which is too warm;
solution: reduce thermostat setting by 11°C.

If our new scenario involves a fridge whose cabinet temperature is 10°C, which is still warmer than it should be, the new solution would be to turn down the thermostat, but by a modified amount (say 6°C).

Reasoning by analogy

This is another structural adaptation technique. If a case history cannot be found in the most appropriate class, then analogous case histories are considered. Given a hierarchically organized database of case histories, the search for analogous cases is relatively straightforward. The search begins by looking at siblings, and then cousins, in a class hierarchy like the one shown in Figure 6.3(a). Some parts of the historical solution may not applicable, as the solution belongs to a different class. Under such circumstances, the inapplicable parts are replaced by referring back to the class of the current problem.

As an example, consider a refrigerator that is found to be excessively noisy. There may not be a case in the database that refers to noisy refrigerators, but there may be a case that describes a noisy washing machine. The solution in that case may have been that the bearings on the washer's drum were worn and needed replacing. This solution is not directly applicable to the refrigerator, as a refrigerator does not have a drum. However, the refrigerator has bearings on the compressor, and so it is concluded that the compressor bearings are worn and are in need of replacement.

Critics

The use of critics has stemmed from work on a planning system called HACKER [7], which can rearrange its planned actions if they are found to be incorrectly ordered (see Chapter 13). The ideas are also applicable to other problems, such as diagnosis (see Chapter 11). Critics are modules that can look at a nearly correct solution and determine what flaws it has, if any, and suggest modifications. In the planning domain, critics would look for unnecessary actions, or actions that make subsequent actions more difficult. Adapting these ideas to diagnosis, critics can be used to "fine-tune" a previous solution so that it fits the current circumstances. For instance, a case-based reasoner might diagnose that the compressor bearings in a refrigerator need replacing. Critics might notice that most compressors have two sets of bearings and that, in this particular refrigerator, one set is fairly new. The modified solution would then be to replace only the older set of bearings.

Reinstantiation

The above adaptation techniques are all structural, i.e., they modify a previous solution. Reinstantiation is a derivational technique, because it involves replaying the *derivation* of the previous solution using the new data. Previously used names, numbers, structures, and components are reinstantiated to the corresponding new ones. Suppose that a case history concerning a central heating system contained the following abduction to explain from a set of a rules why a room felt cold:

```
IF thermostat is set too low
THEN boiler will not switch on

IF boiler is not switched on
THEN radiators stay at ambient temperature

IF radiators are at ambient temperature
THEN room will not warm up

Abductive conclusion: thermostat is set too low
```

By suitable reinstantiation, this case history can be adapted to diagnose why food in a refrigerator is not chilled:

```
IF thermostat is set too high
THEN compressor will not switch on

IF compressor is not switched on
THEN cabinet stays at ambient temperature

IF cabinet is at ambient temperature
THEN food will not be chilled

Abductive conclusion: thermostat is set too high
```

6.3.4 Dealing with mistaken conclusions

Suppose that a system has diagnosed that a particular component is faulty, and that this has caused the failure of an electronic circuit. If it is then discovered that there was in fact nothing wrong with the component, or that replacing it made no difference, then the conclusion needs *repair*. Repair is conceptually very similar to adaptation, and similar techniques can be applied so as to modify the incorrect conclusion to reach a correct one. If this fails, then a completely new solution must be sought. In either case, it is important that the failed conclusion be recorded in the database of case histories with a link to the correct conclusion. If the case is subsequently retrieved in a new scenario, the system will be aware of a possible failure and of a possible way around that failure.

6.4 Summary

Systems that can learn offer a way around the so-called "knowledge acquisition bottleneck." In cases where it is difficult or impossible to extract accurate knowledge about a specific domain, it is clearly an attractive proposition for a computer system to derive its own representation of the knowledge from examples. We have distinguished between two categories of learning systems — symbolic and numerical — with this chapter focusing on the former. Inductive and cased-based methods are two particularly important types of symbolic learning.

Rule induction involves the generation and refinement of prototype rules from a set of examples. An initial prototype rule can be generated from a template or by hypothesizing a causal link between a pair of observations. The prototype rule can then be refined in the light of new evidence by generalizing, specializing, or rejecting it. Rule induction from numerical systems such as neural networks is also possible; discussion of this is deferred until Chapter 9 on hybrid systems.

Case-based reasoning involves storing the details of every case encountered, successful or not. A stored case can subsequently be retrieved and adapted for new, but related, sets of circumstances. This is arguably a good model of human reasoning. The principal difficulties are in recognizing relevant cases, which necessitates a suitable storage and retrieval system, and in adapting stored cases to the new circumstances. The concepts of CBR have been illustrated here with reference to fault diagnosis, but CBR has also found a wide range of other applications including engineering sales [8], help-desk support [9], and planning [10].

References

1. Minton, S., Carbonell, J. G., Knoblock, C. A., Kuokka, D. R., Etzioni, O., and Gil, Y., "Explanation-based learning: a problem-solving perspective," *Artificial Intelligence*, vol. 40, pp. 63–118, 1989.
2. Mitchell, T. M., Utgoff, P. E., and Banerji, R., "Learning by experimentation: acquiring and refining problem-solving heuristics," in *Machine Learning: an artificial intelligence approach*, vol. 1, Michalski, R., Carbonell, J. G., and Mitchell, T. M. (Eds.), pp. 163–190, 1983.
3. Laird, J. E., Rosenbloom, P. S., and Newell, A., "Chunking in SOAR: the anatomy of a general learning mechanism," *Machine Learning*, vol. 1, pp. 11–46, 1986.

4. Laird, J. E., Newell, A., and Rosenbloom, P. S., "SOAR: an architecture for general intelligence," *Artificial Intelligence*, vol. 33, pp. 1–64, 1987.
5. Riesbeck, C. K. and Schank, R. C., *Inside Case-Based Reasoning*, Lawrence Erlbaum Associates, 1989.
6. Ferguso, A. and Bridge, D., "Options for query revision when interacting with case retrieval systems," *Expert Update*, vol. 3, issue 1, pp. 16–27, Spring 2000.
7. Sussman, G. J., *A Computer Model of Skill Acquisition*, Elsevier, 1975.
8. Watson, I. and Gardingen, D., "A distributed case-based reasoning application for engineering sales support," 16th International Joint Conference on Artificial Intelligence (IJCAI'99), Stockholm, Sweden, vol. 1, pp. 600–605, Morgan Kaufmann, 1999.
9. Kunze, M. and Hübner, A., "Textual CBR case studies of projects performed," Lenz, M. and Ashley, K. (Eds.), AAAI'98 Workshop on Textual Case-Based Reasoning, Menlo Park, CA, pp. 58–61, 1998.
10. Marefat, M. and Britanik, J., "Case-based process planning using an object-oriented model representation," *Robotics and Computer Integrated Manufacturing*, vol. 13, pp. 229–251, 1997.

Further reading

- Kolodner, J., *Case-Based Reasoning*, Morgan Kaufmann, 1993.
- Langley, P., *Elements of Machine Learning*, Morgan Kaufmann, 1995.
- Mitchell, T. M., *Machine Learning*, McGraw-Hill, 1997.
- Watson, I. D., *Applying Case-Based Reasoning: techniques for enterprise systems*, Morgan Kaufmann, 1997.

Chapter seven

Optimization algorithms

7.1 Optimization

We have already seen that symbolic learning by induction is a search process, where the search for the correct rule, relationship, or statement is steered by the examples that are encountered. Numerical learning systems can be viewed in the same light. An initial model is set up, and its parameters are progressively refined in the light of experience. The goal is invariably to determine the maximum or minimum value of some function of one or more variables. This is the process of *optimization*. Often the optimization problem is considered to be one of determining a minimum, and the function that is being minimized is referred to as a *cost function*. The cost function might typically be the difference, or *error*, between a desired output and the actual output. Alternatively, optimization is sometimes viewed as maximizing the value of a function, known then as a *fitness function*. In fact the two approaches are equivalent, because the fitness can simply be taken to be the negation of the cost and vice versa, with the optional addition of a constant value to keep both cost and fitness positive. Similarly, fitness and cost are sometimes taken as the reciprocals of each other. The term *objective function* embraces both fitness and cost. Optimization of the objective function might mean either minimizing the cost or maximizing the fitness.

7.2 The search space

The potential solutions to a search problem constitute the *search space* or *parameter space*. If a value is sought for a single variable, or parameter, the search space is one-dimensional. If simultaneous values of n variables are sought, the search space is n-dimensional. Invalid combinations of parameter values can be either explicitly excluded from the search space, or included on the assumption that they will be rejected by the optimization algorithm. In combinatorial problems, the search space comprises combinations of values, the order of which has no particular significance provided that the meaning of each value is known. For example, in a steel rolling mill the combination of parameters that describe the profiles of the rolls can be optimized to maximize the flatness of the manufactured steel [1]. Here, each possible combination of parameter values represents a point in the search space. The extent of the search space is constrained by any limits that apply to the variables.

In contrast, permutation problems involve the ordering of certain attributes. One of the best known examples is the traveling salesperson problem, where he or she must find the shortest route between cities of known location, visiting each city only once. This sort of problem has many real applications, such as in the routing of electrical connections on a semiconductor chip. For each permutation of cities, known as a tour, we can evaluate the cost function as the sum of distances traveled. Each possible tour represents a point in the search space. Permutation problems are often cyclic, so the tour ABCDE is considered the same as BCDEA.

The metaphor of space relies on the notion that certain points in the search space can be considered closer together than others. In the traveling salesperson example, the tour ABCDE is close to ABDCE, but DACEB is distant from both of them. This separation of patterns can be measured intuitively in terms of the number of pair-wise swaps required to turn one tour into another. In the case of binary patterns, the separation of the patterns is usually measured as the *Hamming distance* between them, i.e., the number of bit positions that contain different values. For instance, the binary patterns 01101 and 11110 have a Hamming separation of 3.

We can associate a fitness value with each point in the search space. By plotting the fitness for a two-dimensional search space, we obtain a *fitness landscape* (Figure 7.1). Here the two search parameters are x and y, constrained within a range of allowed values. For higher dimensions of search space a fitness landscape still exists, but is difficult to visualize. A suitable optimization algorithm would involve finding peaks in the fitness landscape or valleys in the cost landscape. Regardless of the number of dimensions, there is a risk of finding a local optimum rather than the global optimum for the function. A global optimum is the point in the search space with the highest fitness. A local optimum is a point whose fitness is higher than all its near neighbors but lower than that of the global optimum.

If neighboring points in the search space have a similar fitness, the landscape is said to be *smooth* or *correlated.* The fitness of any individual point in the search space is, therefore, representative of the quality of the surrounding region. Where neighboring points have very different fitnesses, the landscape is said to be *rugged.* Rugged landscapes typically have large

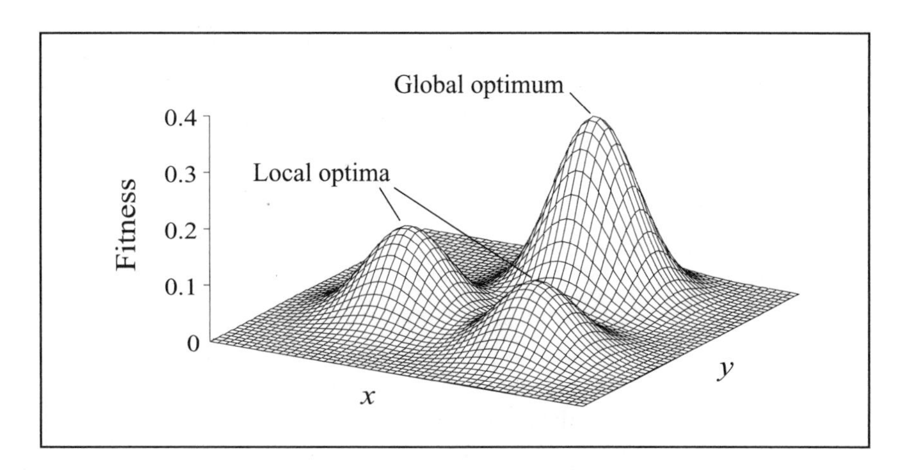

Figure 7.1 A fitness landscape

numbers of local optima and the fitness of an individual point in the search space will not necessarily reflect that of its neighbors.

The idea of a fitness landscape assumes that the function to be optimized remains constant during the optimization process. If this assumption cannot be made, as might be the case in a real-time system, we can think of the problem as finding an optimum in a "fitness seascape" [2].

7.3 Searching the search space

Determining the optimum for an objective function of multiple variables is not straightforward, even when the landscape is static. Although exhaustively evaluating the fitness of each point in the search space will always reveal the optimum, this is usually impracticable because of the enormity of the search space. Thus, the essence of all the numerical optimization techniques is to determine the optimum point in the search space by examining only a fraction of all possible candidates.

The techniques described here are all based upon the idea of choosing a starting point and then altering one or more variables in an attempt to increase the fitness or reduce the cost. The various approaches have the following two key characteristics.

(i) *Whether they are based on a single candidate or a population of candidates*

Some of the methods to be described, such as hill-climbing, maintain a single "best solution so far" which is refined until no further increase in fitness can be achieved. Genetic algorithms, on the other hand, maintain a population of candidate solutions. The overall fitness of the population generally improves with each generation, although some decidedly unfit individual candidates may be added along the way.

(ii) *Whether new candidates can be distant in the search space from the existing ones*
Methods such as hill-climbing take small steps from the start point until they reach either a local or global optimum. To guard against missing the global optimum, it is advisable to repeat the process several times, starting from different points in the search space. An alternative approach, adopted in genetic algorithms and simulated annealing, is to begin with the freedom to roam around the whole of the search space in order to find the regions of highest fitness. This initial *exploration* phase is followed by *exploitation*, i.e., a detailed search of the best regions of the search space identified during exploration. Methods, such as genetic algorithms, that use a population of candidates rather than just one allow several regions to be explored at the same time.

7.4 Hill-climbing and gradient descent algorithms

7.4.1 Hill-climbing

The name *hill-climbing* implies that optimization is viewed as the search for a maximum in a fitness landscape. However, the method can equally be applied to a cost landscape, in which case a better name might be *valley descent*. It is the simplest of the optimization procedures described here. The algorithm is easy to implement, but is inefficient and offers no protection against finding a local minimum rather than the global one. From a randomly selected start point in the search space, i.e., a trial solution, a step is taken in a random direction. If the fitness of the new point is greater than the previous position, it is accepted as the new trial solution. Otherwise the trial solution is unchanged. The process is repeated until the algorithm no longer accepts any steps from the trial solution. At this point the trial solution is assumed to be the optimum. As noted above, one way of guarding against the trap of detecting a local optimum is to repeat the process many times with different starting points.

7.4.2 Steepest gradient descent or ascent

Steepest gradient descent (or ascent) is a refinement of hill-climbing that can speed the convergence toward a minimum cost (or maximum fitness). It is only slightly more sophisticated than hill-climbing, and it offers no protection against finding a local minimum rather than the global one. From a given starting point, i.e., a trial solution, the direction of steepest descent is determined. A point lying a small distance along this direction is then taken as the new trial solution. The process is repeated until it is no longer possible to descend, at which point it is assumed that the optimum has been reached.

If the search space is not continuous but discrete, i.e., it is made up of separate individual points, at each step the new trial solution is the neighbor with the highest fitness or lowest cost. The most extreme form of discrete data is where the search parameters are binary, i.e., they have only two possible values. The parameters can then be placed together so that any point in the search space is represented as a binary string and neighboring points are those at a Hamming distance (see Section 7.2) of 1 from the current trial solution.

7.4.3 Gradient-proportional descent

Gradient-proportional descent, often simply called *gradient descent*, is a variant of steepest gradient descent that can be applied in a cost landscape that is continuous and differentiable, i.e., where the variables can take any value within the allowed range and the cost varies smoothly. Rather than choosing a fixed step size, the size of the steps is allowed to vary in proportion to the local gradient of descent.

7.4.4 Conjugate gradient descent or ascent

Conjugate gradient descent (or ascent) is a simple attempt at avoiding the problem of finding a local, rather than global, optimum in the cost (or fitness) landscape. From a given starting point in the cost landscape, the direction of steepest descent is initially chosen. New trial solutions are then taken by stepping along this direction, with the same direction being retained until the slope begins to curve uphill. When this happens, an alternative direction having a downhill gradient is chosen. When the direction that has been followed curves uphill, and all of the alternative directions are also uphill, it is assumed that the optimum has been reached. As the method does not continually hunt for the sharpest descent, it may be more successful than the steepest gradient descent method in finding the global minimum. However, the technique will never cause a gradient to be climbed, even though this would be necessary in order to escape a local minimum and thereby reach the global minimum.

7.5 Simulated annealing

Simulated annealing [3] owes its name to its similarity to the problem of atoms rearranging themselves in a cooling metal. In the cooling metal, atoms move to form a near-perfect crystal lattice, even though they may have to overcome a localized energy barrier called the activation energy, E_a, in order to do so. The atomic rearrangements within the crystal are probabilistic. The probability P of an atom jumping into a neighboring site is given by:

$$P = \exp(-E_a / kT) \tag{7.1}$$

where k is Boltzmann's constant and T is temperature. At high temperatures, the probability approaches 1, while at $T = 0$ the probability is 0.

In simulated annealing, a trial solution is chosen and the effects of taking a small random step from this position are tested. If the step results in a reduction in the cost function, it replaces the previous solution as the current trial solution. If it does not result in a cost saving, the solution still has a probability P of being accepted as the new trial solution given by:

$$P = \exp(-\Delta E / T) \tag{7.2}$$

This function is shown in Figure 7.2(a). Here, ΔE is the increase in the cost function that would result from the step and is, therefore, analogous to the activation energy in the atomic system. There is no need to include Boltzmann's constant, as ΔE and T no longer represent real energies or temperatures.

The temperature T is simply a numerical value that determines the stability of a trial solution. If T is high, new trial solutions will be generated continually. If T is low, the trial solution will move to a local or global cost minimum — if it is not there already — and will remain there. The value of T is initially set high and is periodically reduced according to a cooling schedule. A commonly used simple cooling schedule is:

$$T_{t+1} = \alpha T_t \tag{7.3}$$

where T_t is the temperature at step number t and α is a constant close to, but below, 1. While T is high, the optimization routine is free to accept many varied solutions, but as it drops, this freedom diminishes. At $T = 0$, the method is equivalent to the hill-climbing algorithm, as shown in Figure 7.2(b).

If the optimization is successful, the final solution will be the global minimum. The success of the technique is dependent upon values chosen for

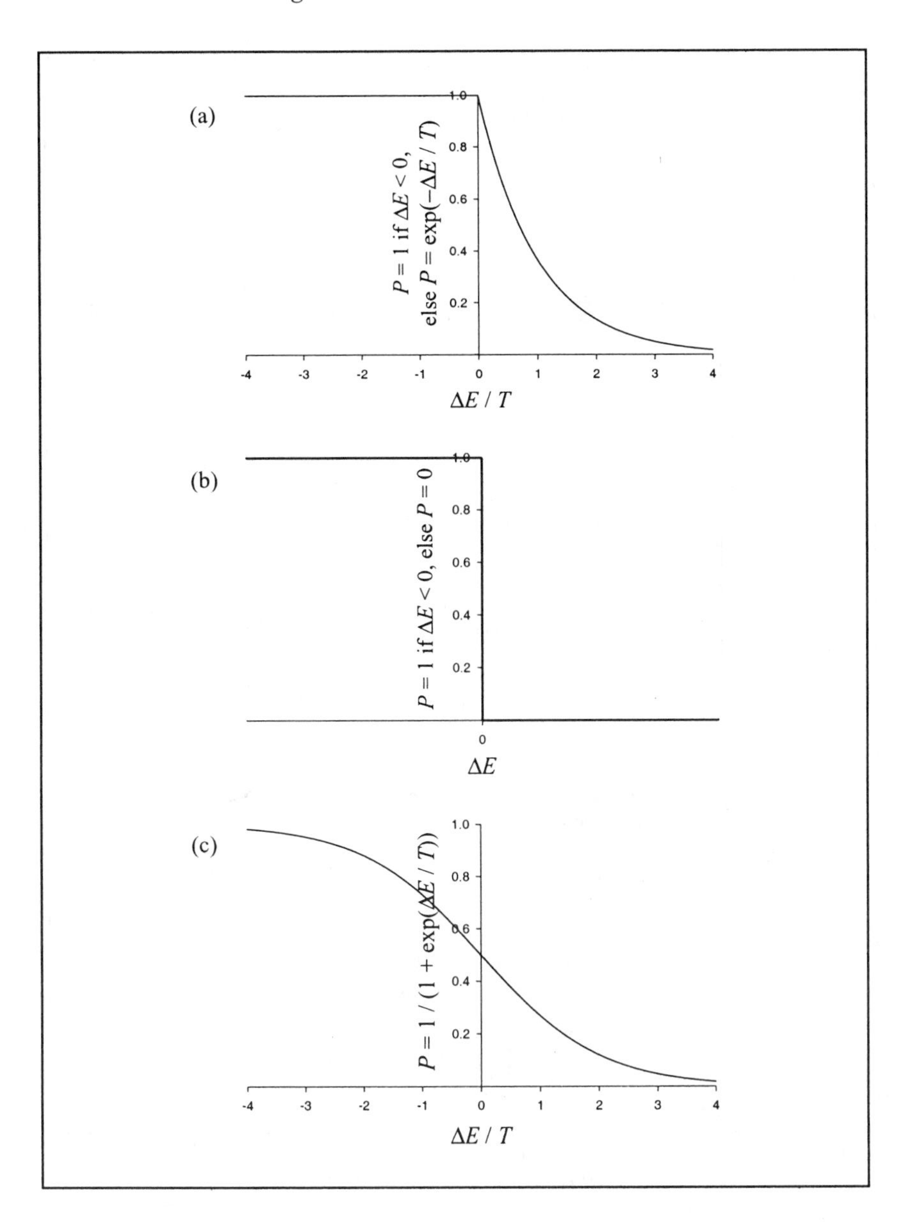

Figure 7.2 Three probability functions

starting temperature, the size and frequency of the temperature decrement, and the size of perturbations applied to the trial solutions. A flowchart for the simulated annealing algorithm is given in Figure 7.3.

Johnson and Picton [4] have described a variant of simulated annealing in which the probability of accepting a trial solution is always probabilistic, even

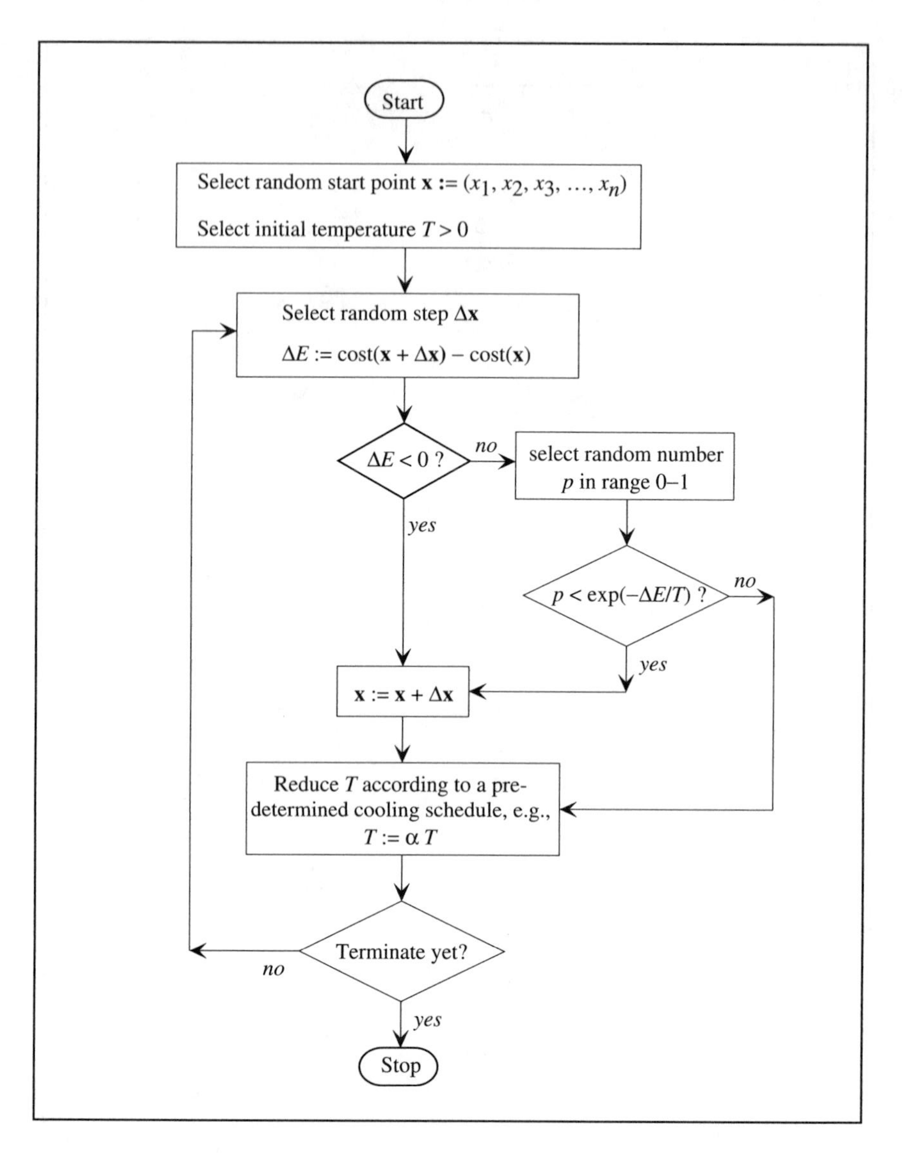

Figure 7.3 Simulated annealing

if it results in a decrease in the cost function (Figure 7.2(c)). Under their scheme, the probability of accepting a trial solution is:

$$P = \frac{1}{1 + \exp(\Delta E / T)} \tag{7.4}$$

where ΔE is positive if the new trial solution increases the cost function, or negative if it decreases it. In the former case P is in the range 0–0.5 and in the latter case it is in the range 0.5–1. At high temperatures, P is close to 0.5 regardless of the fitness of the new trial solution. As with standard simulated annealing, at $T = 0$ the method becomes equivalent to the hill-climbing algorithm, as shown in Figure 7.2(b).

Simulated annealing has been successfully applied to partitioning circuits into electronic chips, positioning chips on printed circuit boards, and circuit layout within chips [3]. It has also been applied to setting the parameters for a finisher mill in the rolling of sheet steel [1].

7.6 Genetic algorithms

Genetic algorithms (GAs) have been inspired by natural evolution, the process by which successive generations of animals and plants are modified so as to approach an optimum form. Each offspring has different features from its parents, i.e., it is not a perfect copy. If the new characteristics are favorable, the offspring is more likely to flourish and pass its characteristics to the next generation. However, an offspring with unfavorable characteristics is likely to die without reproducing. These ideas have been applied to mathematical optimization, where a population of candidate solutions "evolves" toward an optimum [5].

Each cell of a living organism contains a set of chromosomes that define the organism's characteristics. The chromosomes are made up of genes, where each gene determines a particular trait such as eye color. The complete set of genetic material is referred to as the *genome*, and a particular set of gene values constitutes a *genotype*. The resulting set of traits is described as the *phenotype*.

Each individual in the population of candidate solutions is graded according to its fitness. The higher the fitness of a candidate solution, the greater are its chances of reproducing and passing its characteristics to the next generation. In order to implement a GA, the following design decisions need to be made:

- how to use sequences of numbers, known as *chromosomes*, to represent the candidate solutions;
- the size of the population;
- how to evaluate the fitness of each member of the population;
- how to select individuals for reproduction using fitness information (conversely, how to determine which less-fit individuals will not reproduce);

- how to reproduce candidates, i.e., how to create a new generation of candidate solutions from the existing population;
- when to stop the evolutionary process.

These decisions will be addressed in detail in subsequent subsections but, for now, let us look at the most basic form of GA.

7.6.1 The basic GA

All of the other numerical optimization techniques described in the previous sections involved storing just one "best so far" candidate solution. In each case, a new trial solution was generated by taking a small step in a chosen direction. Genetic algorithms are different in both respects. First, a population of several candidate solutions, i.e., chromosomes, is maintained. Second, the members of one generation can be a considerable distance in the search space from the previous generation.

Chromosomes

Each point in the search space can be represented as a unique chromosome, made up of *genes*. Suppose, for example, we are trying to find the maximum value of a fitness function, $f(x, y)$. In this example, the search space variables, x and y, are constrained to the 16 integer values in the range 0–15. A chromosome corresponding to any point in the search space can be represented by two genes:

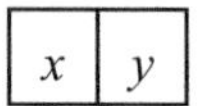

Thus the point (2, 6) in search space would be represented by the following chromosome:

2	6

The possible values for the genes are called *alleles*, so there are 16 alleles for each gene in this example. Each position along the chromosome is known as a *locus*; there are two loci in the above example. The loci are usually constrained to hold only binary values. (The term *evolutionary algorithm* describes the more general case where this constraint is relaxed.) The chromosome could therefore be represented by eight loci comprising the binary numbers 0010 and 0110, which represent the two genes:

0	0	1	0	0	1	1	0

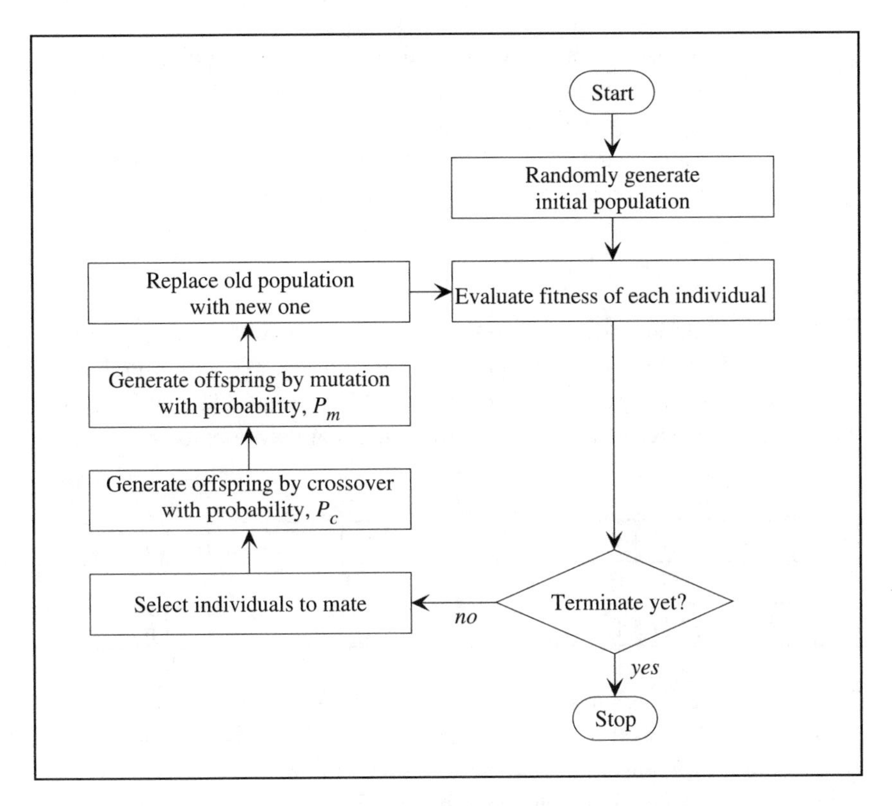

Figure 7.4 The basic GA

Although there are still 16 alleles for the genes, there are now only two possible values (0 and 1) for the loci. The chromosome can be made as long as necessary for problems involving many variables, or where many loci are required for a single gene. In general, there are 2^N alleles for a binary-encoded gene that is N bits wide.

2^{32}

Algorithm outline

A flow chart for the basic GA is shown in Figure 7.4. In the basic algorithm, the following assumptions have been made:

- The initial population is randomly generated.
- Individuals are evaluated according to the fitness function.
- Individuals are selected for reproduction on the basis of fitness; the fitter an individual, the more likely it is to be selected. Further details are given in Section 7.6.2 below.

- Reproduction of chromosomes to produce the next generation is achieved by "breeding" between pairs of chromosomes using the crossover operator and then applying a mutation operator to each of the offspring. The crossover and mutation operators are described below; the balance between them is yet another decision that the GA designer faces.

Crossover

Here, child chromosomes are produced by aligning two parents, picking a random position along their length, and swapping the tails with a probability P_c, known as the crossover probability. An example for an eight-loci chromosome, where the mother and father genes are represented by m_i and f_i respectively, would be:

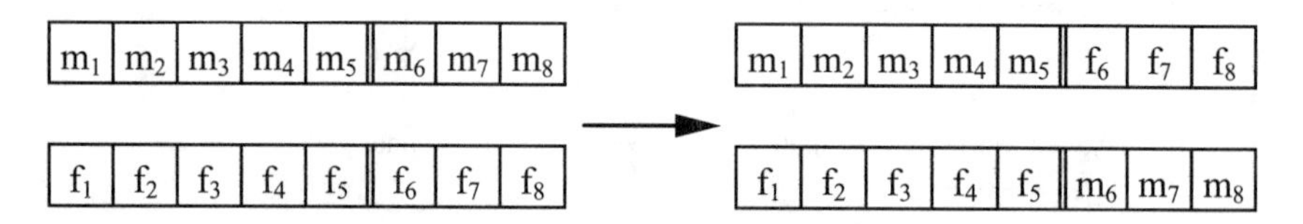

This is known as single-point crossover, as only one position is specified for separating the swapped and unswapped loci. In fact this is a misnomer, as a second cross-over position is always required. In single-point crossover the second crossover position is assumed to be the end of the chromosome. This can be made clearer by considering two-point crossover, where the chromosomes are treated as though they were circular, i.e., m_1 and m_8 are neighboring loci:

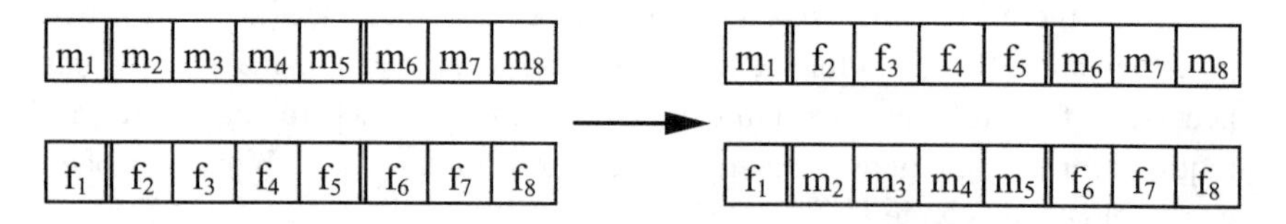

In general, multipoint crossover is also possible, provided there are an even number of crossover points:

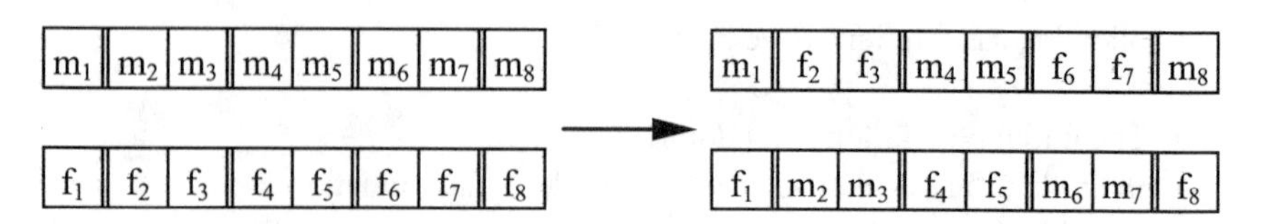

In the extreme case, each locus is considered for crossover, independently of the rest, with crossover probability P_c. This is known as *uniform crossover* [6].

Mutation

Unlike crossover, mutation involves altering the values of one or more loci. This creates new possibilities for gene combinations that can be generated by crossover. Mutation can be carried out in either of two ways:

- The value of a randomly selected gene can be replaced by a randomly generated allele. This works for both binary and nonbinary chromosomes.
- In a binary chromosome, randomly selected loci can be toggled, i.e., 1 becomes 0 and 0 becomes 1.

Individuals are selected randomly for mutation with a probability P_m. The main advantage of mutation is that it puts variety into the gene pool, enabling the GA to explore potentially beneficial regions of the search space that might otherwise be missed. This helps to counter premature convergence, described below.

Validity check

Depending on the optimization problem, an additional check may be required to ensure that the chromosomes in the new generation represent valid points in the search space. Consider, for example, a chromosome comprising four genes, each of which can take three possible values: A, B, or C. The binary representation for each gene would require two bits, where each gene has redundant capacity of one extra value. In general, binary encoding of a gene with n alleles requires X bits, where X is $\log_2 n$ rounded up to the nearest integer. Thus there is redundant capacity of $2^X - n$ values per gene. Using the binary coding $A = 01$, $B = 10$, $C = 11$, a binary chromosome to represent the gene combination BACA would look like this:

1	0	0	1	1	1	0	1

A mutation that toggled the last locus would generate an invalid chromosome, since a gene value of 00 is undefined:

1	0	0	1	1	1	0	0

Similarly defective chromosomes can also be generated by crossover. In each case the problem can be avoided by using *structured* operators, i.e., requiring crossover and mutation to operate at the level of genes rather than loci. Thus, crossover points could be forced to coincide with gene boundaries and mutation could randomly select new values for whole genes. These restrictions would ensure the generation of valid chromosomes, but they also risk producing insufficient variety in the chromosome population.

An alternative approach is to detect and *repair* invalid chromosomes. Once a defective chromosome has been detected, a variety of ways exist to repair it. One approach is to generate "spare" chromosomes in each generation, which can then be randomly selected as replacements for any defective ones.

7.6.2 Selection

It has already been stated that individuals are selected for reproduction on the basis of their fitness, i.e., the fittest chromosomes have the highest likelihood of reproducing. Selection determines not only which individuals will reproduce, but how many offspring they will have. The selection method can have an important impact on the effectiveness of a GA.

Selection is said to be *strong* if the fittest individuals have a much greater probability of reproducing than less fit ones. Selection is said to be *weak* if the fittest individuals have only a slightly greater probability of reproducing than the less fit ones. If the selection method is too strong, the genes of the fittest individuals may dominate the next generation population even though they may be suboptimal. This is known as *premature convergence*, i.e., the exploitation of a small region of the search space before a thorough exploration of the whole space has been achieved. On the other hand, if the selection method is too weak, less fit individuals are given too much opportunity to reproduce and evolution may become too slow. This can be a particular problem during the latter stages of evolution, when the whole population may have congregated within a smooth and fairly flat region of the search space. All individuals in such a region would have similar, relatively high, fitnesses and, thus, it may be difficult to select among them. This can result in *stalled evolution*, i.e., there is insufficient variance in fitness across the population to drive further evolution.

Some alternative methods of selection, given a fitness for each member of the population, are now reviewed. The first approach, fitness-proportionate selection, is prone to both premature convergence and stalled evolution. The other methods are designed to counter these effects.

Fitness-proportionate selection

In this method of selection, an individual's expected number of offspring is proportional to its fitness. The number of times an individual would expect to reproduce is, therefore, equal to its fitness divided by the mean fitness of the population. A method of achieving this, as originally proposed by Holland [5], is *roulette wheel selection with replacement*. The fitness of each individual is first normalized by dividing it by the sum of fitness values for all individuals in

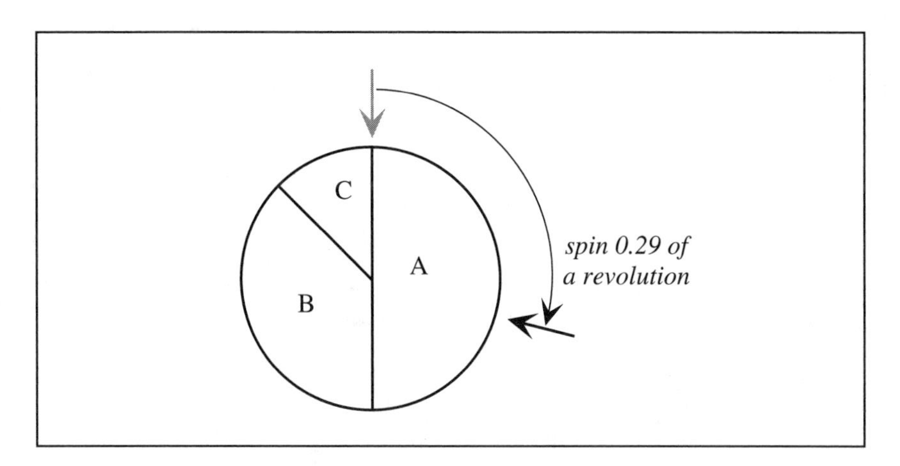

Figure 7.5 Roulette wheel selection. In this example, the normalized fitnesses of A, B, and C are 0.5, 0.375, and 0.125 respectively and the random number is 0.29

the population to yield its selection probability, P_i. Individuals are then imagined on a roulette wheel, with each one allocated a proportion of the circumference equal to their selection probability.

Individuals are selected for reproduction by spinning the notional roulette wheel. This is achieved in software by generating a random number in the range 0–1. From a fixed starting point, a notional pointer moves around the wheel by the fraction of a revolution determined by the random number. In the example in Figure 7.5, the random number 0.29 was spun and individual A selected. The pointer is reset at the origin and spun again to select the next individual. To say that selection proceeds "with replacement" means that previously selected individuals remain available for selection with each spin of the wheel. In all, the wheel is spun N times per generation, where N is the number of individuals in the population.

A perceived drawback of roulette wheel selection is its high degree of variance. An unfit individual could, by chance, reproduce more times than a fitter one. *Stochastic universal selection* is a variant of proportional selection that overcomes this problem. As with proportional selection, individuals are allocated a proportion of the circumference of a wheel according to their fitness value. Rather than a single pointer, there are N equally spaced pointers for a population size N (Figure 7.6). All pointers are moved together around the wheel by the fraction of a revolution determined by the random number. Thus only one spin of the wheel is required in order to select all reproducing individuals and so the method is less computationally demanding than roulette wheel selection.

Both variants of fitness-proportionate selection suffer from the risk of premature convergence and, later on, stalled evolution. Four scaling techniques

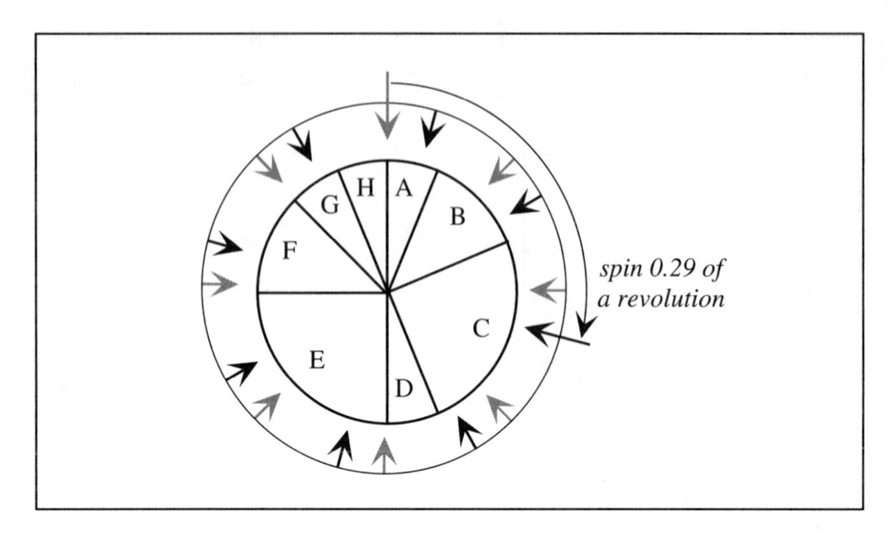

Figure 7.6 Stochastic universal selection, shown here for a population size of eight and a randomly generated spin of 0.29. Gray arrows represent the original pointer positions; black arrows represent the pointer positions after spinning

are described below: linear fitness scaling, Boltzmann fitness scaling, rank selection, and truncation selection. A fifth technique, tournament selection, is also described. These techniques are intended to counter premature convergence by slowing evolution and maintaining diversity in the early generations, and to counter stalled evolution by spreading out the selection rates for the population in the later stages.

Linear fitness scaling

The simplest form of scaling is linear scaling, where the scaled fitness s_i of individual i is scaled linearly with its fitness f_i:

$$s_i = mf_i + c \tag{7.5}$$

where m and c are chosen for each generation to either stretch or contract the range of fitnesses. The raw fitness is sometimes referred to as the *evaluation* or *objective* value. The scaled value, which forms the basis of the selection, is referred to as the *selective* value. Selection can proceed by either roulette wheel selection or stochastic universal selection, where the individuals are now allocated a proportion of the circumference of the wheel according to their selective values instead of their objective values.

The selection probability P_i and the expected number of offspring E_i can be calculated by normalizing the scaled fitness:

$$P_i = \frac{s_i}{\sum_i s_i} \tag{7.6}$$

$$E_i = P_i N \tag{7.7}$$

As defined here, E_i is measured in terms of the amount of genetic material that is passed to the next generation. For instance, an individual of average fitness would have $E_i = 1$, even though this genetic material might be shared among two offspring as a result of crossover.

The key to linear scaling is the selection of suitable values for m and c for each generation. A commonly used method is known as *sigma scaling*, or *sigma truncation* [7]. Here an individual's fitness is scaled according to its deviation from the mean fitness $\bar{f}$ of the population, measured in standard deviations (i.e., 'sigma', σ). The scaled fitness s_i for an individual i is given by:

$$s_i = E_i = \begin{cases} 1 + \dfrac{f_i - \bar{f}}{2\sigma} & \text{if } \sigma \neq 0 \text{ and } \bar{f} - f_i \leq 2\sigma \\ 1 & \text{if } \sigma = 0 \\ 0 & \text{if } \sigma \neq 0 \text{ and } \bar{f} - f_i > 2\sigma \end{cases} \tag{7.8}$$

The main part of Equation 7.8 is the first, which is linear. This can be made explicit by rewriting it in the form of Equation 7.5:

$$s_i = \frac{1}{2\sigma} f_i + \frac{2\sigma - \bar{f}}{2\sigma} \tag{7.9}$$

The last two parts of Equation 7.8 are simply designed to catch anomalous cases. The second part deals with the case of zero variance, and the third part prevents the scaled fitness from becoming negative. In Tanese's implementation as reported in [7], the equation was modified to ensure that even the least fit individuals had a small chance of reproducing:

$$s_i = E_i = \begin{cases} 1 + \dfrac{f_i - \bar{f}}{2\sigma} & \text{if } \sigma \neq 0 \text{ and } \bar{f} - f_i \leq 1.8\sigma \\ 1 & \text{if } \sigma = 0 \\ 0.1 & \text{if } \sigma \neq 0 \text{ and } \bar{f} - f_i > 1.8\sigma \end{cases} \tag{7.10}$$

Sigma scaling tends to maintain a fairly consistent contribution to the gene pool from highly fit individuals. During the early stages of evolution, premature convergence is avoided because even the fittest individuals are only two or three standard deviations above the mean fitness. Later on, when the population has converged, stalled evolution is avoided because the diminished value of σ has the effect of spreading out the scaled fitnesses.

Boltzmann fitness scaling

Boltzmann fitness scaling is a nonlinear method. This technique borrows from simulated annealing the idea of a "temperature" T that drops slowly from generation to generation. The simple Boltzmann scaling function is:

$$s_i = \exp(f_i / T) \tag{7.11}$$

When the temperature is high, all chromosomes have a good chance of reproducing as the fittest are only slightly favored over less fit individuals. This avoids premature convergence and allows extensive exploration of the search space. As the temperature cools according to a preset schedule, less fit individuals are progressively suppressed, allowing exploitation of what should then be the right part of the search space. Stalled evolution is therefore avoided as well.

Rank selection

Rank selection [8] can be thought of as an alternative way of scaling the fitness of the chromosomes in the population. Instead of being derived from the raw fitness values, the scaled fitness is derived from a rank ordering of individuals based on their fitness. The simplest approach is to have the scaled fitness vary linearly with the rank position:

$$s_i = Min + (Max - Min)\frac{rank_i - 1}{N - 1} \tag{7.12}$$

where *Min* is the fitness for the lowest ranking individual ($rank_1$) and *Max* is the fitness for the highest ranking individual ($rank_N$). If this expression is normalized so that $s_i = E_i$, and we require that $Max \geq 0$, then the values of *Max* and *Min* are bounded such that $1 \leq Max \leq 2$ and $Min = 2 - Max$.

The likelihood of being selected for reproduction is dependent only on rank within the current population and not directly on the fitness value, except insofar as this determines rank. Premature convergence and stalled evolution are both avoided because the spread of scaled fitnesses or selective values is maintained, regardless of the distribution of the underlying objective values.

Nolle [9] has proposed the following nonlinear algorithm to produce a stronger form of rank-based selection:

$$s_i = E_i = \frac{1}{c} \ln \left(\frac{N - (rank_i - 1)\left(1 - \frac{1}{e^c}\right)}{N - rank_i \left(1 - \frac{1}{e^c}\right)} \right) \tag{7.13}$$

where c is a constant chosen to control the nonlinearity of the function. In Figure 7.7, this nonlinear algorithm is compared with the standard linear rank-based approach. The nonlinear algorithm stretches the distribution of selection probabilities for the fittest individuals, thereby countering the effects of stalled evolution. Nolle has also experimented with an adaptive algorithm, where c is recalculated at each generation so that it increases as evolution progresses.

Truncation selection

Truncation selection is a variant of rank selection. Here a cut-off rank position $rank_0$ is chosen in advance and the scaled fitness becomes:

$$s_i = \begin{cases} \dfrac{1}{N - rank_0 + 1} & \text{if } rank_i \geq rank_0 \\ 0 & \text{if } rank_i < rank_0 \end{cases} \tag{7.14}$$

The individuals at or above the cutoff have an equal number of expected offspring, while those below the cutoff do not reproduce at all. Premature convergence is avoided as the fittest individuals have the same reproductive chances as any others that meet the threshold requirement, and stalled evolution is avoided as the method always dispenses with the least fit individuals.

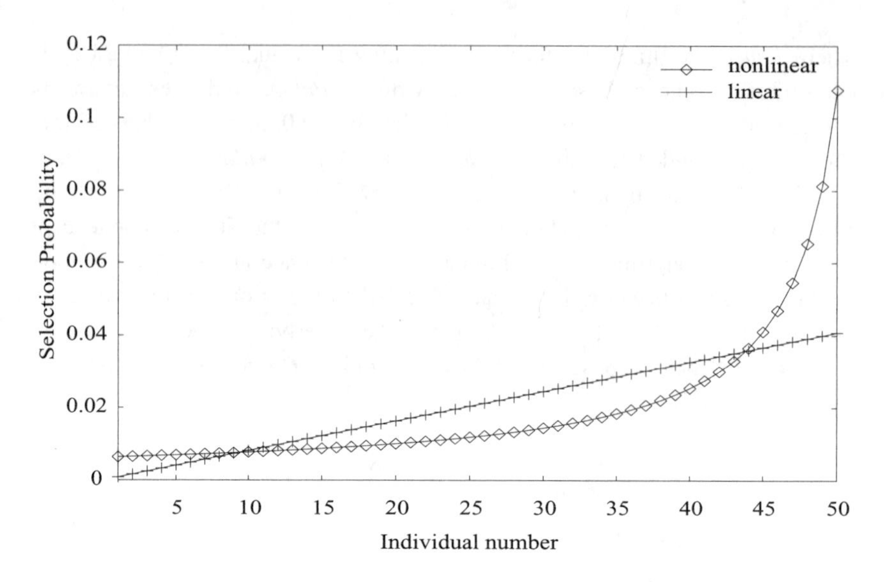

Figure 7.7 Comparison of linear (*Max* = 2.0) and nonlinear (c = 3.0) rank-based selection

Tournament selection

In each of the above techniques for selection, an individual's expected number of offspring is proportional to either its fitness or a scaled version of its fitness. Tournament selection is a different approach in which individuals are selected on the basis of direct competition between them. The most commonly used form is a binary tournament, in which two individuals are selected at random and their fitnesses evaluated. The fitter of the two is then selected for breeding. Further pairs of individuals are then selected at random from the population of size N, which includes previously selected individuals, until N individuals have been selected for reproduction. The fittest individuals would be expected to be selected several times, thereby increasing their probability of reproduction. Although binary tournaments are the norm, tournaments with more than two competitors are possible.

Elitism

The selection methods described so far have all assumed that the whole of the population is replaced with every generation. Such GAs are said to be *generational*. Even in a generational GA, some individuals may, by chance, be identical to members of the previous generation.

Some researchers have found it useful to modify the basic GA so that it is only partially generational through the use of *elitism*. Elitism refers to passing one or more of the fittest individuals unchanged through to the next generation. The fittest solutions found so far are, therefore, preserved within the population. Elitism can be thought of as the allowance of cloning alongside reproduction.

The proportion of new individuals in successive generations is termed the *generation gap*. *Steady-state selection* refers to the extension of elitism so that the generation gap is rather small, with only a few of the least fit individuals being replaced on successive generations. They would typically be replaced by crossover and mutation of the fittest individuals.

Multiobjective optimization

The treatment so far has assumed that there was a single fitness function to optimize, whereas in fact there may be more than one. For instance, Chapter 12 considers the design of a lightweight beam, where both stiffness and strength criteria are considered. A simple approach to the application of a GA to this sort of problem is to combine the fitnesses into a single function. A multi-objective function could be defined as a weighted sum of the individual fitness functions. So, if the strength of a composite beam were considered twice as importance as its stiffness, a suitable function might be (2 × strength + stiffness) / 3.

The difficulty with this approach is that, although the relative weightings are vital, they are likely to be arbitrary. A more sophisticated view is to recognize that there is often no single best solution. Typically there is a trade-off between one fitness measure and the others. Rather than a single optimal solution, a set of solutions exists, from which any improvement in one fitness measure would cause a deterioration in the others. This set is said to be *Pareto optimal*. Ideally, an optimization process will identify all members of the Pareto optimal set, leaving selection of the ultimate solution to a further decision process, possibly knowledge-based.

7.6.3 Gray code

The structure of the chromosome needs to be designed according to the sort of candidate solution we wish to discover. Choosing a good representation is essential if a GA is to work effectively. Section 7.6.1 showed a simple case where the objective was to maximize the value of $f(x,y)$ within a specified range of x and y. The chromosome was defined as two genes, one for each of x and y. Each gene represented a real number and was encoded in a specified number of bits using standard binary encoding. Some researchers have pointed out that this may not be the most suitable encoding, as consecutive values such as 7 (binary 0111) and 8 (binary 1000) often have large Hamming separations [10, 11]. An alternative form of encoding is known as Gray code, shown in

Denary	Binary code	Gray code
0	0000	0000
1	0001	0001
2	0010	0011
3	0011	0010
4	0100	0110
5	0101	0111
6	0110	0101
7	0111	0100
8	1000	1100
9	1001	1101
10	1010	1111
11	1011	1110
12	1100	1010
13	1101	1011
14	1110	1001
15	1111	1000

Table 7.1 Binary and Gray codes for integers 0–15

Table 7.1. Consecutive Gray-coded representations always have a Hamming separation of 1.

7.6.4 Variable length chromosomes

Some researchers have experimented with variable length chromosomes in the hope of achieving more open-ended solutions. The so-called messy GAs are one such approach, using messy chromosomes containing messy genes [12]. In a conventional fixed-length chromosome, the position of a gene on a chromosome determines how it is decoded. This is the *fixed locus assumption.* In a simple GA with 3 binary genes, the string 110 would be decoded as:

gene #1 = 1
gene #2 = 1
gene #3 = 0

In a messy chromosome, the position of the genes does not matter as each gene carries an identifier to indicate what it represents. Instead of containing just a value, each gene comprises two components, shown here bracketed together: (identifier, value). The fixed-length chromosome 110 could be equally represented by any of the following messy chromosomes, all of which are equivalent:

[(1, 1) (2, 1) (3, 0)]
[(1, 1) (3, 0) (2, 1)]
[(2, 1) (1, 1) (3, 0)]
[(2, 1) (3, 0) (1, 1)]
[(3, 0) (1, 1) (2, 1)]
[(3, 0) (2, 1) (1, 1)]

This structure means that messy chromosomes may be *over-specified* or *under-specified.* An over-specified chromosome has some duplication of messy genes. For example:

[(1, 1) (3, 0) (1, 1) (2, 1)]

defines gene #1 twice. Over-specification can result in contradictions, shown here for gene #1:

[(1, 1) (3, 0) (1, 0) (2, 1)]

Such contradictions are normally handled by accepting the first, or leftmost, instance of the gene. So in this specific case, the position of the gene *does* matter.

In an under-specified chromosome, such as [(3, 1) (1, 0)], no value is provided for at least one of the genes, namely gene #2. Chromosome [(1, 1) (3, 0) (1, 0)] is both under-specified and over-specified since it lacks gene #2 but contains multiple occurrences of gene #1. Whether under-specification is problematic depends on the nature of the optimization task. In cases where values must be found for each parameter, under-specified chromosomes must be repaired in some way. The unspecified genes may be set to random values, or to the values used in previously discovered fit chromosomes.

The messy GA itself is broadly similar to the standard GA. Tournament selection (see Section 7.6.2) is usually used to select messy chromosomes for reproduction. Competing individuals are chosen on the basis of their similarity, measured as the number of genes for which both have a value. If the similarity of two individuals is above a specified threshold, a tournament between them is allowed. Mutation is rarely used, and the *cut-and-splice* operator replaces crossover. Each chromosome is cut at a randomly selected point and then chromosome segments from different individuals are spliced together, e.g.:

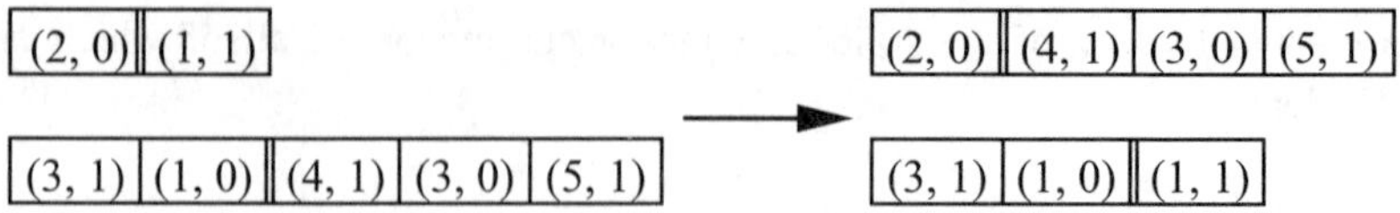

(2, 0)	(1, 1)			
(3, 1)	(1, 0)	(4, 1)	(3, 0)	(5, 1)

→

(2, 0)	(4, 1)	(3, 0)	(5, 1)
(3, 1)	(1, 0)	(1, 1)	

7.6.5 Building block hypothesis

Schema theorem

Holland [5] has proposed the *building block hypothesis*, i.e., that successful chromosomes are made up of good quality blocks of genes. He formalized this idea as *schemata* (plural of *schema*), which are sets of gene values that can be represented by a template. For example, the template 1**0 defines a four-bit gene sequence whose first and fourth positions have defined values, and whose other two positions can take any value.

Holland developed the following equation as the basis of his *schema theorem:*

$$n(H,t+1) \geq n(H,t)\frac{\bar{f}(H,t)}{\bar{f}(t)}\left(1-\frac{P_c d(H)}{l-1}\right)(1-P_m)^{o(H)} \tag{7.15}$$

where H is a schema with at least one instance in the last generation; $d(H)$ is the defining length of H, i.e., the distance between crossover points; l is the chromosome length, $n(H, t)$ is the number of instances of H at time t, $\bar{f}(H, t)$ is the mean fitness of H at time t, $\bar{f}(t)$ is the mean fitness of the population at time t; P_c is the crossover probability; P_m is the mutation probability; and $o(H)$ is the order of H, i.e., the number of defined bits in H. The schema theorem provides a theoretical foundation for genetic algorithms, since it can be used to demonstrate that the number of schemata that are fitter than average increases with each generation, while the number of schemata that are less fit than average diminishes [10].

Inversion

The crossover operator risks breaking up good quality schemata. Holland, therefore, devised a further operator — *inversion* — that may be able to protect good combinations of genes. The process of inversion consists of reversing the order of a short section of loci within a chromosome. It is intended to operate on chromosomes whose interpretation is independent of the ordering of the genes. Although not envisaged by Holland, the messy chromosomes described above provide a good example. The template 1**0 could be represented as the following messy chromosome:

[(1, 1) (2, *) (3, *) (4, 0)]

Rearranging the ordering of the messy genes has no effect on their fitness. The inversion operator could be applied to this messy gene to reverse the order of the last three loci:

(1, 1)	(2, *)	(3, *)	(4, 0)

→

(1, 1)	(4, 0)	(3, *)	(2, *)

The effect has been to bring together the two genes with good quality values, namely the genes with identifier 1 and 4. Although the fitness of the chromosome has been unchanged, these two good quality values are now more likely to be retained as a good quality schema when crossover is applied.

For chromosomes where the order of the genes does affect their interpretation, inversion can be used as a nonsexual means of generating new individuals, additional to mutation. Used in this way, it is particularly applicable to permutation problems such as the traveling salesperson.

7.6.6 Selecting GA parameters

One of the main difficulties in building a practical GA is in choosing suitable values for parameters such as population size, mutation rate, and crossover rate. De Jong's guidelines, as cited in [7], are still widely followed, namely, to start with:

- a relatively high crossover probability (0.6–0.7);
- a relatively low mutation probability (typically set to $1/l$ for chromosomes of length l);
- a moderately sized (50–500) population.

Some of the parameters can be allowed to vary. For example, the crossover rate may be started at an initially high level and then progressively reduced with each generation or in response to particular performance measures.

Given the difficulties in setting the GA parameters, it is unsurprising that many researchers have tried encoding them so that they too might evolve toward optimum values. These *self-adaptive* parameters can be encoded in individual chromosomes, providing values that adapt specifically to the characteristics of the chromosome. Typically, a minimal background mutation rate applies to the population as a whole, and each chromosome includes a gene that encodes a mutation rate to apply to the remaining genes on that chromosome. Self-adaptive parameters do not completely remove the difficulties in choosing parameters, but by deferring the choice to the level of *metaparameters*, i.e., parameters' parameters, it may become less critical.

7.6.7 Monitoring evolution

There is a variety of measures that can be made at run-time in order to monitor the evolutionary progress of a GA, including:

- highest fitness in the current population;
- lowest fitness in the current population;
- mean fitness in the current population;
- standard deviation of fitness in the current population;
- mean Hamming separation between randomly selected sample pairs in the current population;
- bitwise convergence — for each locus the proportion of the population that has the most popular value is calculated and then averaged over the whole chromosome.

The first four all relate to fitness, while the last two provide measures of similarity between chromosomes.

7.6.8 Lamarckian inheritance

Genetic algorithms can be hybridized with local search procedures to help optimize individuals within the population, while maintaining the GA's ability to explore the search space. For instance, a simple search may be carried out in the immediate vicinity of each chromosome to see whether any of its neighbors offers a better solution. Suppose that the task is to find the maximum denary integer represented by a 7-bit binary-encoded chromosome. A chromosome's fitness could be taken as the highest integer represented by itself or by any of its nearest neighbors, i.e., those that have a Hamming separation of 1 from it. For instance, the chromosome 0101100 (denary value 44) would have a fitness of 108, since it has a nearest neighbor 1101100 (denary value 108). The chromosome could either be left unchanged while retaining the fitness value of its fittest neighbor, or it could be replaced by the higher scoring neighbor, in this case 1101100. The latter scheme is known as *Lamarckian inheritance* and is equivalent to including a steepest gradient descent step prior to evaluation.

Lamarckian inheritance is an example of how one optimization algorithm can be enhanced by another. There are many ways in which different artificial intelligence techniques can be mixed within hybrid systems, some of which are reviewed in Chapter 9.

7.6.9 Finding multiple optima

In certain classes of problem, we may want a population to identify several different optima, referred to as *niches*. In such cases, we must provide a mechanism for decomposing the population into several distinct sub-populations, or *species*. These mechanisms fall into one of four broad categories:

- *incest prevention*, in which similar individuals are prevented from mating;
- *speciation*, in which only similar individuals may breed together;
- *crowding*, in which each offspring replaces the member in the current population most similar to it;
- *fitness sharing*, in which an individual's selective value is scaled as a decreasing function of its similarity to other members of the population.

7.6.10 Genetic programming

Since genetic algorithms can evolve numerical populations, Koza et al. reasoned that it ought to be possible to evolve computer programs in a similar way [13]. This concept has led to the thriving research field of genetic programming (GP). The idea is that hierarchically arranged computer programs can be automatically generated to solve a specified problem.

The building blocks of GP are *functions* and *terminals*. Functions are simple pieces of program code that take one or more arguments, perform an operation on them, and return a value. Terminals are constants or variables that can be used as arguments to the functions. An initial population of hierarchically arranged programs is generated by randomly combining functions and terminals. The fitness of these primitive programs for the specified task can be evaluated, and the fittest individuals selected for reproduction. Crossover is achieved by swapping branches of the hierarchical programs between parents. Mutation may occur by changing the terminals, or by replacing a function with another that takes the same number of arguments. If a particularly useful combination of functions and terminals is discovered, it can be nominated as a function in its own right, thereby preserving it within the population.

7.7 Summary

This chapter has reviewed a number of numerical optimization techniques, with particular attention paid to genetic algorithms. All these techniques are based upon minimizing a cost or maximizing a fitness. Often the cost is taken as the error between the output and the desired output.

It is interesting to compare numerical and knowledge-based search techniques. A knowledge-based system is typically used to find a path from a known, current state, such as a set of observations, to a desired state, such as an interpretation or action. In numerical optimization, we know the properties we require of a solution in terms of some fitness measure, but have no knowledge of where it lies in the search space. The problem is of searching for locations within the search space that satisfy a fitness requirement.

All the numerical optimization techniques carry some risk of finding a local optimum rather than the global one. An attempt to overcome this is made with GAs by the inclusion of an exploration phase, in which the algorithm is free to roam the search space seeking good quality regions. Nevertheless, the problem may still remain. A further claimed advantage of GAs is that they result in *robust* optimization, i.e., the fitness of the solutions tends to be insensitive to slight deviations from the precise optimum. Thus, if the precise optimum is on a knife-edge or narrow peak in the fitness landscape, the GA may not find it. Robust optima are preferable in many applications. For instance, design engineers know that their products can be manufactured only to certain tolerances, so a property specified on a fitness knife-edge would be impractical.

It is rarely necessary to code the optimization algorithms from scratch, as a variety of software packages are available, many free of charge. These provide nondomain-specific tools in the same way that an expert system shell does, often as libraries for linking to other software. A GA package typically provides data structures and operators that act on them so that the user has only to define the chromosome structure, fitness function and a set of parameters covering:

- population size,
- mutation rate,
- crossover rate,
- number of generations or other termination condition,
- selection method.

The main problem in applying genetic algorithms to real optimization problems lies in finding a chromosome representation that remains valid after each generation. Nevertheless, the technique has been successfully employed in the automatic scheduling of machining operations [14, 15] and designing communications networks [16].

References

1. Nolle, L., Walters, M., Armstrong, D. A., and Ware, J. A., "Optimum finishing mill set-up in the hot rolling of wide steel strip using simulated

annealing," 5th International Mendel Conference on Soft Computing, Brno, Czech Republic, pp. 100–105, 1999.

2. Hirst, A. J., "Adaptive evolution in static and dynamic environments," PhD thesis, Open University, 1997.
3. Kirkpatrick, S., Gelatt, C. D., and Vecchi, M. P., "Optimization by simulated annealing: quantitative studies," *Science*, vol. 220, pp. 671–680, 1983.
4. Johnson, J. H. and Picton, P. D., *Concepts in Artificial Intelligence*, Butterworth-Heinemann, 1995.
5. Holland, J. H., *Adaptation in Natural and Artificial Systems*, University of Michigan Press, 1975.
6. Syswerda, G., "Uniform crossover in genetic algorithms," Schaffer, J. D. (Ed.), 4th International Conference on Genetic Algorithms, Morgan Kaufmann, 1989.
7. Mitchell, M., *An Introduction to Genetic Algorithms*, MIT Press, 1996.
8. Baker, J. E., "Adaptive selection methods for genetic algorithms," International Conference on Genetic Algorithms and Their Application, pp. 101–111, 1985.
9. Nolle, L., "Application of computational intelligence in the hot rolling of wide steel strip," PhD thesis, Open University, 1999.
10. Goldberg, D. E., *Genetic Algorithms in Search, Optimization, and Machine Learning*, Addison Wesley, 1989.
11. Davis, L. (Ed.), *Handbook of Genetic Algorithms*, Van Nostrand Reinhold, 1991.
12. Goldberg, D. E., Korb, B., and Deb, K., "Messy genetic algorithms: motivation, analysis, and first results.," *Complex Systems*, vol. 3, pp. 493–530, 1989.
13. Koza, J. R., Bennett, I. I. I., H., F., Andre, D., and Keane, M. A. (Eds.), *Genetic Programming III: Darwinian invention and problem solving*, Morgan Kaufmann, 1999.
14. Husbands, P., Mill, F., and Warrington, S., "Generating optimal process plans from first principles," in *Expert Systems for Management and Engineering*, Balagurasamy, E. and Howe, J. (Eds.), pp. 130–152, Ellis Horwood, 1990.
15. Husbands, P. and Mill, F., "Simulated co-evolution as the mechanism for emergent planning and scheduling," Belew, R. and Booker, L. (Eds.), 4th International Conference on Genetic Algorithms, pp. 264–271, Morgan Kaufmann, 1991.

16. Davis, L. and Coombs, S., "Optimizing network link sizes with genetic algorithms," in *Modelling and Simulation Methodology: knowledge systems paradigms*, Elzas, M. S., Oren, T. I., and Zeigler, B. P. (Eds.), North Holland Publishing Co., 1987.

Further reading

- Bäck, T., *Evolutionary Algorithms in Theory and Practice*, Oxford University Press, 1996.
- Davis, L. (Ed.), *Genetic Algorithms and Simulated Annealing*, Pitman/Morgan Kaufmann, 1987.
- Davis, L. (Ed.), *Handbook of Genetic Algorithms*, Van Nostrand Reinhold, 1991.
- Goldberg, D. E., *Genetic Algorithms in Search, Optimization, and Machine Learning*, Addison-Wesley, 1989.
- Johnson, J. H. and Picton, P. D., *Concepts in Artificial Intelligence*, Butterworth-Heinemann, 1995.
- Mitchell, M., *An Introduction to Genetic Algorithms*, MIT Press, 1996.

Chapter eight

Neural networks

8.1 Introduction

Artificial neural networks are a family of techniques for numerical learning, like the optimization algorithms reviewed in Chapter 7, but in contrast to the symbolic learning techniques reviewed in Chapter 6. They consist of many nonlinear computational elements which form the network nodes or *neurons*, linked by weighted interconnections. They are analogous in structure to the neurological system in animals, which is made up of real rather than artificial neural networks. Practical artificial neural networks are much simpler than biological ones, so it is unrealistic to expect them to produce the sophisticated behavior of humans or animals. Nevertheless, they can perform certain tasks, particularly classification, most effectively. Throughout the rest of this book we will use the expression *neural network* to mean an artificial neural network. The technique of using neural networks is described as *connectionism*.

Each node in a neural network may have several inputs, each of which has an associated weighting. The node performs a simple computation on its input values, which are single integers or real numbers, to produce a single numerical value as its output. The output from a node can either form an input to other nodes or be part of the output from the network as a whole. The overall effect is that a neural network generates a pattern of numbers at its outputs in response to a pattern of numbers at its inputs. These patterns of numbers are one-dimensional arrays known as *vectors*, e.g., (0.1, 1.0, 0.2).

Each neuron performs its computation independently of the other neurons, except that the outputs from some neurons may form the inputs to others. Thus, neural networks have a highly parallel structure, allowing them to explore many competing hypotheses simultaneously. This parallelism allows neural networks to take advantage of parallel processing computers. They can also run on conventional serial computers — they just take longer to run that way. Neural networks are tolerant of the failure of individual neurons or interconnections. The performance of the network is said to *degrade gracefully* if these localized failures within the network should occur.

The weights on the node interconnections, together with the overall topology, define the output vector that is derived by the network from a given input vector. The weights do not need to be known in advance, but can be learned by adjusting them automatically using a training algorithm. In the case of *supervised* learning, the weights are derived by repeatedly presenting to the network a set of example input vectors along with the corresponding desired output vector for each of them. The weights are adjusted with each iteration until the actual output for each input is close to the desired vector. In the case of *unsupervised* learning, the examples are presented without any corresponding desired output vectors. With a suitable training algorithm, the network adjusts its weights in accordance with naturally occurring patterns in the data. The output vector then represents the position of the input vector within the discovered patterns of the data.

Part of the appeal of neural networks is that when presented with noisy or incomplete data, they will produce an approximate answer rather than one that is incorrect. This is another aspect of the graceful degradation of neural networks mentioned above. Similarly, when presented with unfamiliar data that lie within the range of its previously seen examples, the network will generally produce an output that is a reasonable interpolation between the example outputs. Neural networks are, however, unable to extrapolate reliably beyond the range of the previously seen examples. Interpolation can also be achieved by fuzzy logic (see Chapter 3). Thus, neural networks and fuzzy logic often represent alternative solutions to a particular engineering problem and may be combined in a hybrid system (see Chapter 9).

8.2 Neural network applications

Neural networks can be applied to a diversity of tasks. In general, the network associates a given input vector $(x_1, x_2, \ldots x_n)$ with a particular output vector $(y_1, y_2, \ldots y_m)$, although the function linking the two may be unknown and may be highly nonlinear. (A linear function is one that can be represented as $f(x) = mx + c$, where m and c are constants; a nonlinear one may include higher order terms for x, or trigonometric or logarithmic functions of x.)

8.2.1 Nonlinear estimation

Neural networks provide a useful technique for determining the values of variables that cannot be measured easily, but which are known to depend in some complex way on other more accessible variables. The measurable variables form the network input vector and the unknown variables constitute the output vector. We can call this use *nonlinear estimation*. The network is initially trained using a set of examples known as the training data. Supervised learning is used, so each example in the training comprises two vectors: an input vector and its corresponding desired output vector. (This assumes that some values for the less accessible variable have been obtained to form the desired outputs.) During training, the network learns to associate the example input vectors with their desired output vectors. When it is subsequently presented with a previously unseen input vector, the network is able to interpolate between similar examples in the training data to generate an output vector.

8.2.2 Classification

Often the output vector from a neural network is used to represent one of a set of known possible outcomes, i.e., the network acts as a *classifier*. For example, a speech recognition system could be devised to recognize three different words: *yes*, *no*, and *maybe*. The digitized sound of the words would be preprocessed in some way to form the input vector. The desired output vector would then be either (0, 0, 1), (0, 1, 0), or (1, 0, 0), representing the three classes of word.

Such a network would be trained using a set of examples known as the *training data*. Each example would comprise a digitized utterance of one of the words as the input vector, using a range of different voices, together with the corresponding desired output vector. During training, the network learns to associate similar input vectors with a particular output vector. When it is subsequently presented with a previously unseen input vector, the network selects the output vector that offers the closest match. This type of classification would not be straightforward using non-connectionist techniques, as the input data rarely correspond exactly to any one example in the training data.

8.2.3 Clustering

Clustering is a form of unsupervised learning, i.e., the training data comprises a set of example input vectors without any corresponding desired output vectors. As successive input vectors are presented, they are clustered into N groups, where the integer N may be prespecified or may be allowed to grow according to the diversity of the data. For instance, digitized preprocessed spoken words could be presented to the network. The network would learn to cluster together the examples that it considered to be in some sense similar to each other. In this example, the clusters might correspond to different words or different voices.

Once the clusters have formed, a second neural network can be trained to associate each cluster with a particular desired output. The overall system then becomes a classifier, where the first network is unsupervised and the second one is supervised. Clustering is useful for data compression and is an important aspect of *data mining*, i.e., finding patterns in complex data.

8.2.4 Content-addressable memory

The use of a neural network as a content-addressable memory is another form of unsupervised learning, so again there are no desired output vectors associated with the training data. During training, each example input vector becomes stored in a dispersed form through the network.

When a previously unseen vector is subsequently presented to the network, it is treated as though it were an incomplete or error-ridden version of one of the stored examples. So the network regenerates the stored example that most closely resembles the presented vector. This can be thought of as a type of classification, where each of the examples in the training data belongs to a separate class, and each represents the ideal vector for that class. It is useful when classes can be characterized by an ideal or perfect example. For example, printed text that is subsequently scanned to form a digitized image will contain noisy and imperfect examples of printed characters. For a given font, an ideal version of each character can be stored in a content-addressable memory and produced as its output whenever an imperfect version is presented as its input.

8.3 Nodes and interconnections

Each node, or neuron, in a neural network is a simple computing element having an input side and an output side. Each node may have directional connections to many other nodes at both its input and output sides. Each input x_i is multiplied by its associated weight w_i. Typically, the node's role is to sum each of its weighted inputs and add a bias term w_0 to form an intermediate quantity called the *activation*, a. It then passes the activation through a nonlinear function f_t known as the *transfer function* or *activation function*. Figure 8.1 shows the function of a single neuron.

The behavior of a neural network depends on its topology, the weights, the bias terms, and the transfer function. The weights and biases can be learned, and the learning behavior of a network depends on the chosen training algorithm. Typically a sigmoid function is used as the transfer function, as shown in Figure 8.2(a). The *sigmoid* function is given by:

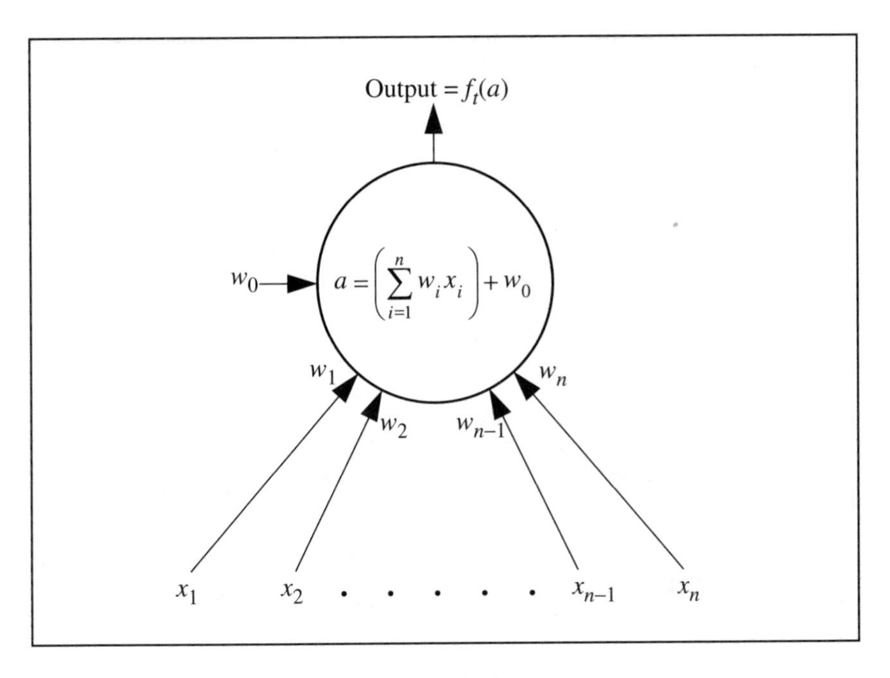

Figure 8.1 A single neuron

$$f_t(a) = \frac{1}{1+e^{-a}} \tag{8.1}$$

For a neuron, the activation a is given by:

$$a = \left(\sum_{i=1}^{n} w_i x_i\right) + w_0 \tag{8.2}$$

where n is the number of inputs and the bias term w_0 is defined separately for each node. Figures 8.2(b) and (c) show the ramp and step functions, which are alternative nonlinear functions sometimes used as transfer functions.

Many network topologies are possible, but we will concentrate on a selection which illustrates some of the different applications for neural networks. We will start by looking at single and multilayer perceptrons, which can be used for categorization or, more generally, for nonlinear mapping. The others are the Hopfield network for use as a content-addressable memory; the MAXNET for selecting the maximum among its inputs; the Hamming network for classification; and finally the ART1 network, the Kohonen self-organizing network, and the radial basis function network, all used for clustering.

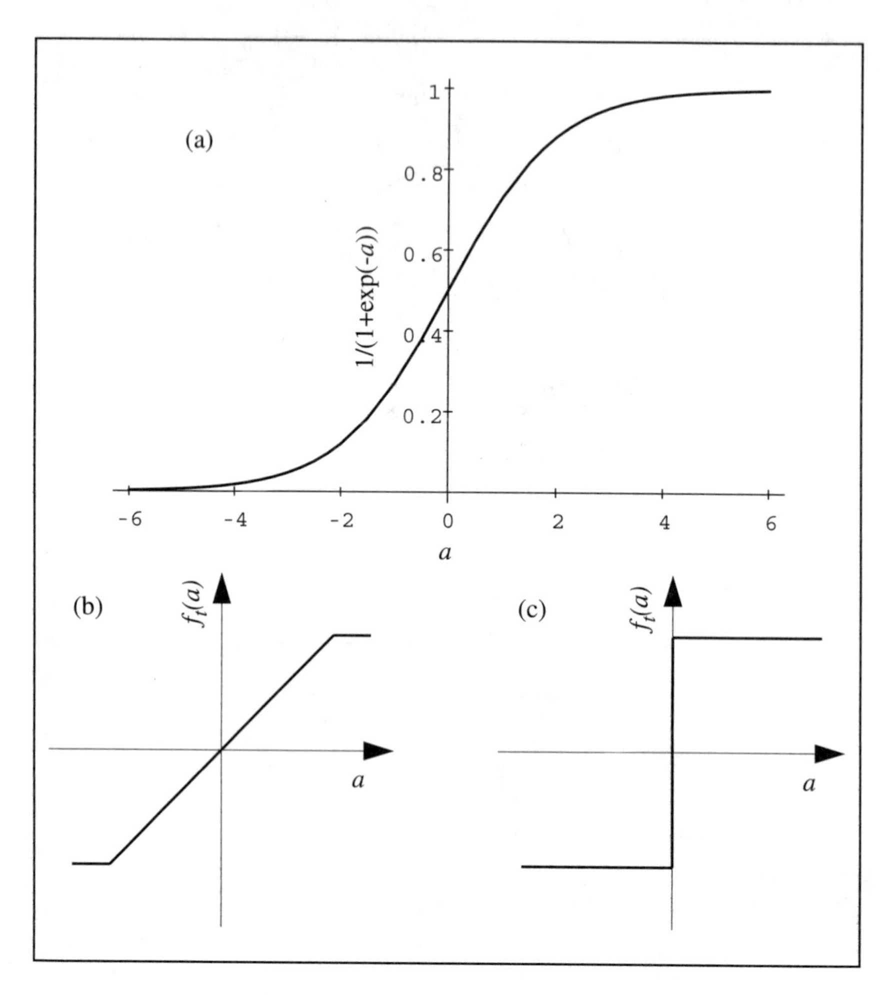

Figure 8.2 Nonlinear transfer functions:
(a) a sigmoid function; (b) a ramp function; (c) a step function

8.4 Single and multilayer perceptrons

8.4.1 Network topology

The topology of a *multilayer perceptron* (MLP) is shown in Figure 8.3. The neurons are organized in layers, such that each neuron is totally connected to the neurons in the layers above and below, but not to the neurons in the same layer. These networks are also called *feedforward networks*, although this term

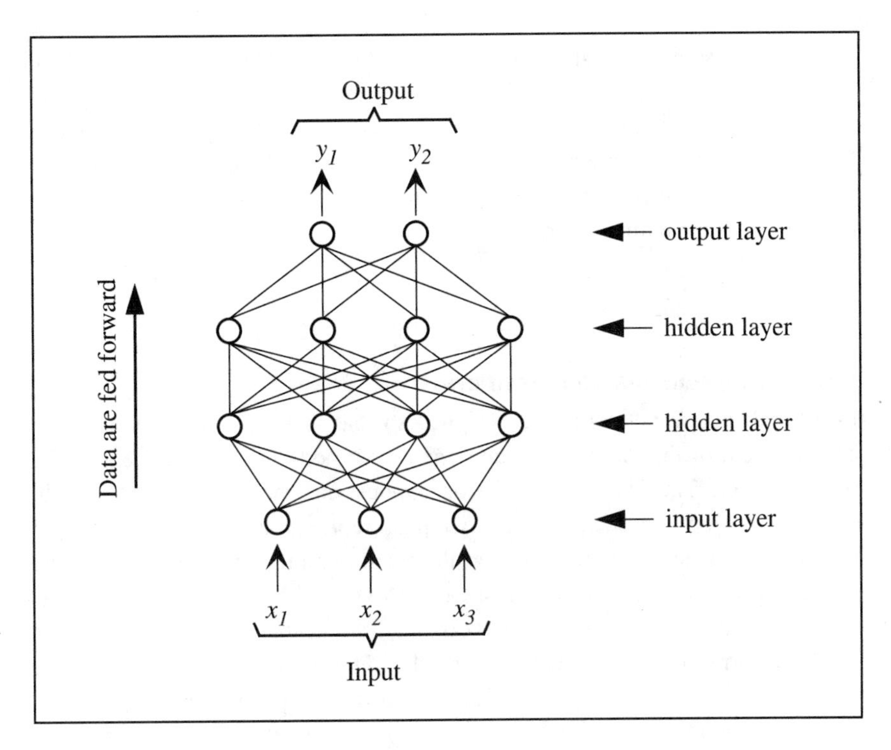

Figure 8.3 A 3–4–4–2 MLP

could be applied more generally to any network where the direction of data flow is always "forwards," i.e., toward the output. MLPs can be used either for classification or as nonlinear estimators. The number of nodes in each layer and the number of layers are determined by the network builder, often on a trial-and-error basis. There is always an input layer and an output layer; the number of nodes in each is determined by the number of inputs and outputs being considered. There may be any number of layers between these two layers. Unlike the input and output layers, the layers between often have no obvious meaning associated with them, and they are known as *hidden layers*. If there are no hidden layers, the network is a *single layer perceptron* (SLP). The network shown in Figure 8.3 has three input nodes, two hidden layers with four nodes each and an output layer of two nodes. It can, therefore, be described as a 3–4–4–2 MLP.

An MLP operates by feeding data forwards along the interconnections from the input layer, through the hidden layers, to the output layer. With the exception of the nodes in the input layer, the inputs to a node are the outputs from each node in the previous layer. At each node apart from those in the input layer, the data are weighted, summed, added to the bias, and then passed through the transfer function.

There is some inconsistency in the literature over the counting of layers, arising from the fact that the input nodes do not perform any processing, but simply feed the input data into the nodes above. Thus although the network in Figure 8.3 is clearly a four-layer network, it only has three processing layers. An SLP has two layers (the input and output layers) but only one processing layer, namely the output layer.

8.4.2 Perceptrons as classifiers

In general, neural networks are designed so that there is one input node for each element of the input vector and one output node for each element of the output vector. Thus in a classification application, each output node would usually represent a particular class. A typical representation for a class would be for a value close to 1 to appear at the corresponding output node, with the remaining output nodes generating a value close to 0. A simple decision rule is needed in conjunction with the network, e.g., the *winner takes all* rule selects the class corresponding to the node with the highest output. If the input vector does not fall into any of the classes, none of the output values may be very high. For this reason, a more sophisticated decision rule might be used, e.g., one that specifies that the output from the winning node must also exceed a predetermined threshold such as 0.5.

More compact representations are also possible for classification problems. Hallam et al. [1] have used just two output nodes to represent four classes. This was achieved by treating both outputs together, so that the four possibilities corresponding to four classes are (0,0), (0,1), (1,0), and (1,1). One drawback of this approach is that it is more difficult to interpret an output which does not closely match one of these possibilities (e.g., what would an output of (0.5, 0.5) represent?).

Let us now return to the more usual case where each output node represents a distinct class. If the input vector has two elements, it can be represented as a point in two-dimensional *state space*, sometimes called the *pattern space*. The process of classification is then one of drawing dividing lines between regions. A single layer perceptron, with two neurons in the input layer and the same number of neurons in the output layer as there are classes, can associate with each class a single straight dividing line, as shown in Figure 8.4(a). Classes that can be separated in this way are said to be *linearly separable*. More generally, n-dimensional input vectors are points in n-dimensional hyperspace. If the classes can be separated by $(n-1)$-dimensional hyperplanes, they are linearly separable.

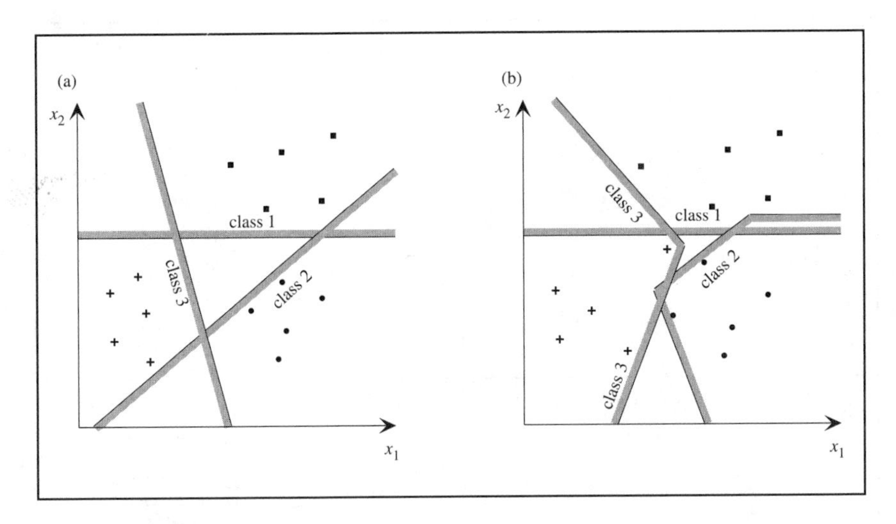

Figure 8.4 Dividing up the pattern space: (a) linearly separable classes; (b) nonlinearly separable classes. Data points belonging to classes 1, 2, and 3 are respectively represented by ■, •, and +

To see how an SLP divides up the pattern space with hyperplanes, consider a single processing neuron of an SLP. Its output, prior to application of the transfer function, is a real number given by Equation 8.2. Regions of the pattern space that clearly belong to the class represented by the neuron will produce a strong positive value, and regions that clearly do not belong to the class will produce a strong negative value. The classification becomes increasingly uncertain as the activation a becomes close to zero, and the dividing criterion is usually assumed to be $a = 0$. This would correspond to an output of 0.5 after the application of the sigmoid transfer function (Figure 8.2(a)). Thus the hyperplane that separates the two regions is given by:

$$\left(\sum_{i=1}^{n} w_i x_i\right) + w_0 = 0 \tag{8.3}$$

In the case of two inputs, Equation 8.3 becomes simply the equation of a straight line, since it can be rearranged as:

$$x_2 = \frac{-w_1}{w_2} x_1 - \frac{w_0}{w_2} \tag{8.4}$$

where $-w_1/w_2$ is the gradient and $-w_0/w_2$ is the intercept on the x_2 axis.

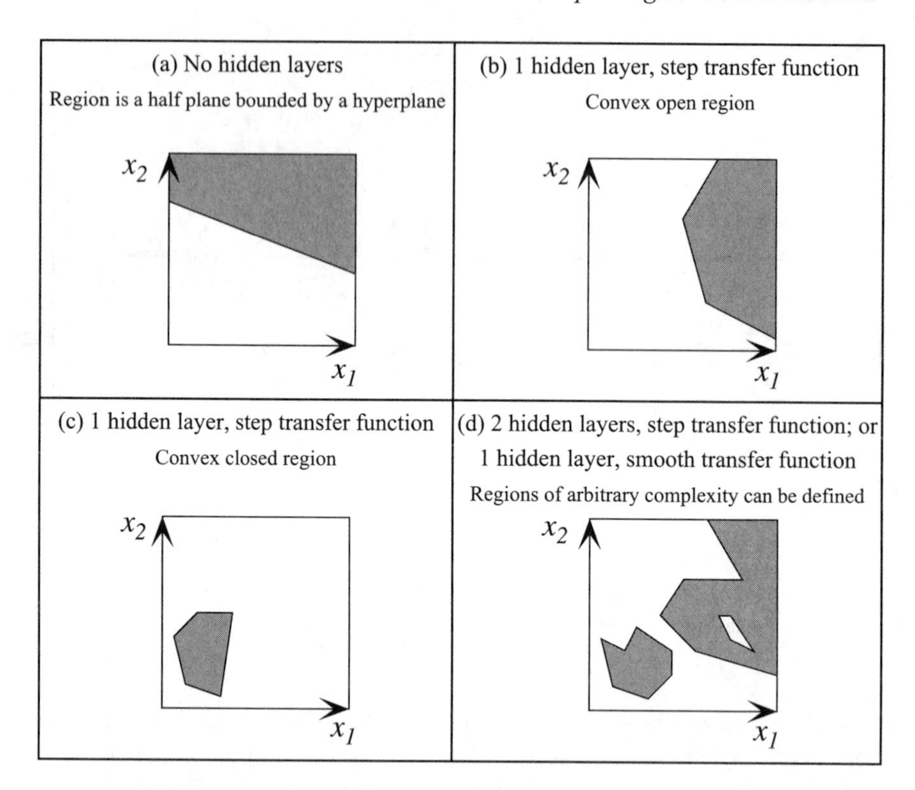
(a) No hidden layers
Region is a half plane bounded by a hyperplane
(b) 1 hidden layer, step transfer function
Convex open region
(c) 1 hidden layer, step transfer function
Convex closed region
(d) 2 hidden layers, step transfer function; or
1 hidden layer, smooth transfer function
Regions of arbitrary complexity can be defined
x_2
x_1

Figure 8.5 Regions in state space distinguished by a perceptron (adapted from [3]). A convex region has the property that a line joining points on the boundary passes only through that region.

For problems that are not linearly separable, as in Figure 8.4(b), regions of arbitrary complexity can be drawn in the state space by a multilayer perceptron with one hidden layer and a differentiable, i.e., smooth, transfer function such as the sigmoid function (Figure 8.2(a)). The first processing layer of the MLP can be thought of as dividing up the state space with straight lines (or hyperplanes), and the second processing layer forms multifaceted regions by Boolean combinations (AND, OR, and NOT) of the linearly separated regions. It is therefore generally accepted that only one hidden layer is necessary to perform any nonlinear mapping or classification with an MLP that uses a sigmoid transfer function (Figure 8.5). This is *Kolmogorov's Existence Theorem* [2]. Similarly, no more than two hidden layers are required if a step transfer function is used. However, the ability to learn from a set of examples cannot be guaranteed and, therefore, the detailed topology of a network inevitably involves a certain amount of trial and error. A pragmatic approach to network design is to start with a small network and expand the number of nodes or layers as necessary.

8.4.3 Training a perceptron

During training, a multilayer perceptron learns to separate the regions in state space by adjusting its weights and bias terms. Appropriate values are learned from a set of examples comprising input vectors and their corresponding desired output vectors. An input vector is applied to the input layer, and the output vector produced at the output layer is compared with the desired output. For each neuron in the output layer, the difference between the generated value and the desired value is the *error*. The overall error for the neural network is expressed as the square root of the mean of the squares of the errors. This is the root-mean-squared (RMS) value, designed to take equal account of both negative and positive errors. The RMS error is minimized by altering the weights and bias terms, which may take many passes through the training data. The search for the combination of weights and biases that produces the minimum RMS error is an optimization problem like those considered in Chapter 7, where the cost function is the RMS error. When the RMS error has become acceptably low for each example vector, the network is said to have *converged* and the weights and bias terms are retained for application of the network to new input data.

One of the most commonly used training algorithms is the *back-error propagation algorithm*, sometimes called the *generalized delta rule* [3, 4]. This is a gradient-proportional descent technique (see Chapter 7), and it relies upon the transfer function being continuous and differentiable. The sigmoid function (Figure 8.2(a)) is a particularly suitable choice since its derivative is simply given by:

$$f_t'(a) = f_t(a)(1 - f_t(a)) \tag{8.5}$$

The use of the back-error propagation algorithm for optimizing weights and bias terms can be made clearer by treating the biases as weights on the interconnections from dummy nodes, whose output is always 1, as shown in Figure 8.6. A flowchart describing the back-error propagation algorithm is presented in Figure 8.7, using the nomenclature shown in Figure 8.6.

At the core of the algorithm is the *delta rule* that determines the modifications to the weights, Δw_{Bij}:

$$\Delta w_{Bij} = \eta \delta_{Bi} y_{Aj} + \alpha(w_{Bij}) \tag{8.6}$$

for all nodes j in layer A and all nodes i in layer B, where $A = B - 1$. Neurons in the output layer and in the hidden layers have an associated error term, δ (pronounced *delta*). When the sigmoid transfer function is used, δ_{Aj} is given by:

$$\left.\begin{aligned}&\text{output layer:}\\&\delta_{Ai} = f_t'(y_{Ai})(d_i - y_{Ai}) = y_{Ai}(1 - y_{Ai})(d_i - y_{Ai})\\&\text{hidden layers:}\\&\delta_{Aj} = f_t'(y_{Aj})\sum_i \delta_{Bi} w_{Bij} = y_{Aj}(1 - y_{Aj})\sum_i \delta_{Bi} w_{Bij}\end{aligned}\right\} \quad (8.7)$$

The *learning rate*, η, is applied to the calculated values for δ_{Aj}. Knight [5] suggests a value for η of about 0.35. As written in Equation 8.6, the delta rule includes a *momentum* coefficient, α, although this is sometimes omitted. Gradient-proportional descent techniques can be inefficient, especially close to a minimum in the cost function, which in this case is the RMS error of the output. To address this, a momentum term forces changes in weight to be dependent on previous weight changes. The value of the momentum coefficient must be in the range 0–1. Knight [5] suggests that α be set to 0.0 for the first few training passes and then increased to 0.9.

Other training algorithms have also been successfully applied to perceptrons. For instance, Willis et al. [6] favor the chemotaxis algorithm, which incorporates a random statistical element in a similar fashion to simulated annealing (Chapter 7).

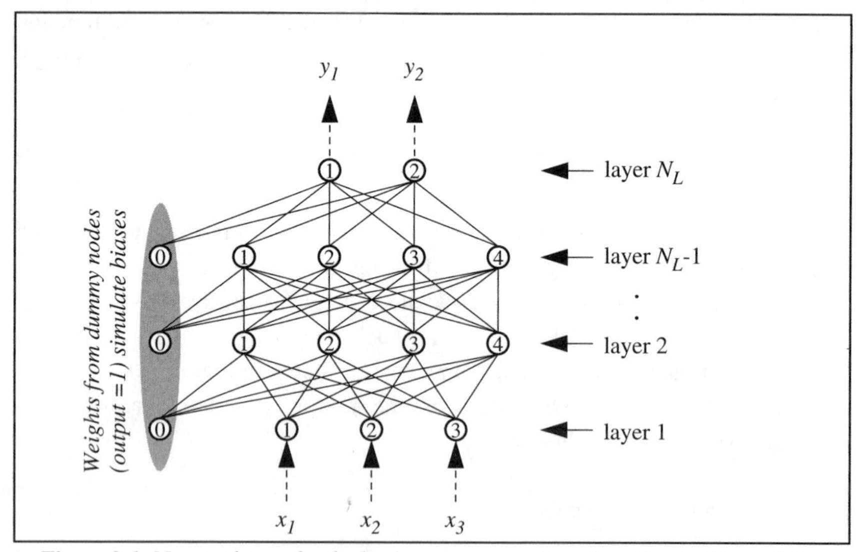

Figure 8.6 Nomenclature for the back-error propagation algorithm in Figure 8.7:
N_L = number of layers (4 in this example)
w_{Aij} = weight between node i on level A and node j on level A–1 ($N_L \geq A \geq 2$)
y_{Ai} = output from node i on level A
δ_{Ai} = an error term associated with node i on level A

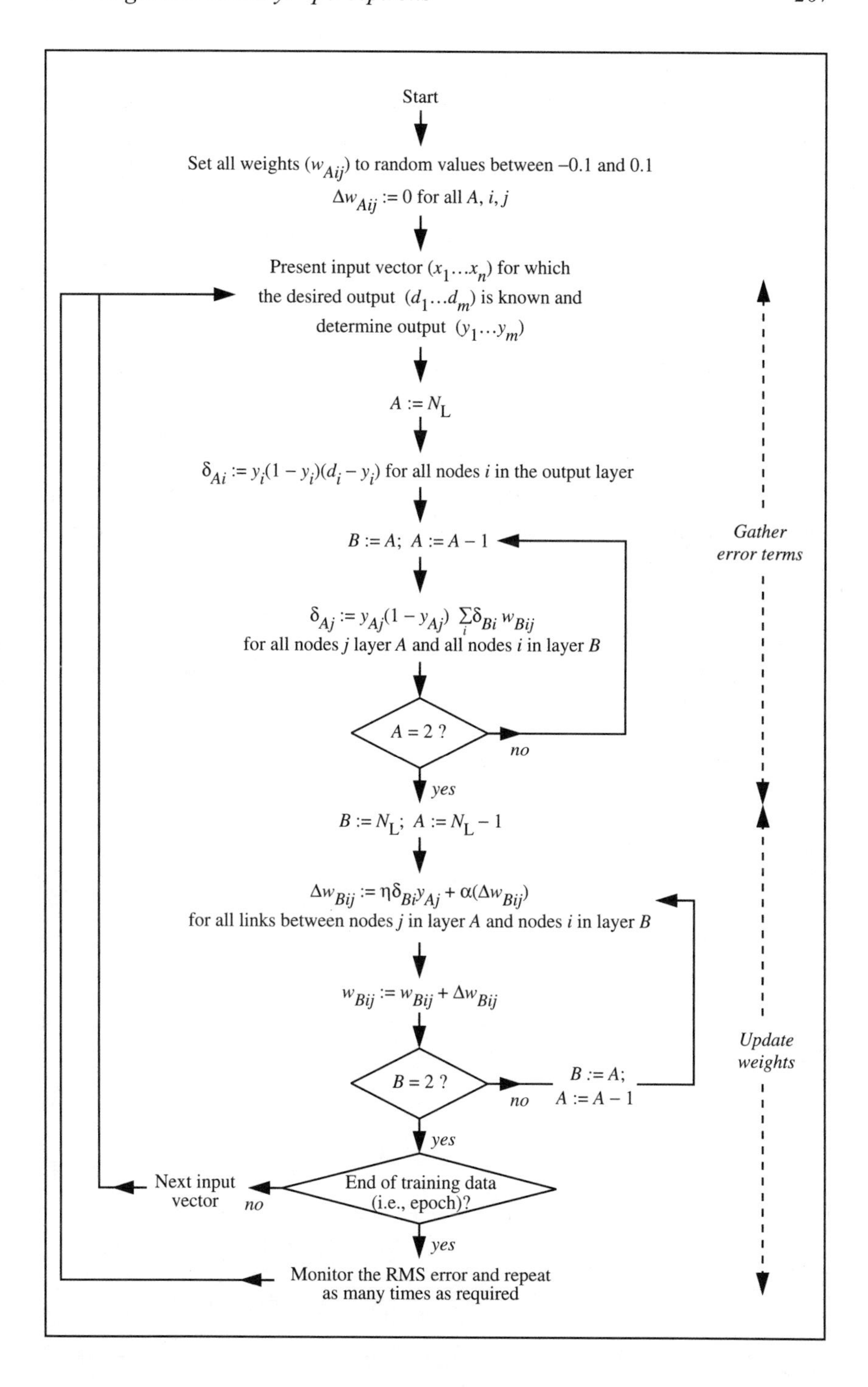
Start
Set all weights (w_{Aij}) to random values between –0.1 and 0.1
$\Delta w_{Aij} := 0$ for all A, i, j
Present input vector $(x_1 \ldots x_n)$ for which the desired output $(d_1 \ldots d_m)$ is known and determine output $(y_1 \ldots y_m)$
$A := N_L$
$\delta_{Ai} := y_i(1 - y_i)(d_i - y_i)$ for all nodes i in the output layer
$B := A;\ A := A - 1$
$\delta_{Aj} := y_{Aj}(1 - y_{Aj}) \sum_i \delta_{Bi} w_{Bij}$
for all nodes j layer A and all nodes i in layer B
$A = 2$?
no
yes
$B := N_L;\ A := N_L - 1$
$\Delta w_{Bij} := \eta \delta_{Bi} y_{Aj} + \alpha(\Delta w_{Bij})$
for all links between nodes j in layer A and nodes i in layer B
$w_{Bij} := w_{Bij} + \Delta w_{Bij}$
$B = 2$?
no
$B := A;$
$A := A - 1$
yes
End of training data (i.e., epoch)?
no
Next input vector
yes
Monitor the RMS error and repeat as many times as required
Gather error terms
Update weights

Figure 8.7 The back-error propagation algorithm (derived from [7] and [5])

8.4.4 Hierarchical perceptrons

In complex problems involving many inputs, some researchers recommend dividing an MLP into several smaller MLPs arranged in a hierarchy, as shown in Figure 8.8. In this example, the hierarchy comprises two levels. The inputs are shared among the MLPs at level 1, and the outputs from these networks form the inputs to an MLP at level 2. This approach is often useful if meaningful intermediate variables can be identified as the outputs from the level 1 MLPs. For example, if the inputs are measurements from sensors for monitoring equipment, the level 1 outputs could represent diagnosis of any faults, and the level 2 outputs could represent the recommended control actions [8]. In this example, a single large MLP could, in principle, be used to map directly from the sensor measurements to the recommended control actions. However, convergence of the smaller networks in the hierarchical MLP is likely to be achieved more easily. Furthermore, as the constituent networks in the hierarchical MLP are independent from each other, they can be trained either separately or in parallel.

8.4.5 Some practical considerations

Sometimes it is appropriate to stop the training process before the point where no further reductions in the RMS error are possible. This is because it is

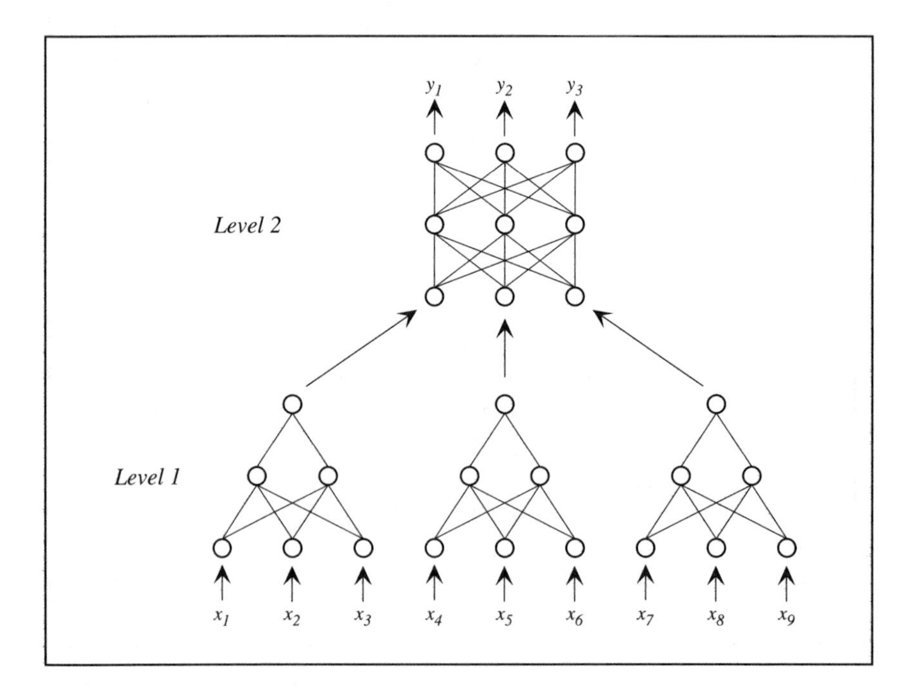

Figure 8.8 A hierarchical MLP

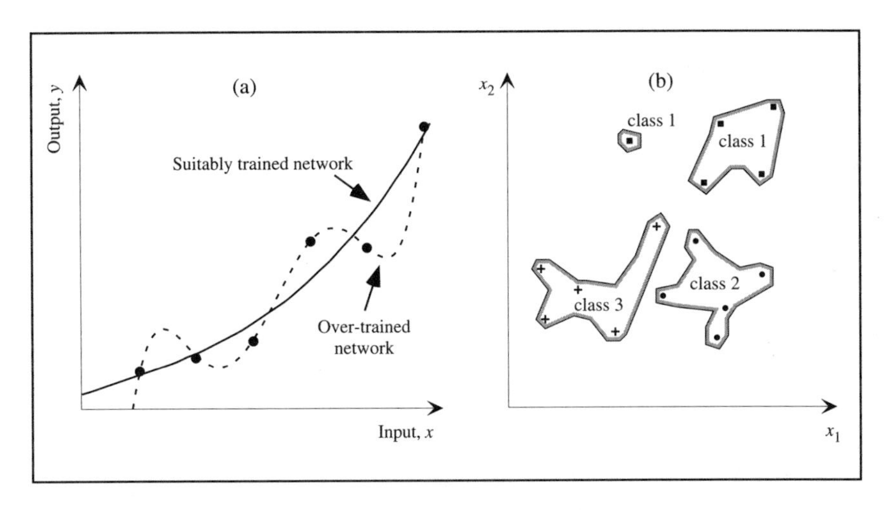

Figure 8.9 The effect of over-training: (a) nonlinear estimation; (b) classification (•, ▪, and + are data points used for training)

possible to *over-train* the network, so that it becomes expert at giving the correct output for the training data, but less expert at dealing with new data. This is likely to be a problem if the network has been trained for too many cycles or if the network is over-complex for the task in hand. For instance, the inclusion of additional hidden layers or large numbers of neurons within the hidden layers will tend to promote over-training. The effect of over-training is shown in Figure 8.9(a) for a nonlinear mapping of a single input parameter onto a single output parameter, and Figure 8.9(b) shows the effect of over-training using the nonlinearly separable classification data from Figure 8.4(b).

One way of avoiding over-training is to divide the data into three sets, known as the *training*, *testing*, and *validation* data. Training takes place using the training data, and the RMS error with these data is monitored. However, at predetermined intervals the training is paused and the current weights saved. At these points, before training resumes, the network is presented with the test data and an RMS error calculated. The RMS error for the training data decreases steadily until it stabilizes. However, the RMS error for the test data may pass through a minimum and then start to increase again because of the effect of over-training, as shown in Figure 8.10. As soon as the RMS error for the test data starts to increase, the network is over-trained, but the previously stored set of weights would be close to the optimum. Finally, the performance of the network can be evaluated by testing it using the previously unseen validation data.

A problem that is frequently encountered in real applications is a shortage of suitable data for training and testing a neural network. Hopgood et al. [9] describe a classification problem where there were only 20 suitable examples,

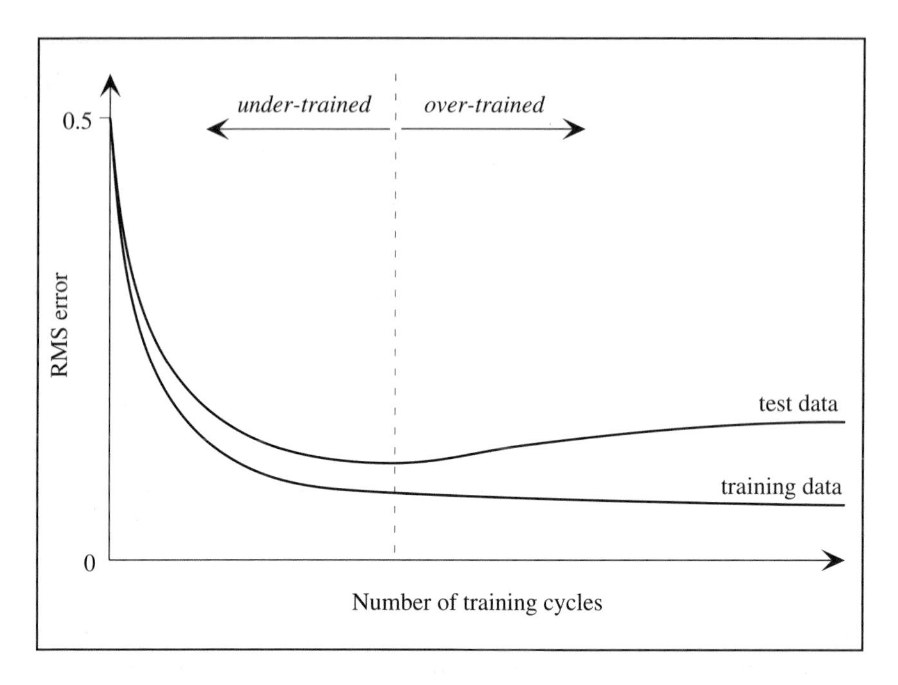

Figure 8.10 RMS error during training

which needed to be shared between the training and testing data. They used a technique called *leave-one-out* as a way of reducing the effect of this problem. The technique involves repeatedly training on all but one of the examples and testing on the missing one. So, in this case, the network would initially be trained on 19 of the examples and tested on the remaining one. This procedure is repeated a further 19 times: omitting a different example each time from the training data, resetting the weights to random values, retraining, and then testing on the omitted example. The leave-one-out technique is clearly time-consuming as it involves resetting the weights, training, testing, and scoring the network many times — 20 times in this example. Its advantage is that the performance of the network can be evaluated using every available example as though it were previously unseen test data.

Neural networks that accept real numbers are only effective if the input values are constrained to suitable ranges, typically between 0 and 1 or between −1 and 1. The range of the outputs depends on the chosen transfer function, e.g., the output range is between 0 and 1 if the sigmoid function is used. In real applications, the actual input and output values may fall outside these ranges or may be constrained to a narrow band within them. In either case the data will need to be scaled, usually linearly, before being presented to the neural network. Some neural network packages perform the scaling automatically.

8.5 The Hopfield network

The Hopfield network has only one layer, and the nodes are used for both input and output. The network topology is shown in Figure 8.11(a). The network is normally used as a content-addressable memory where each training example is treated as a model vector or *exemplar*, to be stored by the network. The Hopfield network uses binary input values, typically 1 and −1. By using the step nonlinearity shown in Figure 8.2(c) as the transfer function f_t the output is forced to remain binary, too. If the activation a is 0, i.e., on the step, the output is indeterminate, so a convention is needed to yield an output of either 1 or −1.

If the network has N_n nodes, then the input and output would both comprise a vector of N_n binary digits. If there are N_e exemplars to be stored, the network weights and biases are set according to the following equations:

$$w_{ij} = \begin{cases} \sum_{k=1}^{N_e} x_{ik} x_{jk} & \text{if } i \neq j \\ 0 & \text{if } i = j \end{cases} \tag{8.8}$$

$$w_{i0} = \sum_{k=1}^{N_e} x_{ik} \tag{8.9}$$

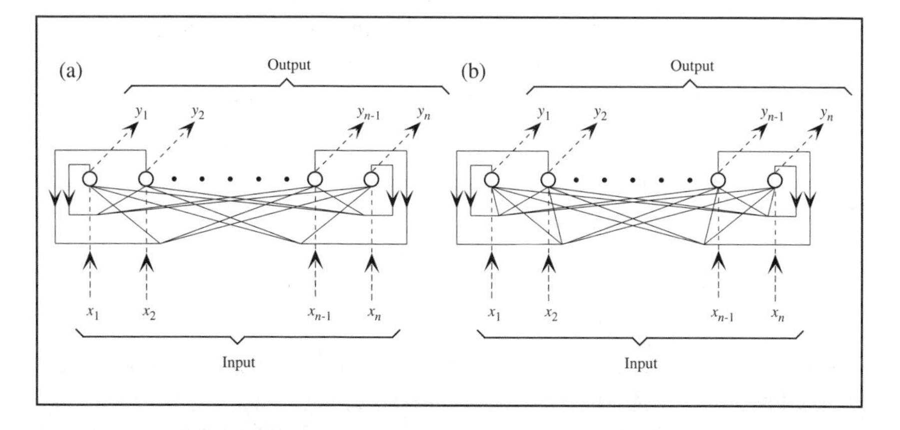

Figure 8.11 The topology of: (a) the Hopfield network, (b) the MAXNET. Circular connections from a node to itself are allowed in the MAXNET, but are disallowed in the Hopfield network

where w_{ij} is the weighting on the connection from node i to node j, w_{i0} is the bias on node i, and x_{ik} is the ith digit of example k. There are no circular connections from a node to itself, hence $w_{ij} = 0$ where $i = j$.

Setting weights in this way constitutes the learning phase, and results in the exemplars being stored in a distributed fashion in the network. If a new vector is subsequently presented as the input, then this vector is initially the output, too, as nodes are used for both input and output. The node function (Figure 8.1) is then performed on each node in parallel. If this is repeated many times, the output will be progressively modified and will converge on the exemplar that most closely resembles the initial input vector. In order to store reliably at least half the exemplars, Hopfield estimated that the number of exemplars (N_e) should not exceed approximately $0.15N_n$ [10].

If the network is to be used as for classification, a further stage is needed in which the result is compared with the exemplars to pick out the one that matches.

8.6 MAXNET

The MAXNET (Figure 8.11(b)) has an identical topology to the Hopfield network, except that the weights on the circular interconnections, w_{ii}, are not always zero as they are in the Hopfield network. The MAXNET is used to recognize which of its inputs has the highest value. In this role it is sometimes used in conjunction with other networks, such as a multilayer perceptron, to select the output node that generates the highest value. Suppose that the multilayer perceptron has four output nodes, corresponding to four different categories. A MAXNET to determine the maximum output value (and, hence, the solution to the classification task) would have four nodes and four alternative output patterns after convergence, i.e., four exemplars:

$$\begin{matrix} X & 0 & 0 & 0 \\ 0 & X & 0 & 0 \\ 0 & 0 & X & 0 \\ 0 & 0 & 0 & X \end{matrix}$$

where $X > 0$. The MAXNET would adjust its highest input value to X, and reduce the others to 0. Note that the MAXNET is constructed to have the same number of nodes (N_n) as the number of exemplars (N_e). Contrast this with the Hopfield network, which needs approximately seven times as many nodes as exemplars.

Using the same notation as Equations 8.8 and 8.9, interconnection weights are set as follows:

$$w_{ij} = \begin{cases} -\varepsilon & \text{where } \varepsilon < \dfrac{1}{N_n} \quad \text{if } i \neq j \\ 1 & \text{if } i = j \end{cases} \tag{8.10}$$

8.7 The Hamming network

The *Hamming network* has two parts — a twin layered feedforward network and a MAXNET, as shown in Figure 8.12. The feedforward network is used to compare the input vector with each of the examples, awarding a matching score to each example. The MAXNET is then used to pick out the example that has attained the highest score. The overall effect is that the network can categorize its input vector.

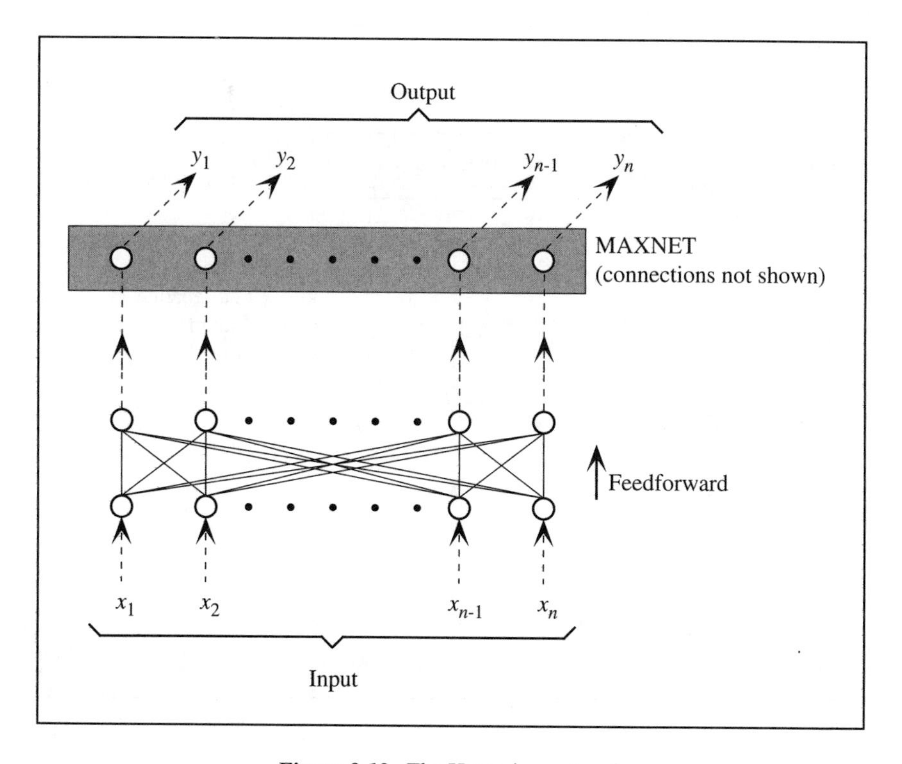

Figure 8.12 The Hamming network

8.8 Adaptive Resonance Theory (ART) networks

The Adaptive Resonance Theory networks (ART1 and ART2) of Carpenter and Grossberg [11] are worthy of mention because they are examples of networks that learn without supervision. The ART1 network topology comprises bidirectional interconnections between a set of input nodes and a MAXNET, as shown in Figure 8.13.

The network classifies the incoming data into clusters. When the first example is presented to the network, it becomes stored as an exemplar or model pattern. The second example is then compared with the exemplar and is either considered sufficiently similar to belong to the same cluster or stored as a new exemplar. If an example is considered to belong to a previously defined cluster, the exemplar for that cluster is modified to take account of the new member. The performance of the network is dependent on the way in which the differences are measured, i.e., the *closeness* measure, and the threshold or

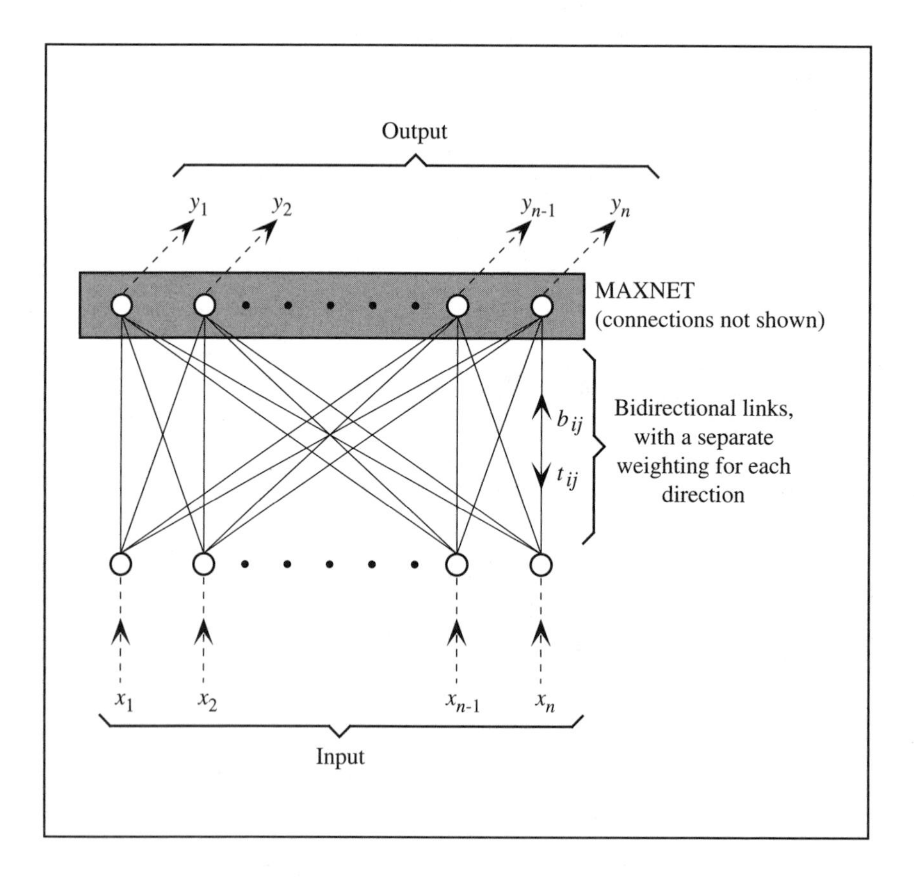

Figure 8.13 The ART1 network

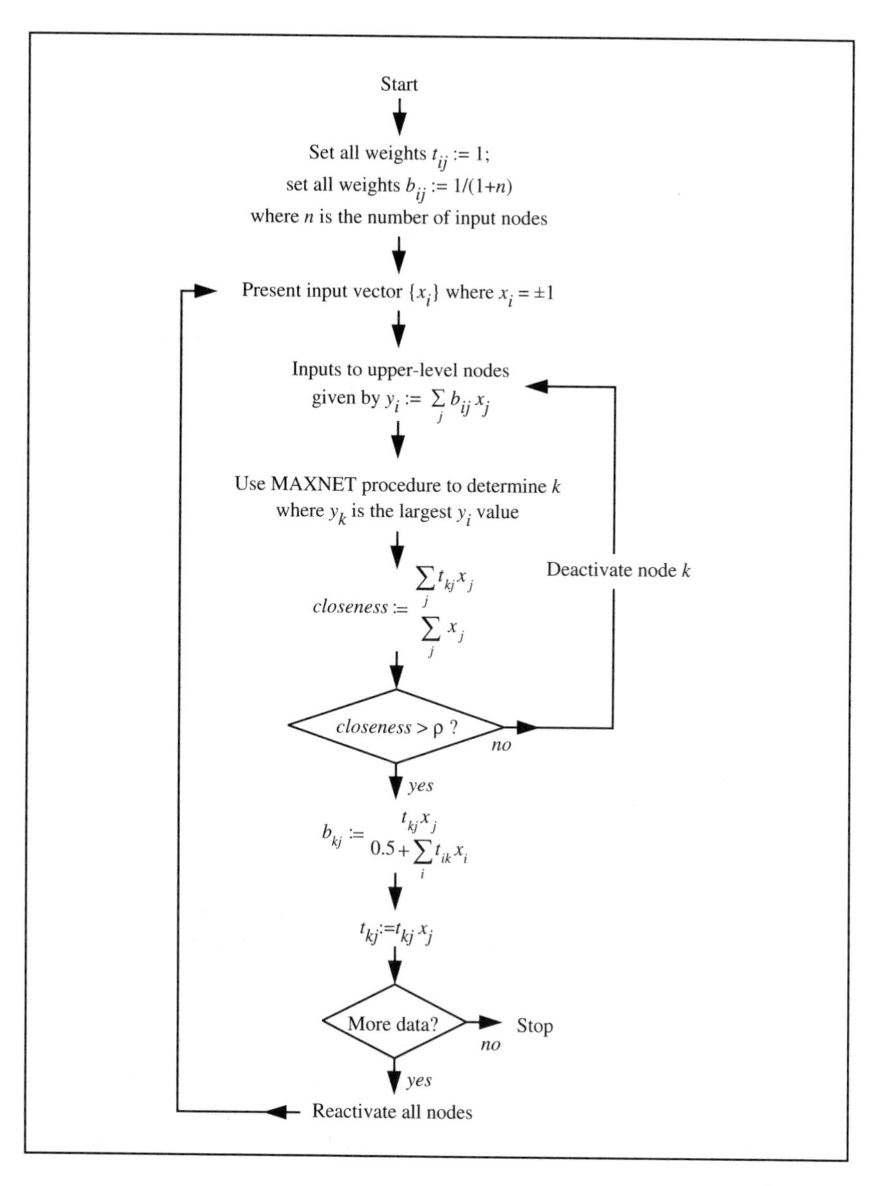

Figure 8.14 Unsupervised learning in ART1

vigilance, ρ, beyond which a new exemplar is stored. As each new vector is presented, it is compared with all of the current exemplars in parallel. The number of exemplars grows as the network is used, i.e., the network learns new patterns. The operation of the ART1 network, which takes binary inputs, is summarized in Figure 8.14. ART2 is similar but takes continuously varying inputs.

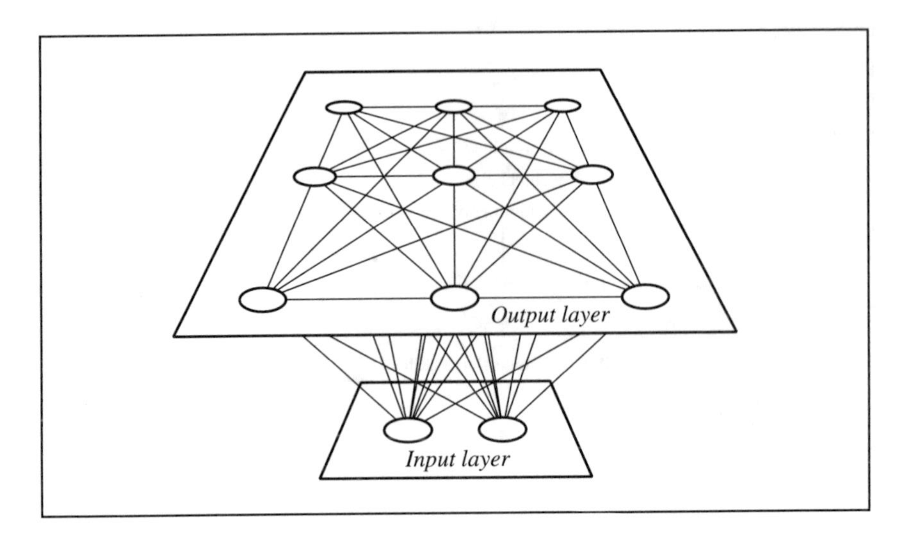

Figure 8.15 A Kohonen self-organizing network

8.9 Kohonen self-organizing networks

Kohonen self-organizing networks, sometimes called self-organizing maps (SOMs), provide another example of networks that can learn without supervision. The processing nodes can be imagined to be arranged in a two-dimensional array, known as the Kohonen layer (Figure 8.15). There is also a separate one-dimensional layer of input nodes, where each input node is connected to each node in the Kohonen layer. As in the MLP, the input neurons perform no processing but simply pass their input values to the processing neurons, where a weighting is applied.

As in the ART networks, the Kohonen network learns to cluster together similar patterns. The learning mechanism, described in [7, 12, 13, 14], involves competition between the neurons to respond to a particular input vector. The "winner" has its weightings set so as to generate a high output, approaching 1. The weightings on nearby neurons are also adjusted so as to produce a high value, but the weights on the "losers" are left alone. The neurons that are nearby the winner constitute a *neighborhood.*

When the trained network is presented with an input pattern, one neuron in the Kohonen layer will produce an output larger than the others, and is said to have fired. When a second similar pattern is presented, the same neuron or one in its neighborhood will fire. As similar patterns cause topologically close neurons to fire, clustering of similar patterns is achieved. The effect can be

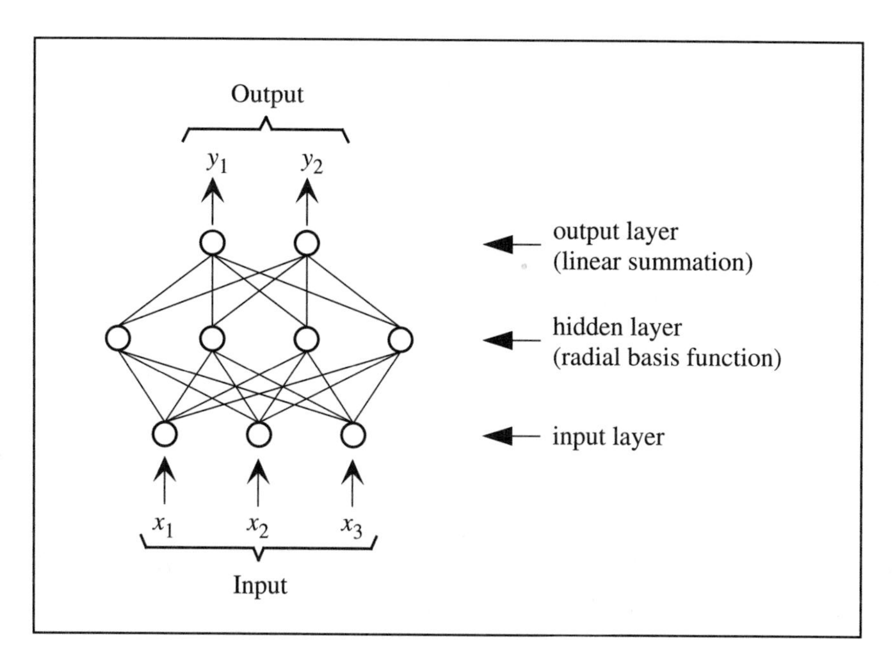

Figure 8.16 A radial basis function network

demonstrated by training the network using pairs of Cartesian coordinates. The trained network has the property that the distribution of the firing neurons corresponds with the Cartesian coordinates represented by the input vector. Thus, if the input elements fall in the range between –1 and 1, then an input vector of (–0.9, 0.9) will cause a neuron close to one corner of the Kohonen layer to fire, while an input vector of (0.9, –0.9) would cause a neuron close to the opposite corner to fire.

Although these networks are unsupervised, they can form part of a hybrid network for supervised learning. This is achieved by passing the coordinates of the firing neuron to an MLP. In this arrangement, learning takes place in two distinct phases. First, the Kohonen self-organizing network learns, without supervision, to associate regions in the pattern space with clusters of neurons in the Kohonen layer. Second, an MLP learns to associate the coordinates of the firing neuron in the Kohonen layer with the desired class.

8.10 Radial basis function networks

Radial basis function (RBF) networks offer another alternative method of unsupervised learning. They are feedforward networks, the overall architecture of which is similar to that of a three-layered perceptron, i.e., an MLP with one

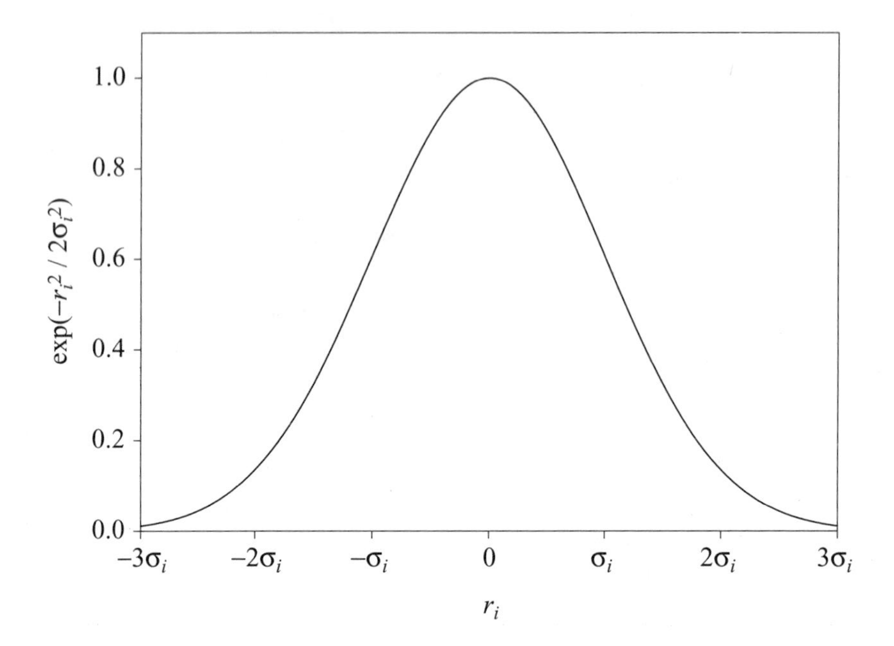

Figure 8.17 Gaussian RBF with standard deviation σ_i

hidden layer. The RBF network architecture is shown in Figure 8.16. The input and output neurons are similar to those of a perceptron, but the neurons in the hidden layer are completely different. The input neurons do not perform any processing, but simply feed the input data into the nodes above. The neurons in the output layer produce the weighted sum of their inputs, which is usually passed through a linear transfer function, in contrast to the nonlinear transfer functions used with perceptrons.

The processing neurons considered so far in this chapter produce an output that is the weighted sum of their inputs, passed through a transfer function. However, in an RBF network the neurons in the hidden layer, sometimes called the *prototype* layer, behave differently. For an input vector (x_1, x_2, … x_n), a neuron i in the hidden layer produces an output, y_i, given by:

$$y_i = f_r(r_i) \tag{8.11}$$

$$r_i = \sqrt{\sum_{j=1}^{n} \left(x_j - w_{ij}\right)^2} \tag{8.12}$$

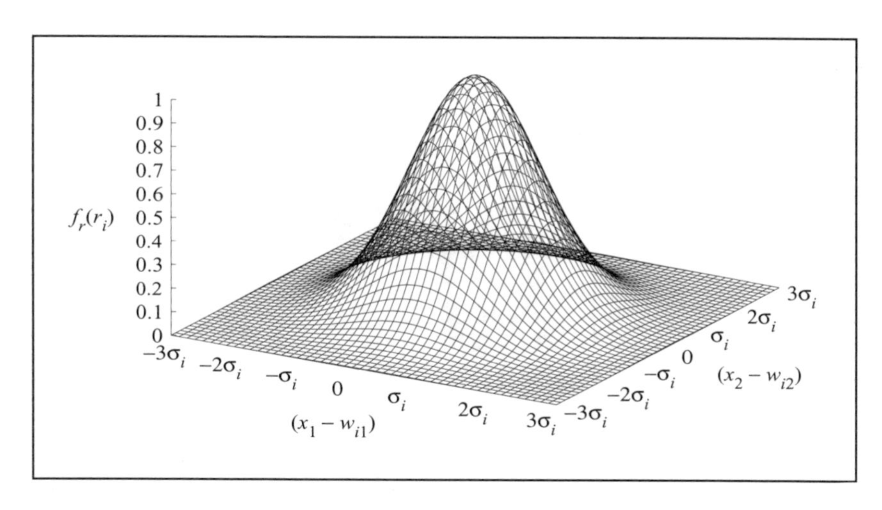

Figure 8.18 Gaussian RBF with standard deviation σ_i applied to two input variables

where w_{ij} are the weights on the inputs to neuron i, and f_r is a symmetrical function known as the *radial basis function* (RBF). The most commonly used RBF is a Gaussian function:

$$f_r(r_i) = \exp\left(\frac{-r_i^2}{2\sigma_i^2}\right) \tag{8.13}$$

where σ_i is the standard deviation of a distribution described by the function (Figure 8.17). Each neuron, i, in the hidden layer has its own separate value for σ_i.

The Euclidean distance between two points is the length of a line drawn between them. If the set of weights (w_{i1}, w_{i2}, … w_{in}) on a given neuron i is treated as the coordinates of a point in pattern space, then r_i is the Euclidean distance from there to the point represented by the input vector (x_1, x_2, … x_n). During unsupervised learning, the network adjusts the weights — more correctly called *centers* in an RBF network — so that each point (w_{i1}, w_{i2}, … w_{in}) represents the center of a cluster of data points in pattern space. Similarly it defines the sizes of the clusters by adjusting the variables σ_i (or equivalent variables if an RBF other than the Gaussian is used). Data points within a certain range, e.g., $2\sigma_i$, from a cluster center might be deemed members of the cluster. Therefore, just as a single-layered perceptron can be thought of as dividing up two-dimensional pattern space by lines, or n-dimensional pattern space by hyperplanes, so the RBF network can be thought of as drawing circles around clusters in two-dimensional pattern space, or hypersheres in n-dimensional pattern space. One such cluster can be identified for each neuron

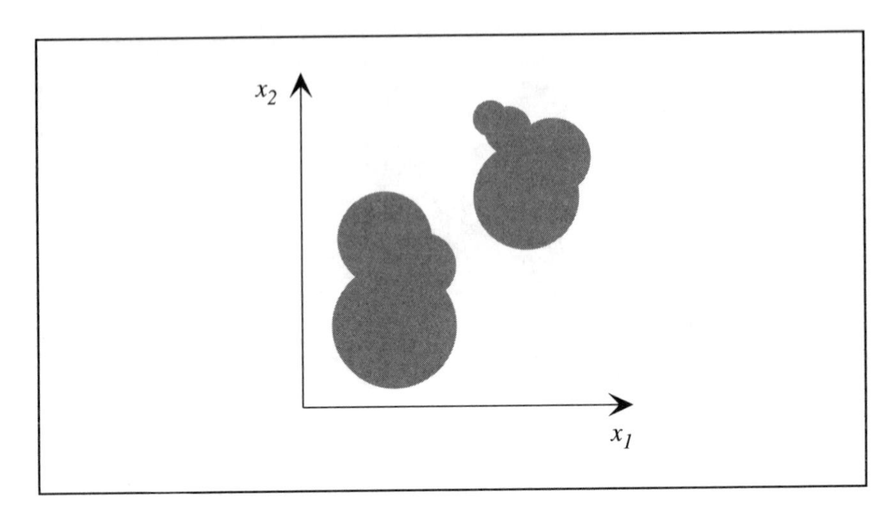

Figure 8.19 RBF networks can define arbitrary shapes for regions in the pattern space

in the hidden layer. Figure 8.18 shows a Gaussian function in two-dimensional pattern space, from which it can be seen that a fixed output value (e.g., 0.5) defines a circle in the pattern space.

The unsupervised learning in the hidden layer is followed by a separate supervised learning phase in which the output neurons learn to associate each cluster with a particular class. By associating several circular clusters of varying center and size with a single class, arbitrary shapes for class regions can be defined (Figure 8.19).

8.11 Summary

Numerical learning is based upon adapting numerical parameters in order to achieve a close match between a desired output and the actual output. Neural networks are an important type of numerical learning technique. They can be used to model any nonlinear mapping between variables, and are frequently used in classification tasks. When presented with data that lie between previously encountered data, neural networks will generally interpolate to produce an output between those generated previously. Neural networks may therefore be a substitute for fuzzy logic in some applications. The parallelism of neural networks makes them ideally suited to parallel processing computers.

A neural network may be used to solve a problem in its own right, or as a component of a larger system. Designing a suitable network for a given application can involve a large amount of trial and error. Whether a network will converge, i.e., learn suitable weightings, will depend on the topology, the transfer function of the nodes, the values of the parameters in the training algorithm, and the training data. It may even depend on the order in which the training data are presented. Although neural networks can alleviate the "knowledge acquisition bottleneck" described in Chapter 6, they can also introduce a "network parameter bottleneck" as the user struggles to configure a network for a particular problem [15].

Another drawback of neural networks is that their reasoning is opaque. The learned weights can rarely be understood in a meaningful way, although rules can, in principle, be extracted from them. Thus, the neural network is often regarded as a "black box" which simply generates an output from a given input. However, by confining the use of the neural network to subtasks within a problem, the overall problem-solving strategy can remain clear.

References

1. Hallam, N. J., Hopgood, A. A., and Woodcock, N., "Defect classification in welds using a feedforward network within a blackboard system," International Neural Network Conference (INNC'90), Paris, vol. 1, pp. 353–356, 1990.
2. Hornick, K., Stinchcombe, M., and White, H., "Multilayer feedforward networks are universal approximators," *Neural Networks*, vol. 2, pp. 359–366, 1989.
3. Rumelhart, D. E., Hinton, G. E., and Williams, R. J., "Learning representations by back-propagating errors," *Nature*, vol. 323, pp. 533–536, 1986.
4. Rumelhart, D. E., Hinton, G. E., and Williams, R. J., "Learning internal representations by error propagation," in *Parallel Distributed Processing: explorations in the microstructures of cognition*, vol. 1, Rumelhart, D. E. and McClelland, J. L. (Eds.), MIT Press, 1986.
5. Knight, K., "Connectionist ideas and algorithms," *Communications of the ACM*, vol. 33, issue 11, pp. 59–74, November 1990.
6. Willis, M. J., Di Massimo, C., Montague, G. A., Tham, M. T., and Morris, A. J., "Artificial neural networks in process engineering," *IEE Proceedings-D*, vol. 138, pp. 256–266, 1991.
7. Lippmann, R. P., "An introduction to computing with neural nets," *IEEE ASSP Magazine*, pp. 4–22, April 1987.
8. Kim, K. H. and Park, J. K., "Application of herarchical neural networks to fault diagnosis of power systems," *International Journal of Electrical Power and Energy Systems*, vol. 15, pp. 65–70, 1993.

9. Hopgood, A. A., Woodcock, N., Hallam, N. J., and Picton, P. D., "Interpreting ultrasonic images using rules, algorithms and neural networks," *European Journal of Nondestructive Testing*, vol. 2, pp. 135–149, 1993.
10. Hopfield, J. J., "Neural networks and physical systems with emergent collective computational abilities," *Proceedings of the National Academy of Science*, vol. 79, pp. 2554–2558, 1982.
11. Carpenter, G. A. and Grossberg, S., "ART2: Self-organization of stable category recognition codes for analog input patterns," *Applied Optics*, vol. 26, pp. 4919–4930, 1987.
12. Kohonen, T., "Adaptive, associative, and self-organising functions in neural computing," *Applied Optics*, vol. 26, pp. 4910–4918, 1987.
13. Kohonen, T., *Self-organization and Associative Memory*, 2nd ed., Springer-Verlag, 1988.
14. Hecht-Nielson, R., *Neurocomputing*, Addison-Wesley, 1990.
15. Woodcock, N., Hallam, N. J., Picton, P., and Hopgood, A. A., "Interpretation of ultrasonic images of weld defects using a hybrid system," International Conference on Neural Networks and their Applications, Nimes, France, 1991.

Further reading

- Beale, R. and Jackson, T., *Neural Computing: an introduction*, IOP Publishing, 1990.
- Bishop, C. M., *Neural Networks for Pattern Recognition*, Clarendon Press, 1995.
- Gurney, K., *An Introduction to Neural Networks*, UCL Press, 1997.
- Hecht-Nielson, R., *Neurocomputing*, Addison-Wesley, 1990.
- Pao, Y.-H., *Adaptive Pattern Recognition and Neural Networks*, Addison-Wesley, 1989.
- Patterson, D. W., *Artificial Neural Networks*, Prentice Hall, 1998.
- Picton, P. D., *Neural Networks*, 2nd ed., Macmillan, 2000.

Chapter nine

Hybrid systems

9.1 Convergence of techniques

One of the aims of this book is to demonstrate that there is a wide variety of computing techniques that can be applied to particular problems. These include symbolic representations such as knowledge-based systems, computational intelligence methods, and conventional programs. In many cases the techniques need not be exclusive alternatives to each other but can be seen as complementary tools that can be brought together within a hybrid system. All of the techniques reviewed in this book can, in principle, be mixed with other techniques. There are several ways in which different computational techniques can be complementary:

Dealing with multifaceted problems

Most real-life problems are complex and have many facets, where each facet may be best suited to a different technique. Therefore, many practical systems are designed as hybrids, incorporating several specialized modules, each of which uses the most suitable tools for its specific task. One way of allowing the modules to communicate is by designing the hybrid as a blackboard system, described in Section 9.2 below.

Capability enhancement

One technique may be used within another to enhance the latter's capabilities. We met an example in Chapter 7, where Lamarckian inheritance involves the inclusion of a steepest gradient descent step within a genetic algorithm. In this example, the aim is to raise the fitness of individual chromosomes and speed convergence of the genetic algorithm toward an optimum.

Parameter setting

Several of the techniques described below, such as neuro-fuzzy, genetic-fuzzy, and genetic-neural systems, are based on the idea of using one technique to set the parameters of another.

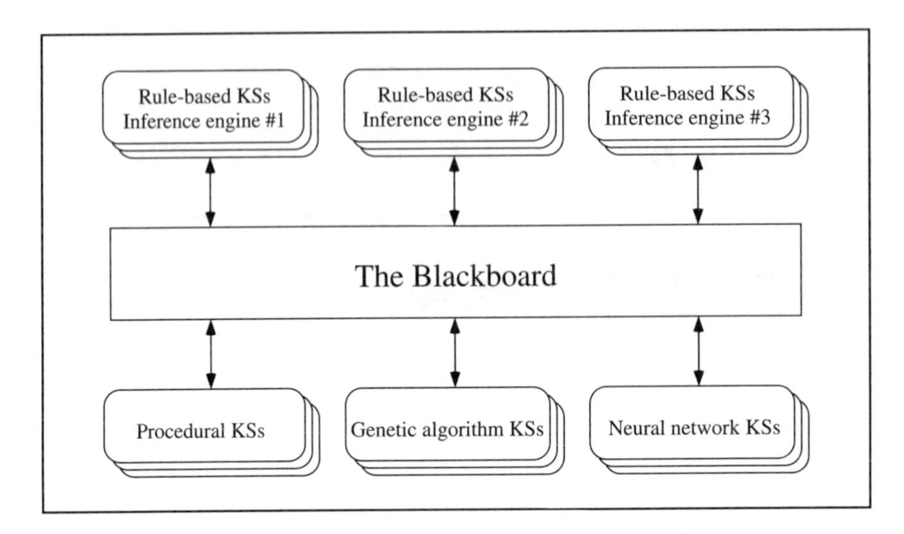

Figure 9.1 The blackboard model. The modules are called knowledge sources (KSs)

Clarification and verification

Neural networks have the ability to learn associations between input vectors and associated outputs. However, the underlying reasons for the associations are opaque, as they are effectively encoded in the weightings on the interconnections between the neurons. Efforts have been made to extract automatically from the network equivalent rules that can be readily understood, and to write verification rules to check the validity of the network's output.

9.2 Blackboard systems

The *blackboard model* or *blackboard architecture*, shown in Figure 9.1, provides a software structure that is well suited to multifaceted tasks. Systems that have this kind of structure are called *blackboard systems*. In such a system, knowledge of the application domain is divided into modules, referred to as *knowledge sources* (or KSs), each of which is designed to tackle a particular subtask. KSs are independent and can communicate only by reading from or writing to the *blackboard*, a globally accessible working memory where the current state of understanding is represented. Knowledge sources can also delete unwanted information from the blackboard.

A blackboard system is analogous to a team of experts who communicate their ideas via a physical blackboard, by adding or deleting items in response to the information that they find there. Each knowledge source represents such an expert having a specialized area of knowledge. As each KS can be encoded in the most suitable form for its particular task, blackboard systems offer a mechanism for the collaborative use of different computational techniques such as rules, neural networks, and fuzzy logic. The inherent modularity of a blackboard system is also helpful for maintenance. Each rule-based knowledge source can use a suitable reasoning strategy for its particular task, e.g., backward- or forward-chaining, and can be thought of as a rule-based system in microcosm.

Knowledge sources are applied in response to information on the blackboard, when they have some contribution to make. This leads to increased efficiency since the detailed knowledge within a knowledge source is only applied when that knowledge source becomes relevant. In the idealized blackboard system, the KSs are said to be *opportunistic*, activating themselves whenever they can contribute to the global solution. However, this is difficult to achieve in practice as it may involve interrupting another knowledge source that is currently active. One approach to KS scheduling is to use a control module that determines the order of KS activation on the basis of applicability and past use of the KSs. As an extension of this idea, a separate blackboard system could select knowledge sources based upon explicit rule-based knowledge of the alternatives. This level of sophistication may, however, result in a slow response.

In the interests of efficiency and clarity, some degree of structure is usually imposed on the blackboard by dividing it into panels. A knowledge source then only needs to look at a small number of panels rather than the whole blackboard. Typically the blackboard panels correspond to different levels of analysis of the problem, progressing from detailed information to more abstract concepts. In the *Hearsay-II* blackboard system for computerized understanding of natural speech, the levels of analysis include those of syllable, word, and phrase [1]. In ultrasonic image interpretation using *ARBS*, described in more detail in Chapter 11, the levels progress from raw signal data, via a description of the significant image features, to a description of the defects in the component [2].

The key advantages of the blackboard architecture, adapted from Feigenbaum [3] can be summarized as follows:

(i) Many and varied sources of knowledge can participate in the development of a solution to a problem.

(ii) Since each knowledge source has access to the blackboard, each can be applied as soon as it becomes appropriate. This is opportunism, i.e., application of the right knowledge at the right time.

(iii) For many types of problem, especially those involving large amounts of numerical processing, the characteristic style of incremental solution development is particularly efficient.

(iv) Different types of reasoning strategy (e.g., data- and goal-driven) can be mixed as appropriate in order to reach a solution.

(v) Hypotheses can be posted onto the blackboard for testing by other knowledge sources. A complete test solution does not have to be built before deciding to modify or abandon the underlying hypothesis.

(vi) In the event that the system is unable to arrive at a complete solution to a problem, the partial solutions appearing on the blackboard are available and may be of some value.

9.3 Genetic-fuzzy systems

The performance of a fuzzy system depends on the definitions of the fuzzy sets and on the fuzzy rules. As these parameters can all be expressed numerically, it is possible to devise a system whereby they are learned automatically using genetic algorithms. A chromosome can be devised that represents the complete set of parameters for a given fuzzy system. The cost function could then be defined as the total error when the fuzzy system is presented with a number of different inputs with known desired outputs.

Often, a set of fuzzy rules for a given problem can be drawn up fairly easily, but defining the most suitable membership functions remains a difficult task. Karr [4, 5] has performed a series of experiments to demonstrate the viability of using genetic algorithms to specify the membership functions. In Karr's scheme, all membership functions are triangular. The variables are constrained to lie within a fixed range, so the fuzzy sets `low` and `high` are both right-angle triangles (Figure 9.2). The slope of these triangles can be altered by moving their intercepts on the abscissa, marked max_1 and min_4 in Figure 9.2. All intermediate fuzzy sets are assumed to have membership functions that are isosceles triangles. Each is defined by two points, max_i and min_i, where i labels the fuzzy set. The chromosome is then a list of all the points max_i and min_i that determine the complete set of membership functions. In several demonstrator systems, Karr's GA-modified fuzzy controller outperformed a fuzzy controller whose membership functions had been set manually. This is perhaps not surprising, since tuning the fuzzy sets is an optimization problem.

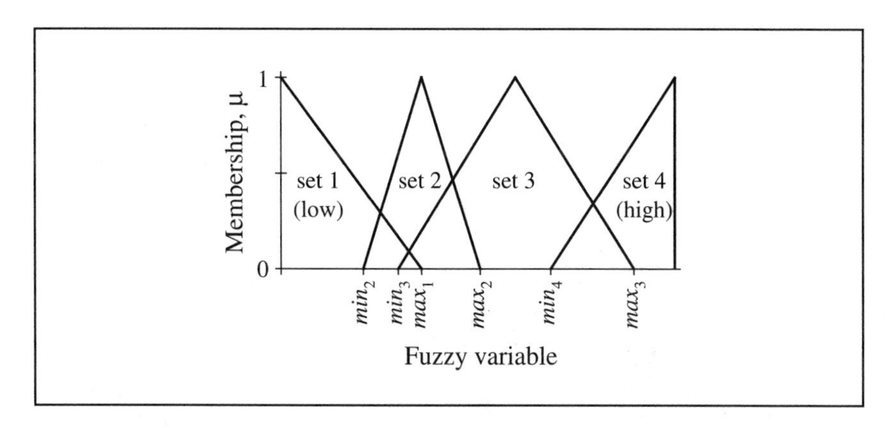

Figure 9.2 Defining triangular membership functions by their intercepts on the abscissa

9.4 Neuro-fuzzy systems

Section 9.3 above shows how a genetic algorithm can be used to optimize the parameters of a fuzzy system. In such a scheme, the genetic algorithm for parameter setting and the fuzzy system that uses those parameters are distinct and separate. The parameters for a fuzzy system can also be learned using neural networks, but here much closer integration is possible between the neural network and the fuzzy system that it represents. A neuro-fuzzy system is a fuzzy system, the parameters of which are derived by a neural network learning technique. It can equally be viewed as a neural network that represents a fuzzy system. The two views are equivalent and it is possible to express a neuro-fuzzy system in either form.

Consider the following fuzzy rules, based on the example used in Chapter 3:

```
/* Rule 9.1f */
IF temperature is high OR water level is high
THEN pressure is high

/* Rule 9.2f */
IF temperature is medium OR water level is medium
THEN pressure is medium

/* Rule 9.3f */
IF temperature is low OR water level is low
THEN pressure is low
```

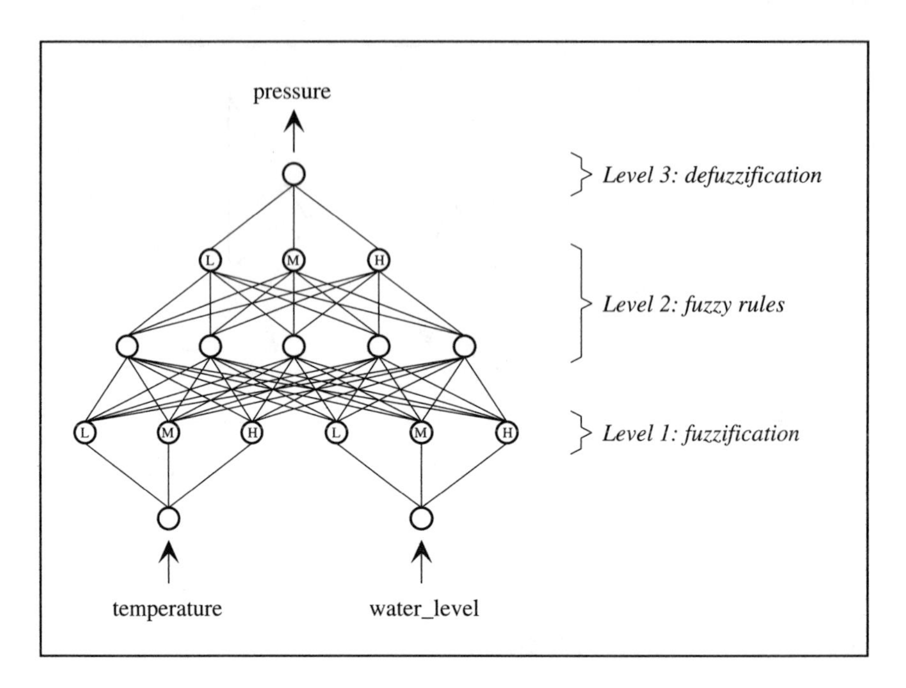

Figure 9.3 A neuro-fuzzy network

These fuzzy rules and the corresponding membership functions can be represented by the neural network shown in Figure 9.3. The first stage is fuzzification, in which any given input value for `temperature` is given a membership value for `low`, `medium`, and `high`. A single layer perceptron, designated level 1 in Figure 9.3, can achieve this because it is a linear classification task. The only difference from other classifications met previously is that the desired output values are not just 0 and 1, but any value in the range between 0 and 1. A similar network is required at level 1 for the other input variable, `water level`. The neurons whose outputs correspond to the `low`, `medium`, and `high` memberships are marked L, M, and H, respectively, in Figure 9.3.

Level 2 of the neuro-fuzzy network performs the role of the fuzzy rules, taking the six membership values as its inputs and generating as its outputs the memberships for `low`, `medium`, and `high` of the fuzzy variable `pressure`. The final stage, at level 3, involves combining these membership values to produce a defuzzified value for the output variable.

The definitions of the fuzzy sets and the fuzzy rules are implicit in the connections and weights of the neuro-fuzzy network. Using a suitable learning mechanism, the weights can be learned from a series of examples. The network can then be used on previously unseen inputs to generate defuzzified output values. In principle, the fuzzy sets and rules can be inferred from the network and run as a fuzzy rule-based system, of the type described in Chapter 4, to produce identical results [6, 7].

9.5 Genetic-neural systems

Part of the practical challenge in building a practical neural network is to choose the right architecture and the right learning parameters. We saw in Chapter 8 that, according to Kolmogorov's Existence Theorem, an MLP with one hidden layer, using the sigmoid transfer function, could perform any mapping from a set of inputs to the desired outputs. Unfortunately, this theorem tells us nothing about the learning parameters, the necessary number of neurons, or whether additional layers would be beneficial. It is, however, possible to use a genetic algorithm to optimize the network design. A suitable cost function might combine the RMS error with duration of training.

Supervised training of a neural network involves adjusting its weights until the output patterns obtained for a range of input patterns are as close as possible to the desired patterns. The different network topologies use different training algorithms for achieving this weight adjustment, typically through back-propagation or errors. However, it is also possible to use a genetic algorithm to train the network. This can be achieved by letting each gene represent a network weight, so that a complete set of network weights is mapped onto an individual chromosome. Each chromosome can be evaluated by testing a neural network with the corresponding weights against a series of test patterns. A fitness value can be assigned according to the error, so that the weights represented by the fittest generated individual correspond to a trained neural network.

9.6 Clarifying and verifying neural networks

Neural networks are often referred to as a type of "black box," with an internal state that conveys no readily useful information to an observer. This metaphor implies the inputs and outputs have significance for an observer, but the weights on the interconnections between the neurons do not. This contrasts with a transparent system, such as a KBS, where the internal state, for example, the value of a variable, does have meaning for an observer. There has been an intensive research effort into *rule extraction* to produce rules which are equivalent to the trained neural network from which they have been extracted [8]. A variety of methods have been reported for extracting different types of rules, including production rules and fuzzy rules.

In safety-critical systems, reliance on the output from a neural network without any means of verification is not acceptable. It has, therefore, been proposed that rules be used to verify that the neural network output is consistent with its input [9]. The use of rules for verification implies that at least some of the domain knowledge can be expressed in rule form. Johnson et al. [10] suggest that an *adjudicator* module be used to decide whether a set of rules or a neural network is likely to provide the more reliable output for a given input. The adjudicator would have access to information relating to the extent of the neural network's training data and could determine whether a neural network would have to interpolate between, or extrapolate from, examples in the training set. Neural networks are good at interpolation but poor at extrapolation. The adjudicator may, therefore, call upon rules to handle the exceptional cases which would otherwise require a neural network to extrapolate from its training data. If heuristic rules are also available for the less exceptional cases, then they could be used to provide an explanation for the neural network's findings. A supervisory rule-based module could dictate the training of a neural network, deciding how many nodes are required, adjusting the learning rate as training proceeds, and deciding when training should terminate.

9.7 Learning classifier systems

Holland's learning classifier systems (LCSs) combine genetic algorithms with rule-based systems to provide a mechanism for rule discovery [11, 12]. The rules are simple production rules (see Section 2.1), coded as a fixed-length mixture of binary numbers and wild-card, i.e., "don't care," characters. Their simple structure makes it possible to generate new rules by means of a genetic algorithm.

The overall LCS is illustrated in Figure 9.4. At the heart of the system is the message list, which fulfils a similar role to the blackboard in a blackboard system. Information from the environment is placed here, along with rule deductions and instructions for the actuators, which act on the environment.

A *credit-apportionment* system, known as the *bucket-brigade algorithm*, is used to maintain a credit balance for each rule. The genetic algorithm uses a rule's credit balance as the measure of its fitness. Conflict resolution (see Section 2.8) between rules in the conflict set is achieved via an auction, in which the rule with the most credits is chosen to fire. In doing so, it must pay some of its credits to the rules that led to its conditions being satisfied. If the fired rule leads to some benefit in the environment, it receives additional credits.

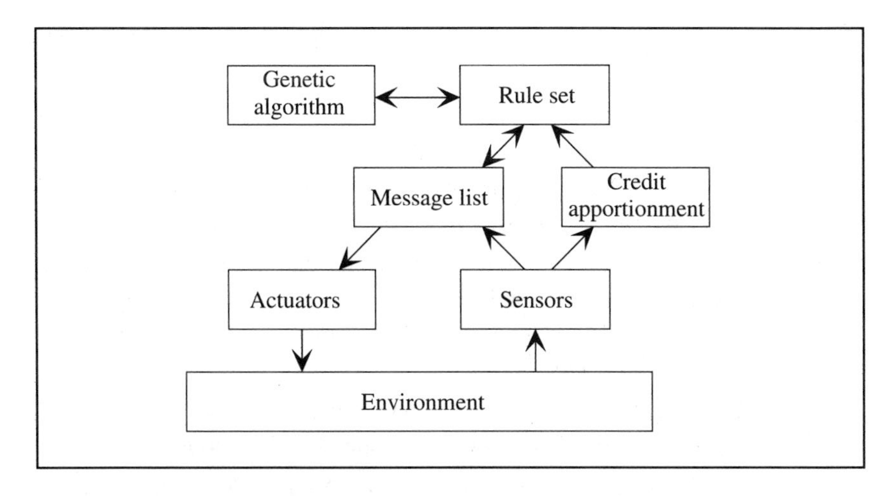

Figure 9.4 Learning classifier system

9.8 Summary

This section has looked at just some of the ways in which different computation techniques — including intelligent systems and conventional programs — can work together within hybrid systems. Chapters 11 through 14 deal with applications of intelligent systems in engineering and science, most of which involve hybrids of some sort.

References

1. Erman, L. D., Hayes-Roth, F., Lesser, V. R., and Reddy, D. R., "The Hearsay-II speech understanding system: integrating knowledge to resolve uncertainty," *ACM Computing Surveys*, vol. 12, pp. 213–253, 1980.
2. Hopgood, A. A., Woodcock, N., Hallam, N. J., and Picton, P. D., "Interpreting ultrasonic images using rules, algorithms and neural networks," *European Journal of Nondestructive Testing*, vol. 2, pp. 135–149, 1993.
3. Feigenbaum, E. A., Foreword to *Blackboard systems*, Englemore, R. S. and Morgan, A. J. (Eds.), Addison-Wesley, 1988.
4. Karr, C. L., "Genetic algorithms for fuzzy controllers," *AI Expert*, pp. 26–33, February 1991.

5. Karr, C. L., "Applying genetics to fuzzy logic," *AI Expert*, pp. 38–43, March 1991.

6. Altug, S., Chow, M. Y., and Trussell, H. J., "Heuristic constraints enforcement for training of and rule extraction from a fuzzy/neural architecture — Part II: Implementation and application," *IEEE Transactions on Fuzzy Systems*, vol. 7, pp. 151–159, 1999.

7. Chow, M. Y., Altug, S., and Trussell, H. J., "Heuristic constraints enforcement for training of and knowledge extraction from a fuzzy/neural architecture — Part I: Foundation," *IEEE Transactions on Fuzzy Systems*, vol. 7, pp. 143–150, 1999.

8. Tsukimoto, H., "Extracting rules from trained neural networks," *IEEE Transactions on Neural Networks*, vol. 11, pp. 377–389, 2000.

9. Picton, P. D., Johnson, J. H., and Hallam, N. J., "Neural networks in safety-critical systems," 3rd International Congress on Condition Monitoring and Diagnostic Engineering Management, Southampton, UK, 1991.

10. Johnson, J. H., Hallam, N. J., and Picton, P. D., "Safety critical neurocomputing: explanation and verification in knowledge augmented neural networks," Colloquium on Human–Computer Interaction, London, IEE, 1990.

11. Holland, J. H. and Reitman, J. S., "Cognitive systems based on adaptive algorithms," in *Pattern-directed inference systems*, Waterman, D. A. and Hayes-Roth, F. (Eds.), pp. 313–329, Academic Press, 1978.

12. Goldberg, D. E., *Genetic Algorithms in Search, Optimization, and Machine Learning*, Addison-Wesley, 1989.

Further reading

- Craig, I., *Blackboard Systems*, Intellect, 1995.
- Englemore, R. S. and Morgan, A. J. (Eds.), *Blackboard Systems*, Addison-Wesley, 1988.
- Jain, L. C. and Martin, N. M. (Eds.), *Fusion of Neural Networks, Fuzzy Sets, and Genetic Algorithms: industrial applications*, CRC Press, 1999.
- Medsker, L. R., *Hybrid Intelligent Systems*, Kluwer, 1995.
- Ruan, D. (Ed.), *Intelligent Hybrid Systems: fuzzy logic, neural networks and genetic algorithms*, Kluwer, 1997.

Chapter ten

Tools and languages

10.1 A range of intelligent systems tools

The previous chapters have introduced a range of intelligent systems techniques, covering both knowledge-based systems (KBSs) and computational intelligence. The tools available to assist in constructing intelligent systems can be roughly divided into the following categories:

- stand-alone packages, e.g., expert system shells and neural network packages;
- KBS toolkits, e.g., Goldworks, Kappa, and Flex;
- libraries, e.g., neural network libraries for MatLab and C++;
- AI programming languages for processing words, symbols, and relations, e.g., Lisp and Prolog;
- object-oriented programming languages, e.g., Smalltalk, C++, CLOS, and Java;
- conventional programming languages, e.g., C, Pascal, and Fortran.

Expert system shells are designed to allow the rapid design and implementation of rule-based expert systems, but tend to lack flexibility. The programming languages offer much greater flexibility and can be used to build customized tools. Most programming languages are procedural rather than declarative, although the Prolog language incorporates both programming styles. The KBS toolkits aim to provide the best of all worlds. They typically offer a mixture of facilities so the programmer has free access to the underlying language and to tools such as rule-based and object-oriented programming. The provision of these tools can save considerable programming effort, while the ability to access the underlying language gives the freedom to build extra facilities.

10.2 Expert system shells

An expert system shell is an expert system that is complete except for the knowledge base. It includes an inference engine, a user interface for programming, and a user interface for running the system. Typically, the programming interface comprises a specialized editor for creating rules in a predetermined format, and some debugging tools. The shell user enters rules in a declarative fashion and ideally should not need to be concerned with the workings of the inference engine. In practice this ideal is rarely met, and a typical difficulty in using a shell is ensuring that rules are applied when expected. As the user has no direct control over the inference engine, it is usually necessary to gain some insight into its workings and to tailor the rules to achieve the desired effect. This is not necessarily easy and detracts from the advantages of having a separate knowledge base. Nevertheless, shells are easy to use in other respects and allow a simple knowledge-based system to be constructed quickly. They are, therefore, useful for building prototype expert systems. However, their inflexible facilities for knowledge representation and inference tend to limit their general applicability.

10.3 Toolkits and libraries

KBS toolkits usually offer a range of knowledge representation facilities, such as rules, objects or frames, and uncertainty handling. Their flexibility stems from granting the user access to the underlying programming language, so that the standard facilities can be altered or enhanced where necessary. They tend to be hungry for processing power and memory, although hardware improvements have made this less of a problem. Many toolkits are based on the AI languages, e.g., Goldworks and Flex are based on Lisp and Prolog, respectively. Toolkits built on the C or C++ languages, e.g., Kappa, have also gained popularity.

Libraries offer similar advantages to toolkits. Instead of supplying the complete programming environment, libraries provide specific functionality on the assumption that the underlying programming environment already exists. There is a range of KBS and computational intelligence libraries available, e.g., for C++, MatLab and Java.

10.4 Artificial intelligence languages

The two main AI programming languages, namely, Lisp and Prolog, will be introduced in subsequent sections of this chapter. A key feature of both languages is the ability to manipulate symbolic data, i.e., characters and words, as well as numerical data. One of the most important structures for this manipulation is lists, introduced below.

10.4.1 Lists

A knowledge base may contain a mixture of numbers, letters, words, punctuation, and complete sentences. Most computer languages can handle such a mixture of characters, provided the general format can be anticipated in advance. Suppose we wanted to represent a simple fact such as:

```
pressure in valve #2 is 12.8 MPa
```

where 1MPa = $1MNm^{-2}$. One possible C++ implementation (see Chapter 4) of this fact is to define a `Valve` class and the instance `valve2`, as follows:

```
class Valve
{
  public:
    Valve(int id, float pres, const char *u);       //constructor
    ~Valve();                                       // destructor
  private:
    int identity_number;
    float pressure;
    char *units;
}

Valve::Valve(int id, float pres, const char *u)
{                                         //constructor definition
    identity_number = id;
    pressure = pres;
    units = (char*) u;
}

Valve::~Valve()
{                                          //destructor definition
    cout << "instance of Valve destroyed" << endl;
}

main()
{
    Valve valve2(2, 12.8, "MPa"); // create an instance of Valve
}
```

The class definition could be specialized to handle different types of valves, but a completely new construction would be needed for a fact or rule with a different format, such as:

```
if pressure is above 10 MPa then close valve
```

Lists are data structures that allow words, numbers, and symbols to be combined in a wide variety of ways. They are useful for symbol manipulation and are a feature of Lisp and Prolog. The above example could be represented as a list in Lisp or Prolog, respectively, as follows:

```
(close_valve (exceeds pressure 10 mpa))

[if, pressure, exceeds, 10, MPa, then, close, valve]
```

Lisp uses round brackets with elements separated by spaces, whereas Prolog uses square brackets with elements separated by commas. The Lisp example includes a list within a list, i.e., a *nested list.* Although the Prolog example looks like a rule, it is really just a list of words and numbers. Separate code would be required to interpret it as a rule.

It should be noted that C++ is a versatile language and that lists can be implemented within it by creating pairs of values and pointers, forming a so-called linked list. However, one strength of the AI languages is the integration of lists into the language, together with the necessary facilities for manipulating those lists.

10.4.2 Other data types

As well as lists, there is a number of other data types available in the AI languages. Unlike most other languages, variables in the AI languages can be used to store any type of data. A list could be assigned to a given variable immediately after assigning a real number to the same variable. The declaration of variables is not always necessary. However, declarations are normally made in Lisp to specify explicitly the scope of a variable, because undeclared variables are assumed to be global (i.e., memory is allocated for the whole duration of a program or interactive session).

Although the assignment of different types of data to variables is transparent to the programmer, the computer needs to know the type of data in order to handle it correctly. There are various techniques for achieving this. Typically, the value associated with a variable includes a tag, hidden from the programmer, that labels its type. Some commonly used data types in the AI languages are shown in Table 10.1.

Lisp also allows the creation of arrays and structures, similar to those used in C. Strings can be made up of any printable characters, including numbers and spaces, enclosed in quotation marks. Words can generally be regarded as a sequence of one or more characters without any spaces, as spaces and certain other characters may act as separators for words, depending on the language.

Type	*Examples*
integer	`0, 23, -15`
real (floating point) number	`3.1415927, -1.24`
string	`"a string in Lisp or Prolog"`
word	`myword, x, z34`
list	`(a list of words in Lisp)` `[a, list, in, Prolog]`

Table 10.1 Some data types

Examples of words include variable names and the elements of the lists shown in Table 10.1. List elements are not always words, as lists can contain nested lists or numbers. The term *atom* is used in both languages to denote a fundamental data type that cannot be made up from other data types. For example, numbers and words are atoms, but lists are not.

During the execution of a program, various data structures may be created. There is, therefore, a need for management of the computer memory, i.e., memory must be allocated when needed and subsequently reclaimed. In languages such as C, the responsibility for memory management rests with the programmer, who must allocate and reclaim memory at appropriate places in the program using the `malloc` and `free` commands (or their equivalent). In the AI languages (and some object-oriented languages, such as Smalltalk — see Chapter 4) the memory is managed automatically. The programming environment must be capable of both dynamically allocating memory and freeing memory that is no longer required. The latter process is called *garbage collection*. This is a useful facility, although it can result in a momentary pause in the computer's response while the garbage collection takes place.

10.4.3 Programming environments

In general, both of the AI languages considered here form part of their own interactive programming environment. Typically, there is a console window into which code can be typed on-line. Such instructions are interpreted and obeyed immediately and the output printed on the screen.

Typing commands in this way is a useful way of inspecting the values of variables and testing ideas, but it is not a practical way of writing a program. Instead, a text editor is used to enter code so that it can subsequently be modified, saved, compiled, and run. In an integrated programming environment, the code can be evaluated or compiled from within the editor.

Debugging tools can stop the program at a selected point or when an error occurs, so the assignments up to that point can be examined.

Code that has been written in an AI language will, in general, run more slowly than a compiled C program. However, the appeal of AI languages is their power in terms of flexibility for the programmer, rather than their computational or memory efficiency.

10.5 Lisp

10.5.1 Background

It has already been noted that a feature of the AI languages is the integration of lists and list manipulation into the language. This is particularly so in the case of Lisp, as a Lisp program is itself a list made up of many lists. Indeed the name *Lisp* is derived from the phrase *list processing.*

Historically, Lisp has developed in an unregulated fashion. Different syntax and features were introduced into different implementations, partly as a consequence of the existing hardware and software environment. As a result, many different Lisp dialects such as Interlisp, Franzlisp, Maclisp, Zetalisp, and Scheme were developed. A standard was subsequently produced — Common Lisp — that aims to combine the most useful and portable features of the dialects into one machine-independent language [1]. This standard form of the language is also the basis of CLOS (Common Lisp Object Standard), an object-oriented extension of Common Lisp. All the examples introduced here are based upon the definition of Common Lisp and should work on any Common Lisp system.

A list is a collection of words, numbers, strings, functions, and further lists, enclosed in parentheses. The following are all examples of lists:

```
(a b c d)

(My car is 10 years old)

(a list (a b c) followed by an empty list ())
```

Lisp uses the key word `nil` to denote an empty list, so that `()` and `nil` are equivalent. Lisp extends the idea of a language based upon list manipulation to the point where everything is either a list or an element of a list. There are only a few basic rules to remember in order to understand how Lisp works. However, because its structure is so different from other languages, Lisp may seem rather strange to the novice.

10.5.2 Lisp functions

While it is valid to describe Lisp as a procedural language (i.e., the computer is told exactly what to do), a more precise description would be that it is a *functional* language. This is because a Lisp program is made up of lists that are interpreted as functions that, by definition, return a single value.

The three key rules to understanding Lisp are:

- each item of a list is evaluated;
- the first item is interpreted as a function name;
- the remaining items are the parameters (or arguments) of the function.

Some of the functions with which we will be dealing are strictly speaking *macros*, which are predefined combinations of functions. However, the distinction need not concern us here. With a few exceptions, the parameters to a function are always evaluated before the function itself. The parameters themselves may also be functions and can even be the same function (thereby permitting recursion). The following example would be a valid Lisp call to the function `print`. The computer's prompt, that precedes user input, varies between implementations but is shown here as `<cl>`, for "Common Lisp":

```
<cl> (print "hello world")
hello world
hello world
```

The first element in the list was interpreted as a function name, and the second item as its argument. The argument evaluates to the string "hello world." It might seem surprising that `hello world` is printed twice. It is first printed because we instructed Lisp to do so. It is then printed again because Lisp always prints out the value of the function that it is given. In this example the function `print` returned as its value the item that it had printed. Consider what will happen if we type the following:

```
<cl> (print hello)
Error: Unbound variable: HELLO.
```

This error message has come about because Lisp evaluates every item in the list. In this example, the interpreter tried to evaluate `hello`, but found it to be undefined. A defined variable is one that has a value (perhaps another function) assigned to it. However, in this case we didn't really want `hello` to be evaluated. There are many instances in writing Lisp code when we would like to suppress Lisp's habit of evaluating every argument. A special function, `quote`, is provided for this specific purpose.

The quote function takes only one argument, which it does not evaluate but simply returns in the same form that it is typed:

```
<cl> (quote hello)
hello

<cl> (print (quote hello))
hello
hello
```

The quote function is used so often that a shorthand form has been made available. The following are equivalent:

```
(quote hello)
```

and

```
'hello
```

We would also wish to suppress the habit of functions evaluating their arguments when making assignments. Here is an assignment in C++:

```
my_variable=2.5;     /* assign value 2.5 to my_variable in C++ */
```

A value of 2.5 is assigned to my_variable, which would need to have been declared as type float. Lisp provides various functions for achieving this, one of which is setf:

```
<cl> (setf my_variable 2.5)
2.5
```

Since the first argument is a variable name, only the second argument to setf is evaluated. This is because we want to make an assignment to the variable name and not to whatever was previously assigned to it. The second parameter of setf, namely 2.5, evaluates to itself, and this value is then assigned to my_variable. Like all Lisp functions, setf returns a value, in this case 2.5, and this value is printed on the screen by the Lisp interpreter.

As we noted earlier, Lisp usually tries to evaluate the parameters of a function before attempting to evaluate the function itself. The parameters may be further functions with their own parameters. There is no practical limit to the embedding of functions within functions in this manner, so complex composite functions can easily be built up and perhaps given names so that they can be reused. The important rules for reading or writing Lisp code are:

- list elements are interpreted as a function name and its parameters (unless the list is the argument to `quote`, `setf`, or a similar function);
- a function always returns a value.

Lisp code is constructed entirely from lists, and lists also represent an important form of data. Therefore, it is hardly surprising that built-in functions for manipulating lists form an important part of the Lisp language. Two basic functions for list manipulation are `first` and `rest`. For historical reasons, the functions `car` and `cdr` are also available to perform the same tasks. The `first` function returns the first item of that list, while `rest` returns the list but with the first item removed. Here are some examples:

```
<cl>(first '(a b c d))
a

<cl>(rest '(a b c d))
(b c d)

<cl>(first (rest '(a b c d)))
b

<cl>(first '((a b)(c d)))
(a b)
```

Used together, `first` and `rest` can find any element in a list. However, two convenient functions for finding parts of a list are `nth` and `nthcdr`:

`(nth 0 x)` finds the 1st element of list `x`
`(nth 1 x)` finds the 2nd element of list `x`
`(nth 2 x)` finds the 3rd element of list `x`
`(nthcdr 2 x)` is the same as `(rest (rest x))`
`(nthcdr 3 x)` is the same as `(rest (rest (rest x)))`

Note that `rest` and `nthcdr` both return a list, while `first` and `nth` may return a list or an atom. When `rest` is applied to a list that contains only one element, an empty list is returned, which is written as `()` or `nil`. When either `first` or `rest` is applied to an empty list, `nil` is returned.

There are many other functions provided in Lisp for list manipulation and program control. It is not intended that this overview of Lisp should introduce them all. The purpose of this section is not to replace the many texts on Lisp, but to give the reader a feel for the unusual syntax and structure of Lisp programs. We will see, by means of a worked example, how these programs can be constructed and, in so doing, will meet some of the important functions in Lisp.

In Chapter 12, the problem of selecting materials will be discussed. One approach involves (among other things) finding those materials that meet a specification or list of specifications. This task is used here as an illustration of the features of the two main AI languages. The problem is to define a Lisp function or Prolog relation called `accept`. When a materials specification or set of specifications are passed as parameters to `accept`, it should return a list of materials in the database that meet the specification(s). The list returned should contain pairs of material names and types. Thus, in Lisp syntax, the following would be a valid result from `accept`:

```
((POLYPROPYLENE . THERMOPLASTIC)
 (POLYURETHANE_FOAM . THERMOSET))
```

In this case, the list returned contains two elements, corresponding to the two materials that meet the specification. Each of the two elements is itself a list of two elements: a material name and its type. Here, each sublist is a special kind of list in Lisp called a *dotted pair* (see Section 10.5.3).

The specification that is passed to `accept` as a parameter will have a predefined format. If just one specification is to be applied, this will be expressed as a list of the form:

```
(property_name  minimum_value  tolerance)
```

An example specification in Lisp would be:

```
(flexural_modulus 1.0 0.1)
```

The units of flexural modulus are assumed to be GNm^{-2}. For a material to meet this specification, its flexural modulus must be at least 1.0 – 0.1, i.e., $0.9GNm^{-2}$. If more than one specification is to be applied, the specifications will be grouped together into a list of the form:

```
((property1  minimum_value1  tolerance1)
 (property2  minimum_value2  tolerance2)
 (property3  minimum_value3  tolerance3))
```

When presented with a list of this type, `accept` should find those materials in the database that meet all of the specifications.

Box 10.1 Problem definition: finding materials that meet some specifications

10.5.3 A worked example

We will discuss in Chapter 12 the application of intelligent systems to problems of selection. One of the selection tasks that will be discussed in some detail is the selection of an appropriate material from which to manufacture a product or component. For the purposes of this worked example, let us assume that we wish to build a shortlist of polymers that meet some numerical specifications. We will endeavor to solve this problem using Lisp, and later in this chapter will encode the same example in Prolog. The problem is specified in detail in Box 10.1.

The first part of our program will be the setting up of some appropriate materials data. There are a number of ways of doing this, including the creation of a list called `materials_database`:

```
(defvar materials_database nil)
(setf materials_database '(
    (abs thermoplastic
        (impact_resistance 0.2)
        (flexural_modulus 2.7)
        (maximum_temperature 70))
    (polypropylene thermoplastic
        (impact_resistance 0.07)
        (flexural_modulus 1.5)
        (maximum_temperature 100))
    (polystyrene thermoplastic
        (impact_resistance 0.02)
        (flexural_modulus 3.0)
        (maximum_temperature 50))
    (polyurethane_foam thermoset
        (impact_resistance 1.06)
        (flexural_modulus 0.9)
        (maximum_temperature 80))
    (pvc thermoplastic
        (impact_resistance 1.06)
        (flexural_modulus 0.007)
        (maximum_temperature 50))
    (silicone thermoset
        (impact_resistance 0.02)
        (flexural_modulus 3.5)
        (maximum_temperature 240))))
```

The function `defvar` is used to declare the variable `materials_database` and to assign to it an initial value, in this case `nil`. The function `setf` is used to assign the list of materials properties to `materials_database`. In a real application we would be more likely to read this information from a file than to have it "hard-coded" into our program, but this representation will suffice for the moment. The variable `materials_database` is used to store a list, each element of which is also a list. Each of these nested lists contains information about one polymer. The information is in the form of a name and category, followed by further lists in which property names and values are stored. Since `materials_database` does not have a predeclared size, materials and materials properties can be added or removed with ease.

Our task is to define a Lisp function, to be called `accept`, which takes a set of specifications as its parameters and returns a list of polymers (both their names and types) that meet the specifications. To define `accept`, we will use the function `defun`, whose purpose is to define a function. Like `quote`, `defun` does not evaluate its arguments. We will define `accept` as taking a single parameter, `spec_list`, that is a list of specifications to be met. Comments are indicated by a semi-colon:

```
(defun accept (spec_list)
  (let ((shortlist (setup)))
    (if (atom (first spec_list))
;if the first element of spec_list is an atom then
;consider the specification
      (setf shortlist (meets_one spec_list shortlist))
;else consider each specification in turn
      (dolist (each_spec spec_list)
        (setf shortlist (meets_one each_spec shortlist))))
      shortlist))                                ;return shortlist
```

In defining `accept`, we have assumed the form that the specification `spec_list` will take. It will be either a list of the form:

```
(property minimum_value tolerance)
```

or a list made up of several such specifications:

```
((property1 minimum_value1 tolerance1) (property2 minimum_value2
tolerance2) (property3 minimum_value3 tolerance3)).
```

Where more than one specification is given within `spec_list`, they must all be met.

The first function of `accept` is `let`, a function that allows us to declare and initialize local variables. In our example the variable `shortlist` is declared and assigned the value returned by the function `setup`, which we have still to define. The variable `shortlist` is local in the sense that it exists only between `(let` and the matching closing bracket.

There then follows a Lisp conditional statement. The `if` function is used to test whether `spec_list` comprises one or many specifications. It does this by ascertaining whether the first element of spec_list is an atom. If it is, then spec_list contains only one specification, and the first element of spec_list is expected to be the name of the property to be considered. If, on the other hand, spec_list contains more than one specification, then its first element will be a list representing the first specification, so the test for whether the first element of spec_list is an atom will return nil (meaning "false").

The general form of the if function is:

```
(if condition function1 function2)
```

which is interpreted as:

```
IF condition is true THEN do function1 ELSE do function2
```

In our example, function1 involves setting the value of shortlist to the value returned by the function meets_one (yet to be defined), which is passed the single specification and the current shortlist of materials as its parameters. The "else" part of the conditional function (function2) is more complicated, and comprises the dolist control function. In our example, dolist initially assigns the first element of spec_list to the local variable each_spec. It then evaluates all functions between (dolist and its associated closing bracket. In our example this is only one function, which sets shortlist to the value returned from the function meets_one. Since dolist is an iterative control function, it will repeat the process with the second element of spec_list, and so on until there are no elements left.

It is our intention that the function accept should return a list of polymers that meet the specification or specifications. When a function is made up of many functions, the value returned is the last value to be evaluated. To ensure that the value returned by accept is the final shortlist of polymers, shortlist is evaluated as the last line of the definition of accept. Note that the bracketing is such that this evaluation takes place within the scope of the let function, within which shortlist is a local variable.

We have seen that the first task performed within the function accept was to establish the initial shortlist of polymers by evaluating the function setup. This function produces a list of all polymers (and their types) that are known to the system through the definition of materials_database. The setup function is defined as follows:

```
(defun setup ()
  (let ((shortlist nil))
    (dolist (material materials_database)
      (setf shortlist (cons (cons (first material) (nth 1
                            material)) shortlist)))
    shortlist))
```

The empty brackets at the end of the first line of the function definition signify that setup takes no parameters. As in the definition of accept, let is used to declare a local variable and to give it an initial value of nil, i.e., the empty list. Then dolist is used to consider each element of materials_database in turn and to assign it to the local variable material. Each successive value of material is a list comprising the polymer name, its type, and lists of properties and values. The intention is to extract from this a list comprising a name and type only, and to collect all such two-element lists together into the list shortlist. The Lisp function cons adds an element to a list, and can therefore be used to build up shortlist.

Assuming that its second parameter is a list, cons will return the result of adding its first parameter to the front of that list. This can be illustrated by the following example:

```
<cl>(cons 'a '(b c d))
(a b c d)
```

However, the use of cons is still valid even if the second parameter is not a list. In such circumstances, a special type of two-element list, called a dotted pair, is produced:

```
<cl>(cons 'a 'b)
(a.b)
```

In our definition of setup we have used two embedded calls to cons, one of which produces a dotted pair while the other produces an ordinary list. The most deeply embedded (or *nested*) call to cons is called first, as Lisp always evaluates parameters to a function before evaluating the function itself. So the first cons to be evaluated returns a dotted pair comprising the first element of material (which is a polymer name) and the second element of material (which is the polymer type). The second cons to be evaluated returns the result of adding the dotted pair to the front of shortlist. Then shortlist is updated to this value by the call of setf. As in the definition of accept, shortlist is evaluated after the dolist loop has terminated, to ensure that setup returns the last value of shortlist.

As we have already seen, accept passes a single specification, together with the current shortlist, as parameters to the function meets_one. It is this function which performs the most important task in our program, namely, deciding which materials in the shortlist meet the specification:

```
(defun meets_one (spec shortlist)
  (dolist (material shortlist)
```

```
    (let ((actual (get_database_value (first spec) (first
                 material))))
      (if (< actual (- (nth 1 spec) (nth 2 spec)))
        (setf shortlist (remove material shortlist)))))
          ;;pseudo-C equivalent:
          ;;if actual < (value - tolerance)
          ;;shortlist=shortlist without material
shortlist)
```

We have already met most of the Lisp functions that make up `meets_one`. In order to consider each material–type pair in the shortlist, `dolist` is used. The actual value of a property (such as maximum operating temperature) for a given material is found by passing the property name and the material name as arguments to the function `get_database_value`, which has yet to be defined. The value returned from `get_database_value` is stored in the local variable `actual`. The value of `actual` is then compared with the result of subtracting the specified tolerance from the specified target value. In our example, the tolerance is used simply to make the specification less severe. It may be surprising at first to see that subtraction and arithmetic comparison are both handled as functions within Lisp. Nevertheless, this treatment is consistent with other Lisp operations. The code includes a comment showing the conventional positioning of the operators in other languages such as C. Note that `nth` is used in order to extract the second and third elements of `spec`, which represent the specification value and tolerance, respectively.

If the database value, `actual`, of the property in question is less than the specification value minus the tolerance, then the material is removed from the shortlist. The Lisp function `remove` is provided for this purpose. Because the arguments to the function are evaluated before the function itself, it follows that `remove` is not able to alter `shortlist` itself. Rather it returns the list that would be produced if `material` were removed from a copy of `shortlist`. Therefore, `setf` has to be used to set `shortlist` to this new value.

As in our other functions, `shortlist` is evaluated, so its value will be returned as the value of the function `meets_one`. At a glance, it may appear that this is unnecessary, as `setf` would return the value of `shortlist`. However, it is important to remember that this function is embedded within other functions. The last function to be evaluated is in fact `dolist`. When `dolist` terminates by reaching the end of `shortlist`, it returns the empty list, `()`, which is clearly not the desired result.

There now remains only one more function to define in order to make our Lisp program work, namely, `get_database_value`. As mentioned previously, `get_database_value` should return the actual value of a property for a material, when the property and material names are passed as arguments. Common Lisp provides us with the function `find`, which is ideally suited to this task. The syntax of `find` is best illustrated by example. The following function call will search the list `materials_database` until it finds a list whose `first` element is identically equal (`eq`) to the value of `material`:

```
(find material materials_database :test #'eq :key #'first)
```

Other tests and keys can be used in conjunction with `find`, as defined in [1]. Having found the data corresponding to the material of interest, the name and type can be removed using `nthcdr`, and `find` can be called again in order to find the list corresponding to the property of interest. The function `get_database_value` can therefore be written as follows:

```
(defun get_database_value (prop_name material)
  (nth 1 (find prop_name
    (nthcdr 2
      (find material materials_database :test #'eq :key
      #'first))
     :test #'eq :key #'first)))
```

We are now in a position to put our program together, as shown in Box 10.2, and to ask Lisp some questions about polymers:

```
<cl> (accept'(maximum_temperature 100 5))
((SILICONE . THERMOSET) (POLYPROPYLENE . THERMOPLASTIC))

<cl> (accept '((maximum_temperature 100 5)(impact_resistance
      0.05 0)))
((POLYPROPYLENE . THERMOPLASTIC))
```

Once the function definitions have been evaluated, each function (such as `accept`) becomes known to Lisp in the same way as all of the predefined functions such as `setf` and `dolist`. Therefore, we have extended the Lisp language so that it has become specialized for our own specific purposes. It is this ability that makes Lisp such a powerful and flexible language. If a programmer does not like the facilities that Lisp offers, he or she can alter the syntax by defining his or her own functions, thereby producing an alternative language. This is the basis upon which the Lisp-based KBS toolkits described in Sections 10.1 and 10.3 are built. These environments offer the programmer not only a Lisp interpreter, but also a vast number of Lisp functions which together provide a language for object-oriented programming (Chapter 4) and rule-based programming (Chapter 2). In addition, an attractive user interface is normally provided, and this too is implemented as an extension of the Lisp language.

```
(defvar materials_database nil)
(setf materials_database '(
                             (abs thermoplastic
                                     (impact_resistance 0.2)
                                     (flexural_modulus 2.7)
                                     (maximum_temperature 70))
                             (polypropylene thermoplastic
                                     (impact_resistance 0.07)
                                     (flexural_modulus 1.5)
                                     (maximum_temperature 100))
                             (polystyrene thermoplastic
                                     (impact_resistance 0.02)
                                     (flexural_modulus 3.0)
                                     (maximum_temperature 50))
                             (polyurethane_foam thermoset
                                     (impact_resistance 1.06)
                                     (flexural_modulus 0.9)
                                     (maximum_temperature 80))
                             (pvc thermoplastic
                                     (impact_resistance 1.06)
                                     (flexural_modulus 0.007)
                                     (maximum_temperature 50))
                             (silicone thermoset
                                     (impact_resistance 0.02)
                                     (flexural_modulus 3.5)
                                     (maximum_temperature 240))))

(defun accept (spec_list)
  (let ((shortlist (setup)))
    (if (atom (first spec_list))
;;if the first element of spec_list is an atom then consider the
;;specification
      (setf shortlist (meets_one spec_list shortlist))
;;else consider each specification in turn
      (dolist (each_spec spec_list)
        (setf shortlist (meets_one each_spec shortlist))))
      shortlist))                                           ;return shortlist

(defun setup ()
  (let ((shortlist nil))
    (dolist (material materials_database)
      (setf shortlist (cons (cons (first material) (nth 1 material))
shortlist)))
    shortlist))

(defun meets_one (spec shortlist)
  (dolist (material shortlist)
    (let ((actual (get_database_value (first spec) (first material))))
      (if (< actual (- (nth 1 spec) (nth 2 spec)))
        (setf shortlist (remove material shortlist)))))
           ;;pseudo-C equivalent:
           ;;if actual< (value - tolerance)
           ;;shortlist=shortlist without material
shortlist)

(defun get_database_value (prop_name material)
  (nth 1 (find prop_name
           (nthcdr 2
             (find material materials_database :test #'eq :key
               #'first))
           :test #'eq :key #'first)))
```

Box 10.2 A worked example in Lisp

10.6 Prolog

10.6.1 Background

Prolog is an AI language that can be programmed declaratively. It is, therefore, very different from Lisp, which is a procedural (or, more precisely, functional) language that can be used to build declarative applications such as expert system shells. As we will see, although Prolog can be used declaratively, an appreciation of the procedural behavior of the language is needed. In other words, programmers need to understand how Prolog uses the declarative information that they supply.

Prolog is suited to symbolic (rather than numerical) problems, particularly logical problems involving relationships between items. It is also suitable for tasks that involve data lookup and retrieval, as pattern-matching is fundamental to the functionality of the language. Because Prolog is so different from other languages in its underlying concepts, many newcomers find it a difficult language. Whereas most languages can be learned rapidly by someone with computing experience, Prolog is perhaps more easily learned by someone who has never programmed.

10.6.2 A worked example

The main building blocks of Prolog are lists (as in Lisp) and *relations*, which can be used to construct *clauses*. We will demonstrate the declarative nature of Prolog programs by constructing a small program for selecting polymers from a database of polymer properties. The task will be identical to that used to illustrate Lisp, namely selecting from a database those polymers which meet a numerical specification or set of specifications. The problem is described in more detail in Box 10.1. We have already said that Prolog is good for data lookup, so let us begin by creating a small database containing some properties of materials. Our database will comprise a number of clauses such as this one involving the relation `materials_database`:

```
materials_database(polypropylene, thermoplastic,
[maximum_temperature, 100]).
```

The above clause means that the three items in parentheses are related through the relation called `materials_database`. The third argument of the clause is a list (denoted by square brackets), while the first two arguments are atoms. The clause is our first piece of Prolog code, and it is purely declarative. We have given the computer some information about polypropylene, and this is sufficient to produce a working (though rather trivial) program. Even though we have not given Prolog any procedural information (i.e., we haven't told it how to use the information about polypropylene), we can still ask it some questions. Having typed the above line of Prolog code, not forgetting the period, we can ask Prolog the question:

"What type of material is polypropylene?"

Depending on the Prolog implementation, the screen prompt is usually `?-`, indicating that what follows is treated as a query. Our query to Prolog could be expressed as:

```
?- materials_database(polypropylene, Family, _).
```

Prolog would respond:

```
Family = thermoplastic
```

This simple example illustrates several features of Prolog. Our program is a single line of code, which states that polypropylene is a material of type thermoplastic and has a maximum operating temperature of 100 (°C assumed). Thus, the program is purely declarative. We have told Prolog what we know about polypropylene, but have given Prolog no procedural instructions about what to do with that information. Nevertheless, we were able to ask a sensible question and receive a sensible reply. Our query includes some distinct data types. As `polypropylene` began with a lower-case letter, it was recognized as a constant, whereas `Family` was recognized as a variable by virtue of beginning with an upper-case letter. These distinctions stem from the following rules, which are always observed:

- variables in Prolog can begin either with an uppercase letter or with an underscore character (e.g., `X`, `My_variable`, `_another` are all valid variable names); and
- constants begin with a lower case letter (e.g., `adrian`, `polypropylene`, `pi` are all valid names for constants).

When presented with our query, Prolog has attempted to *match* the query to the relations (only one relation in our example) that it has stored. In order for any two terms to match, either:

- the two terms must be identical; or

- it must be possible to set (or *instantiate*) any variables in such a way that the two terms become identical.

If Prolog is trying to match two or more clauses and comes across multiple occurrences of the same variable name, it will always instantiate them identically. The only exception to this rule is the underscore character, which has a special meaning when used on its own. Each occurrence of the underscore character's appearing alone means:

I don't care what '_' matches so long as it matches something.

Multiple occurrences of the character can be matched to different values. The '_' character is used when the value of a variable is not needed in the evaluation of a clause. Thus:

```
materials_database(polypropylene, thermoplastic,
[maximum_temperature, 100]).
```

matches:

```
materials_database(polypropylene, Family, _).
```

The relation name, `materials_database`, and its number of arguments (or its *arity*) are the same in each case. The first argument to `materials_database` is identical in each case, and the remaining two can be made identical by instantiating the variable `Family` to `thermoplastic` and the underscore variable to the list `[maximum_temperature, 100]`. We don't care what the underscore variable matches, so long as it matches something.

Now let us see if we can extend our example into a useful program. First we will make our database more useful by adding some more data:

```
materials_database(abs, thermoplastic,
  [[impact_resistance, 0.2],
   [flexural_modulus, 2.7],
   [maximum_temperature, 70]]).
materials_database(polypropylene, thermoplastic,
  [[impact_resistance, 0.07],
   [flexural_modulus, 1.5],
   [maximum_temperature, 100]]).
materials_database(polystyrene, thermoplastic,
  [[impact_resistance, 0.02],
   [flexural_modulus, 3.0],
   [maximum_temperature, 50]]).
materials_database(polyurethane_foam, thermoset,
  [[impact_resistance, 1.06],
```

```
    [flexural_modulus, 0.9],
    [maximum_temperature, 80]]).
materials_database(pvc, thermoplastic,
   [[impact_resistance, 1.06],
    [flexural_modulus, 0.007],
    [maximum_temperature, 50]]).
materials_database(silicone, thermoset,
   [[impact_resistance, 0.02],
    [flexural_modulus, 3.5],
    [maximum_temperature, 240]]).
```

Our aim is to build a program that can select from the database those materials that meet a set of specifications. This requirement can be translated directly into a Prolog rule:

```
accept(Material,Type,Spec_list):-
  materials_database(Material,Type,Stored_data),
  meets_all_specs(Spec_list,Stored_data).
```

The `:-` symbol stands for the word "if" in the rule. Thus, the above rule means:

accept a material, given a list of specifications, if that material is in the database and if the stored data about the material meet the specifications.

We now have to let Prolog know what we mean by a material meeting all of the specifications in the user's specification list. The simplest case is when there are no specifications at all, in other words, the specification list is empty. In this case the (nonexistent) specifications will be met regardless of the stored data. This fact can be simply coded in Prolog as:

```
meets_all_specs([],_).
```

The next most straightforward case to deal with is when there is only one specification, which we can code as follows:

```
meets_all_specs(Spec_list, Data):-
  Spec_list= [Spec1|Rest],
  atom(Spec1),
  meets_one_spec([Spec1|Rest],Data).
```

This rule introduces the list separator `|`, which is used to separate the first element of a list from the rest of the list. As an example, consider the following Prolog query:

```
?- [Spec1|Rest] = [flexural_modulus, 1.0, 0.1].
Spec1 = flexural_modulus, Rest = [1,0.1]
```

The assignments to the variables immediately before and after the list separator are analogous to taking the `first` and `rest` of a list in Lisp. Consistent with this analogy, the item immediately following a `|` symbol will always be instantiated to a list. Returning now to our rule, the first condition requires Prolog to try to match `Spec_list` to the template `[Spec1|Rest]`. If the match is successful, `Spec1` will become instantiated to the first element of `Spec_list` and `Rest` instantiated to `Spec_list` with its first element removed.

We could make our rule more compact by combining the first condition of the rule with the arguments to the goal:

```
meets_all_specs([Spec1|Rest], Data):-
  atom(Spec1),
  meets_one_spec([Spec1|Rest],Data).
```

If the match is successful, we need to establish whether `Spec_list` contains one or many specifications. This can be achieved by testing the type of its first element. If the first element is an atom, the user has supplied a single specification, whereas a list indicates that more than one specification has been supplied. All this assumes that the intended format was used for the query. The built-in Prolog relation `atom` succeeds if its argument is an atom, and otherwise it fails.

We have not yet told Prolog what is meant by the relation called `meets_one_spec`, but we will do so shortly. Next we will consider the general case of the user's supplying several specifications:

```
meets_all_specs([Spec1|Rest],Data):-
  not atom(Spec1),
  meets_one_spec(Spec1,Data),
  meets_all_specs(Rest,Data).
```

An important feature demonstrated by this rule is the use of recursion, that is, the reuse of `meets_all_specs` within its own definition. Our rule says that the stored data meets the user's specification if each of the following is satisfied:

- we can separate the first specification from the remainder;
- the first specification is not an atom;
- the stored data meet the first specification;
- the stored data meet all of the remaining specifications.

When presented with a list of specifications, individual specifications will be stripped off the list one at a time, and the rule will be deemed to have been satisfied if the stored data satisfy each of them.

Having dealt with multiple specifications by breaking down the list into a set of single specifications, it now remains for us to define what we mean by a specification's being met. This is coded as follows:

```
meets_one_spec([Property, Spec_value, Tolerance], List):-
  member([Property, Actual_value], List),
  Actual_value>Spec_value-Tolerance.
```

As in the Lisp example, we explicitly state that a user's specification must be in a fixed format, i.e., the material property name, its target value, and the tolerance of that value must appear in sequence in a list. A new relation called `member` is introduced in order to check that the stored data for a given material include the property being specified, and to assign the stored value for that property to `Actual_value`. If the relation `member` is not built into our particular Prolog implementation, we will have to define it ourselves. Finally, the stored value is deemed to meet the specification if it is greater than the specification minus the tolerance.

The definition of `member` (taken from [2]) is similar in concept to our definition of `meets_all_specs`. The definition is that an item is a member of a list if that item is the first member of the list or if the list may be split so that the item is the first member of the second part of the list. This can be expressed more concisely and elegantly in Prolog than it can in English:

```
member(A,[A|L]).
member(A,[_|L]):-member(A,L).
```

Our program is now complete and ready to be interrogated. The program is shown in full in Box 10.3. In order to run a Prolog program, Prolog must be set a goal that it can try to prove. If successful, it will return all of the sets of instantiations necessary to satisfy that goal. In our case the goal is to find the materials that meet some specifications.

Now let us test our program with some example queries (or goals). First we will determine which materials have a maximum operating temperature of at least 100°C, with a 5°C tolerance:

```
?- accept(M, T, [maximum_temperature, 100, 5]).
M = polypropylene, T = thermoplastic;
M = silicone, T = thermoset;
no
```

The word no at the end indicates that Prolog's final attempt to find a match to the specification, after it had already found two such matches, was unsuccessful. We can now extend our query to find all materials which, as well as meeting the temperature requirement, have an impact resistance of at least 0.05 kJ/m:

```
?- accept(M,T,[[maximum_temperature, 100, 5],
   [impact_resistance, 0.05, 0]]).
M = polypropylene, T = thermoplastic;
no
```

```
materials_database(abs, thermoplastic,
  [[impact_resistance, 0.2],
   [flexural_modulus, 2.7],
   [maximum_temperature, 70]]).
materials_database(polypropylene, thermoplastic,
  [[impact_resistance, 0.07],
   [flexural_modulus, 1.5],
   [maximum_temperature, 100]]).
materials_database(polystyrene, thermoplastic,
  [[impact_resistance, 0.02],
   [flexural_modulus, 3.0],
   [maximum_temperature, 50]]).
materials_database(polyurethane_foam, thermoset,
  [[impact_resistance, 1.06],
   [flexural_modulus, 0.9],
   [maximum_temperature, 80]]).
materials_database(pvc, thermoplastic,
  [[impact_resistance, 1.06],
   [flexural_modulus, 0.007],
   [maximum_temperature, 50]]).
materials_database(silicone, thermoset,
  [[impact_resistance, 0.02],
   [flexural_modulus, 3.5],
   [maximum_temperature, 240]]).

accept(Material,Type,Spec_list):-
  materials_database(Material,Type,Stored_data),
  meets_all_specs(Spec_list,Stored_data).

meets_all_specs([],_).
meets_all_specs([Spec1|Rest],Data):-
  atom(Spec1),
  meets_one_spec([Spec1|Rest],Data).
meets_all_specs([Spec1|Rest],Data):-
  not atom(Spec1),
  meets_one_spec(Spec1,Data),
  meets_all_specs(Rest,Data).

meets_one_spec([Property, Spec_value, Tolerance], List):-
  member([Property, Actual_value], List),
  Actual_value>Spec_value-Tolerance.

member(A,[A|L]).
member(A,[_|L]):-member(A,L).
```

Box 10.3 A worked example in Prolog

10.6.3 Backtracking in Prolog

So far we have seen how to program declaratively in Prolog, without giving any thought to how Prolog uses the declarative program to decide upon a sequential series of actions. In the example shown in the above section, it was not necessary to know how Prolog used the information supplied to arrive at the correct answer. This represents the ideal of declarative programming in Prolog. However, the Prolog programmer invariably needs to have an idea of the procedural behavior of Prolog in order to ensure that a program performs correctly and efficiently. In many circumstances it is possible to type a valid declarative program, but for the program to fail to work as anticipated because the programmer has failed to take into account how Prolog works.

Let us start by considering our last example query to Prolog:

```
?- accept(M,T,[[maximum_temperature, 100, 5],
   [impact_resistance, 0.05, 0]]).
```

Prolog treats this query as a goal, whose truth it attempts to establish. As the goal contains some variables (`M` and `T`), these will need to be instantiated in order to achieve the goal. Prolog's first attempt at achieving the goal is to see whether the program contains any clauses that directly match the query. In our example it does not, but it does find a rule with the `accept` relation as its conclusion:

```
accept(Material,Type,Spec_list):-
  materials_database(Material,Type,Stored_data),
  meets_all_specs(Spec_list,Stored_data).
```

Prolog now knows that if it can establish the two conditions with `M` matched to `Material`, `T` matched to `Type`, and `Spec_list` instantiated to `[[maximum_temperature, 100, 5], [impact_resistance, 0.05, 0]]`, the goal is achieved. The two conditions then become goals in their own right. The first one, involving the relation `materials_database`, is easily achieved, and the second condition:

```
meets_all_specs(Spec_list,Stored_data).
```

becomes the new goal. Prolog's first attempt at satisfying this goal is to look at the relation:

```
meets_all_specs([],_).
```

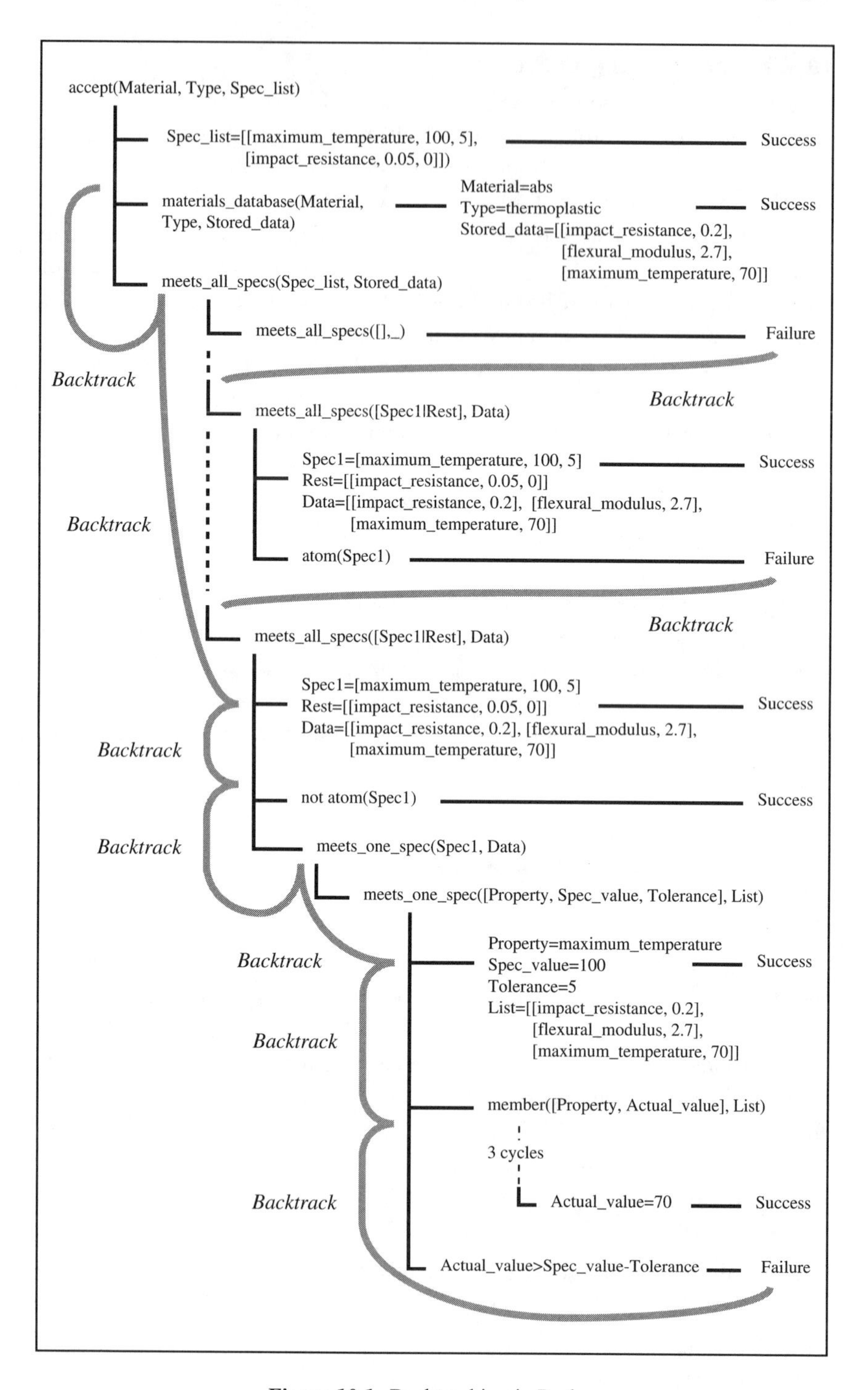

Figure 10.1 Backtracking in Prolog

However, this doesn't help as `Spec_list` is not instantiated to an empty list. Prolog must at this point *backtrack* to try another way of achieving the current sub-goal. In other words, Prolog remembers the stage where it was before the failed attempt and resumes its reasoning along another path from there. Figure 10.1 shows the reasoning followed by Prolog when presented with the goal:

```
?- accept(M,T,[[maximum_temperature, 100, 5],
   [impact_resistance, 0.05, 0]]).
```

The illustration shows Prolog's first attempt at a solution, namely, `M=abs` and `T=thermoplastic`, and the steps that are followed before rejecting these particular instantiations as a solution. The use of backtracking is sensible up until the point where it is discovered that the maximum operating temperature of ABS (acrylonitrile-butadiene-styrene) is too low. When this has been determined, we would ideally like the program to reject ABS as a candidate material and move on to the next contender. However, Prolog does not give up so easily. Instead, it backtracks through every step that it has taken, checking to see whether there may be an alternative solution (or set of instantiations) that could be used. Ultimately it arrives back at the `materials_database` relation, and `Material` and `Type` become reinstantiated.

Prolog provides two facilities for controlling backtracking. They can be used to increase efficiency and to alter the meaning of a program. These facilities are:

- the *order* of Prolog code;
- the use of the *cut* operator.

Prolog tries out possible solutions to a problem in the order in which they are presented. Thus, in our example, Prolog always starts by assuming that the user has supplied a single materials specification. Only when it discovers that this is not the case does Prolog consider that the user may have submitted a list of several specifications. This is an appropriate ordering, as it is sensible to try the simplest solution first. In general, the ordering of code will affect the procedural meaning of a Prolog program (i.e., *how* the problem will be solved), but not its declarative meaning. However, as soon as the Prolog programmer starts to use the second facility, namely, the *cut* operator, the order of Prolog clauses can affect both the procedural and the declarative meaning of programs.

In order to prevent Prolog from carrying out unwanted backtracking, the cut symbol (`!`) can be used. Cuts can be inserted as though they were goals in their own right. When Prolog comes across a cut, backtracking is prevented. Cuts can be used to make our example program more efficient by forcing

```
accept(Material,Type,Spec_list):-
  materials_database(Material,Type,Stored_data),
  meets_all_specs(Spec_list,Stored_data).

meets_all_specs([],_):-!.

meets_all_specs([Spec1|Rest],Data):-
  atom(Spec1),!,
  meets_one_spec([Spec1|Rest],Data).

meets_all_specs([Spec1|Rest],Data):-
  meets_one_spec(Spec1,Data),
  meets_all_specs(Rest,Data).

meets_one_spec([Property, Spec_value, Tolerance], List):-
  member([Property, Actual_value], List),!,
  Actual_value>Spec_value-Tolerance.

member(A,[A|L]).
member(A,[_|L]):-member(A,L).
```

Box 10.4 A Prolog program with cuts

Prolog to immediately try a new material once it has established whether or not a given material meets the specification. The revised program, with cuts included, is shown in Box 10.4 (the setting up of the `materials_database` relations is unchanged and has been omitted).

Although the discussion so far has regarded the cut as a means of improving efficiency by eliminating unwanted backtracking, cuts can also alter the declarative meaning of programs. This can be illustrated by referring once again to our materials selection program. The program contains three alternative means of achieving the `meets_all_specs` goal. The first deals with the case where the first argument is the empty list. The two others take identical arguments, and a distinction is made based upon whether the first element of the first argument (a list) is an atom. If the element is an atom, then the alternative case need not be considered, and this can be achieved using a cut (note that `%` indicates a comment):

```
meets_all_specs([Spec1|Rest],Data):-
  atom(Spec1),!,              % cut placed here
  meets_one_spec([Spec1|Rest],Data).

meets_all_specs([Spec1|Rest],Data):-
  not atom(Spec1),            % this test is now redundant
  meets_one_spec(Spec1,Data),
  meets_all_specs(Rest,Data).
```

Because of the positioning of the cut, if:

```
atom(Spec1)
```

is successful, then the alternative rule will not be considered. Therefore, the test:

```
not atom(Spec1)
```

is now redundant and can be removed. However, this test can only be removed provided that the cut is included in the previous rule. This example shows that a cut can be used to create rules of the form:

```
IF ... THEN ... ELSE
```

While much of the above discussion has concentrated on overcoming the inefficiencies that backtracking can introduce, it is important to remember that backtracking is essential for searching out a solution, and the elegance of Prolog in many applications lies in its ability to backtrack without the programmer's needing to program this behavior explicitly.

10.7 Comparison of AI languages

For each AI language, the worked example gives some feel for the language structure and syntax. However, it does not form the basis for a fair comparison of their merit. The Prolog code is the most compact and elegant solution to the problem of choosing materials which meet a specification, because Prolog is particularly good at tasks that involve pattern matching and retrieval of data. However, the language places a number of constraints on the programmer, particularly in committing him or her to one particular search strategy. As we have seen, the programmer can control this strategy to some extent by judicious ordering of clauses and use of the cut mechanism. Prolog doesn't need a structure for iteration, e.g., `FOR x FROM 1 TO 10`, as recursion can be used to achieve the same effect.

Our Lisp program has a completely different structure from the Prolog example, as it has been programmed procedurally (or, more precisely, functionally). The language provides excellent facilities for manipulating symbols and lists of symbols. It is a powerful language that allows practically any reasoning strategy to be implemented. In fact, it is so flexible that it can be reconfigured by the programmer, although the worked example does not do justice to this flexibility. In particular, the materials database took the form of a long, flat list, whereas there are more structured ways of representing the data. As we have seen in Chapter 4, and will see in Chapter 12, object-oriented programming allows a hierarchical representation of materials properties.

10.8 Summary

Ease of use
The following tools and languages are arranged in order of increasing ease of use for building a *simple* knowledge-based system:

- conventional languages;
- AI languages;
- AI toolkits and libraries;
- expert system shells.

Sophistication
While expert system shells are suitable for some simple problems, most are inflexible and have only limited facilities for knowledge representation. They are difficult to adapt to complex real-world problems, where AI languages or AI toolkits are usually more appropriate.

Characteristics of AI languages
The two AI languages discussed here, Lisp and Prolog, are both well-suited to problems involving the manipulation of symbols. However, they are less suited than conventional languages to numerical problems. In contrast with conventional languages, the same variable can be used to hold a variety of data types. The AI languages allow various types of data to be combined into lists, and they provide facilities for list manipulation.

Prolog can be used declaratively and includes a backtracking mechanism that allows it to explore all possible ways in which a goal might be achieved. The programmer can exercise control over Prolog's backtracking by careful ordering of clauses and through the use of cuts.

In Lisp, lists are not only used for data, but also constitute the programs themselves. Lisp functions are represented as lists containing a function name and its arguments. As every Lisp function returns a value, the arguments themselves can be functions. Both languages are flexible, elegant, and concise.

References

1. Steele, G. L., *Common Lisp: the language*, 2nd ed., Digital Press, 1990.
2. Bratko, I., *Prolog Programming for Artificial Intelligence*, 3rd ed., Longman, 2000.

Further reading

- Bratko, I., *Prolog Programming for Artificial Intelligence*, 3rd ed., Longman, 2000.
- Clocksin, W. F. and Mellish, C. S., *Programming in Prolog*, 4th ed., Springer-Verlag, 1994.
- Covington, M. A., Nute, D., and Vellino, A., *Prolog Programming in Depth*, Prentice Hall, 1996.
- Graham, P., *The ANSI Common Lisp Book*, Prentice Hall, 1995.
- Hasemer, T. and Domingue, J., *Common Lisp Programming for Artificial Intelligence*, Addison-Wesley, 1989.
- Kreiker, P., *Visual Lisp: guide to artful programming*, Delmar, 2000.
- Steele, G. L., *Common Lisp: the language*, 2nd ed., Digital Press, 1990.
- Sterling, L. and Shapiro, E., *The Art of Prolog: advanced programming techniques*, 2nd ed., MIT Press, 1994.

Chapter eleven

Systems for interpretation and diagnosis

11.1 Introduction

Diagnosis is the process of determining the nature of a fault or malfunction, based on a set of symptoms. Input data (i.e., the symptoms) are interpreted and the underlying cause of these symptoms is the output. Diagnosis is, therefore, a special case of the more general problem of interpretation. There are many circumstances in which we may wish to interpret data, other than diagnosing problems. Examples include the interpretation of images (e.g., optical, x-ray, ultrasonic, electron microscopic), meters, gauges, and statistical data (e.g., from surveys of people or from counts registered by a radiation counter). This chapter will examine some of the intelligent systems techniques that are used for diagnosis and for more general interpretation problems. The diagnosis of faults in a refrigerator will be used as an illustrative example, and the interpretation of ultrasonic images from welds in steel plates will be used as a detailed case study.

Since the inception of expert systems in the late 1960s and early 1970s, diagnosis and interpretation have been favorite application areas. Some of these early expert systems were quite successful and became "classics." Three examples of these early successes are outlined below.

MYCIN
This was a medical system for diagnosing infectious diseases and for selecting an antibiotic drug treatment. It is frequently referenced because of its novel (at the time) use of certainty theory (see Chapter 3).

PROSPECTOR
This system interpreted geological data and made recommendations of suitable sites for mineral prospecting. The system made use of Bayesian updating as a means of handling uncertainty (see Chapter 3).

DENDRAL
This system interpreted mass-spectrometry data, and was notable for its use of a three-phase approach to the problem:

Phase 1: *plan*
Suggest molecular substructures that may be present to guide Phase 2.

Phase 2: *generate hypotheses*
Generate all plausible molecular structures.

Phase 3: *test*
For each generated structure, compare the predicted and actual data. Reject poorly matching structures and place the remainder in rank order.

A large number of intelligent systems has been produced more recently, using many different techniques, to tackle a range of diagnosis and interpretation problems in science, technology, and engineering. Rule-based diagnostic systems have been applied to power plants [1], electronic circuits [2], furnaces [3], an oxygen generator for use on Mars [4], and batteries in the Hubble space telescope [5]. Bayesian updating and fuzzy logic have been used in nuclear power generation [6] and automobile assembly [7], respectively. Neural networks have been used for pump diagnosis [8], and a neural network–rule-based system hybrid has been applied to the diagnosis of electrical discharge machines [9]. One of the most important techniques for diagnosis is case-based reasoning (CBR), described in Chapter 6. CBR has been used to diagnose faults in electronic circuits [10], emergency battery backup systems [11], and software [12]. We will also see in this chapter the importance of models of physical systems such as power plants [13]. Applications of intelligent systems for the more general problem of interpretation include the interpretation of drawings [14], seismic data [15], optical spectra [16], and ultrasonic images [17]. The last is a hybrid system, described in a detailed case study in Section 11.5.

11.2 Deduction and abduction for diagnosis

Given some information about the state of the world, we can often infer additional information. If this inference is logically correct (i.e., guaranteed to be true given that the starting information is true), then this process is termed *deduction*. Deduction is used to predict an effect from a given cause (see Chapter 1). Consider, for instance, a domestic refrigerator (Figure 11.1). If a

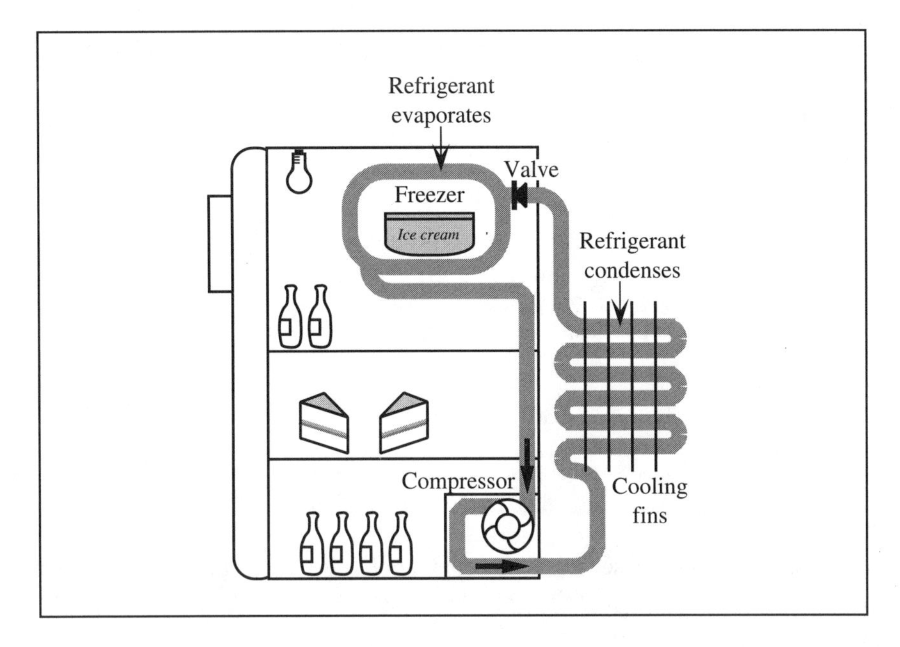

Figure 11.1 A domestic refrigerator

refrigerator is unplugged from the electricity supply, we can confidently predict that, after a few hours, the ice in the freezer will melt and the food in the main compartment will no longer be chilled. The new assertion (that the ice melts) follows logically from the given facts and assertions (that the power is disconnected). Imagine that we have an object representation of a refrigerator, with an attribute `state` that is a list describing the current state of the refrigerator. If my refrigerator is represented by the object instance `my_fridge`, then a simple deductive rule might be as follows, where a question mark denotes a matched variable, as in Section 2.6:

```
/* Rule 11.1 */
IF my_fridge.state contains "unplugged at time ?t"
AND (time_now - ?t) > 5 hours
THEN my_fridge.ice := melted
AND  my_fridge.food_temperature := room_temperature
```

While deductive rules have an important role to play, they are inadequate on their own for problems of diagnosis and interpretation. Instead of determining an effect given a cause, diagnosis and interpretation involve finding a cause given an effect. This is termed *abduction* (see also Chapters 1 and 3) and involves drawing a plausible inference rather than a certain one. Thus, if we observe that the ice in our freezer has melted and the food has warmed to room temperature, we could infer from Rule 11.1 that our refrigerator is unplugged. However this is clearly an unsound inference as there may be several other reasons for the observed effects. For instance, by reference to Rules 11.2 and 11.3 below, we might, respectively, infer that the fuse has blown or that there is a power blackout:

```
/* Rule 11.2 */
IF my_fridge.state contains "fuse blown at time ?t"
AND (time_now - ?t) > 5 hours
THEN my_fridge.ice := melted
AND  my_fridge.food_temperature := room_temperature

/* Rule 11.3 */
IF my_fridge.state contains "power blackout at time ?t"
AND (time_now - ?t) > 5 hours
THEN my_fridge.ice := melted
AND  my_fridge.food_temperature := room_temperature
```

So, given the observed symptoms about the temperature within the refrigerator, the cause might be that the refrigerator is unplugged, or it might be that the fuse has blown, or there might be a power blackout. Three different approaches to tackling the uncertainty of abduction are outlined below.

Exhaustive testing

We could use rules where the condition parts exhaustively test for every eventuality. Rules are, therefore, required of the form:

```
IF my_fridge.state contains "unplugged at time ?t"
AND (time_now - ?t) > 5 hours
AND NOT(a power blackout)
AND NOT(fuse blown)
AND ...
AND ...
THEN my_fridge.ice := melted
AND  my_fridge.food_temperature := room_temperature
```

This is not a practical solution except for trivially simple domains, and the resulting rule base would be difficult to modify or maintain. In the RESCU system [18], those rules which *can* be used with confidence for both deduction and abduction are labeled as *reversible*. Such rules describe a one-to-one mapping between cause and effect, such that no other causes can lead to the same effect.

Explicit modeling of uncertainty

The uncertainty can be explicitly represented using techniques such as those described in Chapter 3. This is the approach that was adopted in MYCIN and

PROSPECTOR. As noted in Chapter 3, many of the techniques for representing uncertainty are founded on assumptions that are not necessarily valid.

Hypothesize-and-test
A tentative hypothesis can be put forward for further investigation. The hypothesis may be subsequently confirmed or abandoned. This hypothesize-and-test approach, which was used in DENDRAL, avoids the pitfalls of finding a valid means of representing and propagating uncertainty.

There may be additional sources of uncertainty, as well as the intrinsic uncertainty associated with abduction. For example, the evidence itself may be uncertain (e.g., we may not be sure that the food isn't cool) or vague (e.g., just what does "cool" or "chilled" mean precisely?).

The production of a hypothesis that may be subsequently accepted, refined, or rejected is similar, but not identical, to *nonmonotonic logic*. Under nonmonotonic logic, an earlier conclusion may be withdrawn in the light of new evidence. Conclusions that can be withdrawn in this way are termed *defeasible*, and a defeasible conclusion is assumed to be valid until such time as it is withdrawn.

In contrast, the hypothesize-and-test approach involves an active search for supporting evidence once a hypothesis has been drawn. If sufficient evidence is found, the hypothesis becomes held as a conclusion. If contrary evidence is found then the hypothesis is rejected. If insufficient evidence is found, the hypothesis remains unconfirmed.

This distinction between nonmonotonic logic and the hypothesize-and-test approach can be illustrated by considering the case of our nonworking refrigerator. Suppose that the refrigerator is plugged in, the compressor is silent, and the light does not come on when the door is opened. Using the hypothesize-and-test approach, we might produce the hypothesis that there is a power blackout. We would then look for supporting evidence by testing whether another appliance is also inoperable. If this supporting evidence were found, the hypothesis would be confirmed. If other appliances were found to be working, the hypothesis would be withdrawn. Under nonmonotonic logic, reasoning would continue based on the assumption of a power blackout until the conclusion is defeated (if it is defeated at all). Confirmation is neither required nor sought in the case of nonmonotonic reasoning. Implicitly, the following default assumption is made:

```
IF an electrical appliance is dead,
AND there is no proof that there is not a power failure
THEN by default, we can infer that there is a power failure
```

The term *default reasoning* describes this kind of implicit rule in nonmonotonic logic.

11.3 Depth of knowledge

11.3.1 Shallow knowledge

The early successes of diagnostic expert systems are largely attributable to the use of shallow knowledge, expressed as rules. This is the knowledge that a human expert might acquire by experience, without regard to the underlying reasons. For instance, a mechanic looking at a broken refrigerator might hypothesize that, if the refrigerator makes a humming noise but does not get cold, then it has lost coolant. While he or she may also know the detailed workings of a refrigerator, this detailed knowledge need not be used. Shallow knowledge can be easily represented:

```
/* Rule 11.4 */
IF refrigerator makes humming sound
AND refrigerator does not get cold
THEN hypothesize loss of coolant
```

Note that we are using the hypothesize-and-test approach to dealing with uncertainty in this example. With the coolant as its focus of attention, an expert system may then progress by seeking further evidence in support of its hypothesis (e.g., the presence of a leak in the pipes). Shallow knowledge is given a variety of names in the literature, including *heuristic*, *experiential*, *empirical*, *compiled*, *surface*, and *low-road.*

Expert systems built upon shallow knowledge may look impressive since they can rapidly move from a set of input data (the symptoms) to some plausible conclusions with a minimal number of intermediate steps, just as a human expert might. However, the limitations of such an expert system are easily exposed by presenting it with a situation that is outside its narrow area of expertise. When it is confronted with a set of data about which it has no explicit rules, the system cannot respond or, worse still, may give wrong answers.

There is a second important deficiency of shallow-reasoning expert systems. Because the knowledge bypasses the causal links between an observation and a deduction, the system has no understanding of its knowledge. Therefore it is unable to give helpful explanations of its reasoning. The best it can do is to regurgitate the chain of heuristic rules that lead from the observations to the conclusions.

11.3.2 Deep knowledge

Deep knowledge is the fundamental building block of understanding. A number of deep rules might make up the causal links underlying a shallow rule. For instance, the effect of the shallow Rule 11.4 (above) may be achieved by the following deep rules:

```
/* Rule 11.5 */
IF current flows in the windings of the compressor
THEN there is an induced rotational force on the windings

/* Rule 11.6 */
IF motor windings are rotating
AND axle is attached to windings and compressor vanes
THEN compressor axle and vanes are rotating

/* Rule 11.7 */
IF a part is moving
THEN it may vibrate or rub against its mounting

/* Rule 11.8 */
IF two surfaces are vibrating or rubbing against each other
THEN mechanical energy is converted to heat and sound

/* Rule 11.9 */
IF compressor vanes rotate
AND coolant is present as a gas at the compressor inlet
THEN coolant is drawn through the compressor and pressurized

/* Rule 11.10 */
IF pressure on a gas exceeds its vapor pressure
THEN the gas will condense to form a liquid

/* Rule 11.11 */
IF a gas is condensing
THEN it will release its latent heat of vaporization
```

Figure 11.2 shows how these deep rules might be used to draw the same hypothesis as the shallow Rule 11.4.

There is no clear distinction between deep and shallow knowledge, but some rules are deeper than others. Thus, while Rule 11.5 is deep in relation to the shallow Rule 11.4, it could be considered shallow compared with knowledge of the flow of electrons in a magnetic field and the origins of the electromotive force that gives rise to the movement of the windings. In recognition of this, Fink and Lusth [19] distinguish *fundamental knowledge*, which is the deepest knowledge that has relevance to the domain. Thus, "unsupported items fall to the ground" may be considered a fundamental rule

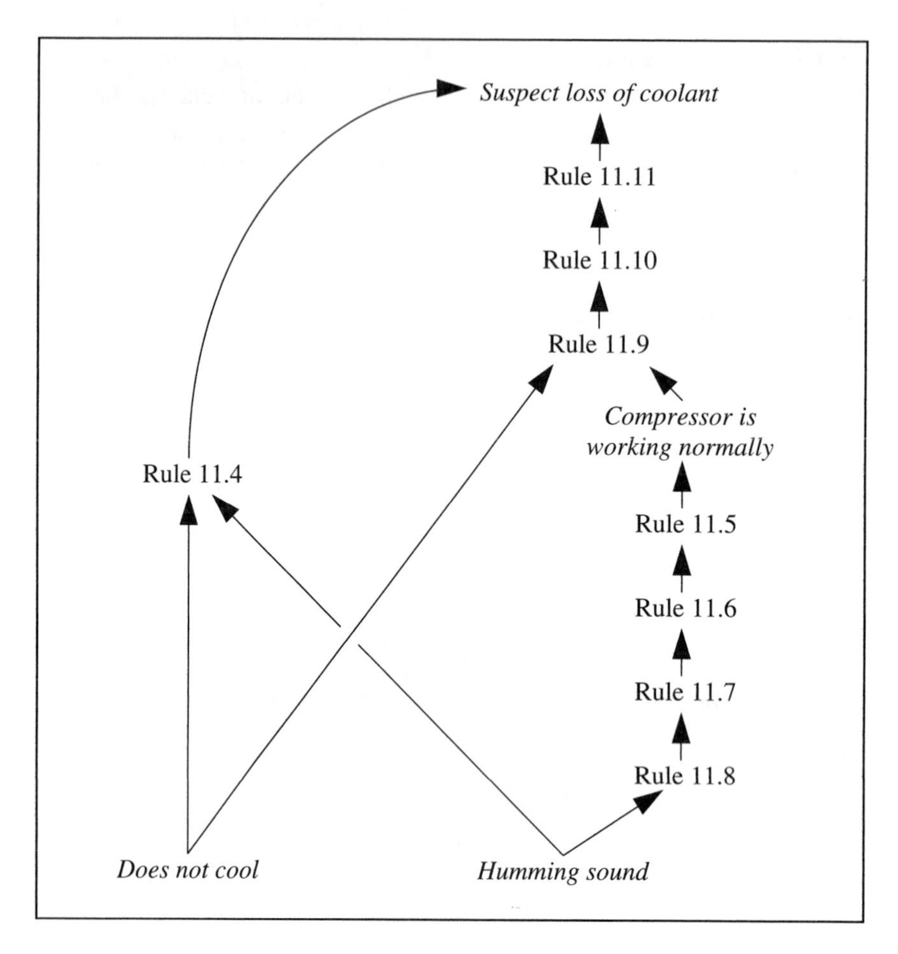

Figure 11.2 Comparing deep and shallow knowledge about a refrigerator

in a domain where details of Newton's laws of gravitation are unnecessary. Similarly Kirchoff's first law (which states that the sum of the input current is equal to the sum of the output current at any point in a circuit) would be considered a deep rule for most electrical or electronic diagnosis problems but is, nevertheless, rather shallow compared with detailed knowledge of the behavior of electrons under the influence of an electric field. Thus, Kirchoff's first law is not fundamental in the broad domain of physics, since it can be derived from deeper knowledge. However, it may be treated as fundamental for most practical problems.

11.3.3 Combining shallow and deep knowledge

There are merits and disadvantages to both deep and shallow knowledge. A system based on shallow knowledge can be very efficient, but it possesses no understanding and, therefore, has no ability to explain its reasoning. It will also fail dismally when confronted with a problem that lies beyond its expertise. The use of deep knowledge can alleviate these limitations, but the knowledge base will be much larger and less efficient. There is a trade-off between the two approaches.

The Integrated Diagnostic Model (IDM) [19] is a system that attempts to integrate deep and shallow knowledge. Knowledge is partitioned into deep and shallow knowledge bases, either one of which is in control at any one time. The controlling knowledge base decides which data to gather, either directly or through dialogue with the user. However, the other knowledge base remains active and is still able to make deductions. For instance, if the shallow knowledge base is in control and has determined that the light comes on when the refrigerator door is opened (as it should), the deep knowledge base can be used to deduce that the power supply is working, that the fuses are OK, that wires from power supply to bulb are OK, and that the bulb is OK. All of this knowledge then becomes available to both knowledge bases.

In IDM, the shallow knowledge base is given control first, and it passes control to the deep knowledge base if it fails to find a solution. The rationale for this is that a quick solution is worth trying first. The shallow knowledge base can be much more efficient at obtaining a solution, but it is more limited in the range of problems that it can handle. If it fails, the information that has been gathered is still of use to the deep knowledge base.

A shallow knowledge base in IDM can be expanded through experience in two ways:

(i) Each solved case can be stored and used to assist in solving similar cases. This is the basis of case-based reasoning (see Chapter 6).

(ii) If a common pattern between symptom and conclusions has emerged over a large number of cases, then a new shallow rule can be created. This is an example of rule induction (see Chapter 6).

11.4 Model-based reasoning

11.4.1 The limitations of rules

The amount of knowledge about a device that can be represented in rules alone is somewhat limited. A deep understanding of how a device works — and what can go wrong with it — is facilitated by a physical model of the device being examined. Fulton and Pepe [20] have highlighted three major inadequacies of a purely rule-based diagnostic system:

(i) Building a complete rule set is a massive task. For every possible failure, the rule-writer must predict a complete set of symptoms. In many cases this information may not even be available, because the failure may never have happened before. It may be possible to overcome the latter hurdle by deliberately causing a fault and observing the sensors. However, this is inappropriate in some circumstances, such as an overheated core in a nuclear power station.

(ii) Symptoms are often in the form of sensor readings, and a large number of rules is needed solely to verify the sensor data [21]. Before an alarm signal can be believed at face value, related sensors must be checked to see whether they are consistent with the alarm. Without this measure, there is no reason to assume that a physical failure has occurred rather than a sensor failure.

(iii) Even supposing that a complete rule set could be built, it would rapidly become obsolete. As there is frequently an interdependence between rules, updating the rule base may require more thought than simply adding new rules.

These difficulties can be circumvented by means of a model of the physical system. Rather than storing a huge collection of symptom–cause pairs in the form of rules, these pairs can be *generated* by applying physical principles to the model.

11.4.2 Modeling function, structure, and state

Practically all physical devices are made up of fundamental components such as tubes, wires, batteries, and valves. As each of these performs a fairly simple role, it also has a simple failure mode. For example, a wire may break and fail to conduct electricity, a tube can spring a leak, a battery can lose its charge, and a valve may be blocked. Given a model of how these components operate and interact to form a device, faults can be diagnosed by determining the effects of local malfunctions on the global view (i.e., on the overall device). Reasoning through consideration of the behavior of the components is sometimes termed reasoning from *second principles*. *First principles* are the basic laws of physics which determine component behavior.

Numerous different techniques and representations have been devised for modeling physical devices. Examples include the Integrated Diagnostic Model (IDM) [19], Knowledge Engineer's Assistant (KEATS) [22], DEDALE [23], FLAME [24], NODAL [25], and GDE [26]. These representations are generally object-oriented (see Chapter 4). The device is made up of a number of components, each of which is represented as an instance of a class of component. The *function* of each component is defined within its class definition. The *structure* of a device is defined by links between the instances of components that make up the device. The device may be in one of several *states*, for example a refrigerator door may be open or closed, and the thermostat may have switched the compressor on or off. These states are defined by setting the values of instance variables.

Object-oriented programming is particularly suited to device modeling because of this clear separation of function, structure, and state. These three aspects of a device are fundamental to understanding its operation and possible malfunctions. A contrast can be drawn with mathematical modeling, where a device can often be modeled more easily by considering its overall activity than by analyzing the component processes.

Function

The function of a component is defined by the methods and attributes of its class. Fink and Lusth [19] define four functional primitives, which are classes of components. All components are considered to be specializations of one of these four classes, although Fink and Lusth hint at the possible need to add further functional primitives in some applications. The four functional primitives are:

- *transformer* — transforms one substance into another;
- *regulator* — alters the output of substance *B*, based upon changes in the input of substance *A*;
- *reservoir* — stores a substance for later output;
- *conduit* — transports substances between other functional primitives.

A fifth class, *sensor*, may be added to this list. A sensor object simply displays the value of its input.

The word "substance" is intended to be interpreted loosely. Thus, water and electricity would both be treated as substances. The scheme was not intended to be completely general purpose, and Fink and Lusth acknowledge that it would need modifications in different domains. However, such modifications may be impractical in many domains, where specialized modeling may be more appropriate. As an illustration of the kind of modification required, Fink and Lusth point out that the behavior of an electrical conduit (a wire) is very different from a water conduit (a pipe), since a break in a wire will stop the flow of electricity, while a broken pipe will cause an out-gush of water.

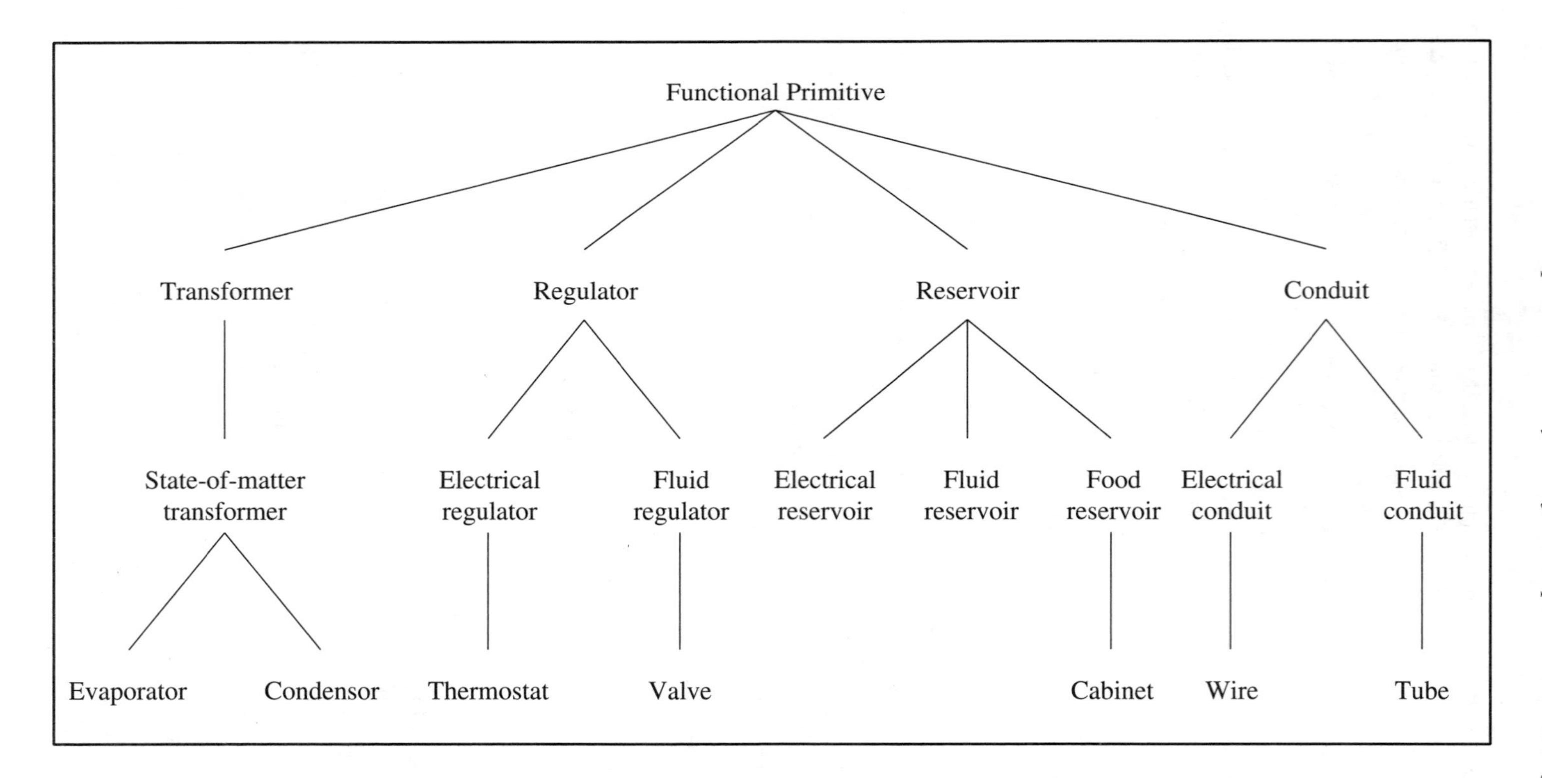

Figure 11.3 Functional hierarchy for some components of a refrigerator

These differences can be recognized by making pipes and wires specializations of the class `conduit` in an object-oriented representation. The `pipe` class requires that a substance be pushed through the conduit, whereas the `wire` class requires that a substance (i.e., electricity) be both pushed and pulled through the conduit. Gas pipes and water pipes would then be two of many possible specializations of the class `pipe`. Figure 11.3 shows a functional hierarchy of classes for some of the components of a refrigerator.

Structure

Links between component instances can be used to represent their physical associations, thereby defining the structure of a device. For example, two resistors connected in series might be represented as two instances of the class `resistor` and one instance of the class `wire`. Each instance of `resistor` would have a link to the instance of `wire`, to represent the electrical contact between them.

Figure 11.4 shows the instances and links that define some of the structure of a refrigerator. This figure illustrates that a compressor can be regarded either as a device made up from several components or as a component of a refrigerator. It is, therefore, an example of a *functional group*. The compressor forms a distinct module in the physical system. However, functional groups in general need not be modular in the physical sense. Thus, the evaporation section of the refrigerator may be regarded as a functional group even though it is not physically separate from the condensation section, and the light system forms a functional group even though the switch is physically removed from the bulb.

Many device-modeling systems can produce a graphical display of the structural layout of a device (similar to Figure 11.4), given a definition of the instances and the links between them. Some systems (e.g., KEATS and IDM) allow the reverse process as well. With these systems, the user draws a structural diagram such as Figure 11.4 on the computer screen, and the instances and links are generated automatically.

In devices where functional groups exist, the device structure is hierarchical. The hierarchical relationship can be represented by means of the composition relationship between objects (see Chapter 4). It is often adequate to consider just three levels of the structural hierarchy:

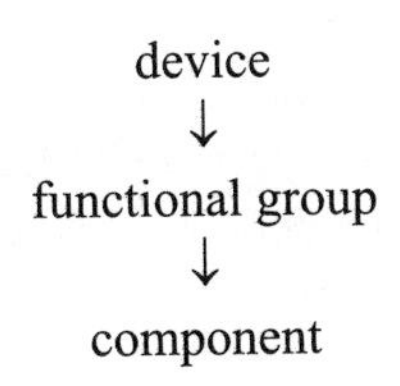

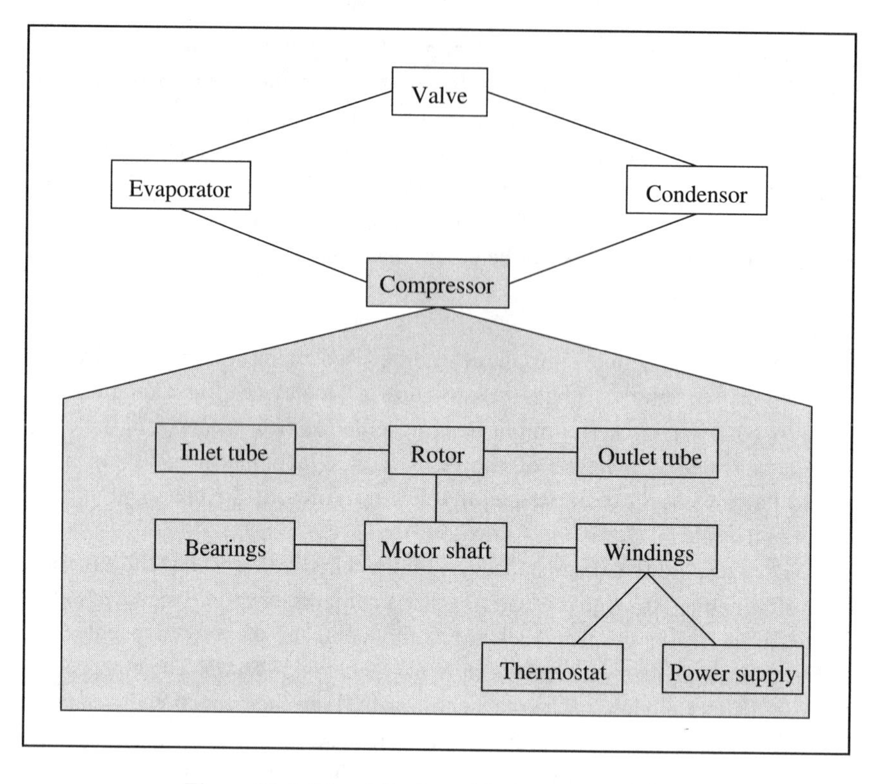

Figure 11.4 Visual display of the instances and links that define the structure of a refrigerator

The application of a three-level hierarchy to the structure of a refrigerator is shown in Figure 11.5.

State

As already noted, a device may be in one of many alternative states. For example, a refrigerator door may be open or closed, and the compressor may be running or stopped. A state can be represented by setting appropriate instance variables on the components or functional groups, and transitions between states can be represented on a state map [27] as shown in Figures 11.6 and 11.7.

The state of some components and functional groups will be dependent on other functional groups or on external factors. Let us consider a refrigerator that is working correctly. The compressor will be in the state *running* only if the thermostat is in the state *closed circuit*. The state of the thermostat will alter according to the cabinet temperature. The cabinet temperature is partially dependent on an external factor, namely, the room temperature, particularly if the refrigerator door is open.

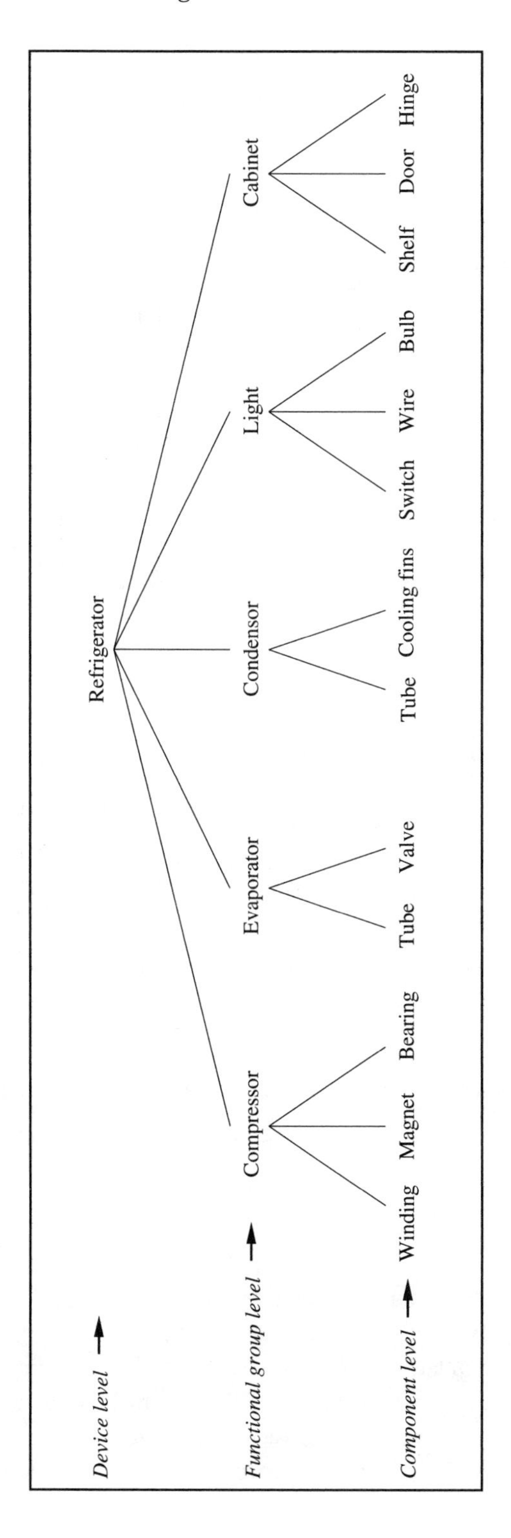

Figure 11.5 Structural hierarchy for some components of a refrigerator

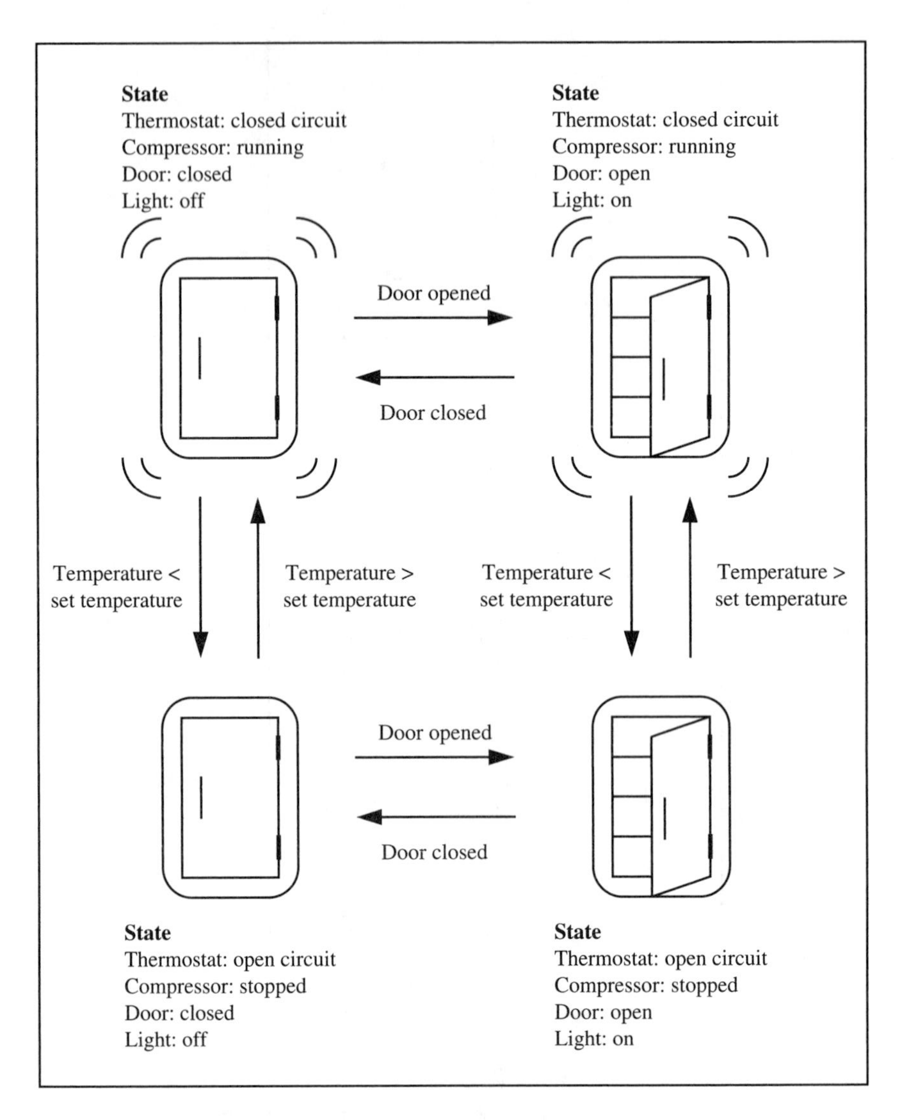

Figure 11.6 State map for a refrigerator working normally

The thermostat behavior can be modeled by making the attribute `temperature` of the `cabinet` object an active value (see Section 4.11.4). The `thermostat` object is updated every time the registered temperature alters by more than a particular amount (say, 0.5°C). A method attached to the thermostat would toggle its state between *open circuit* and *closed circuit* depending on a comparison between the registered temperature and the set

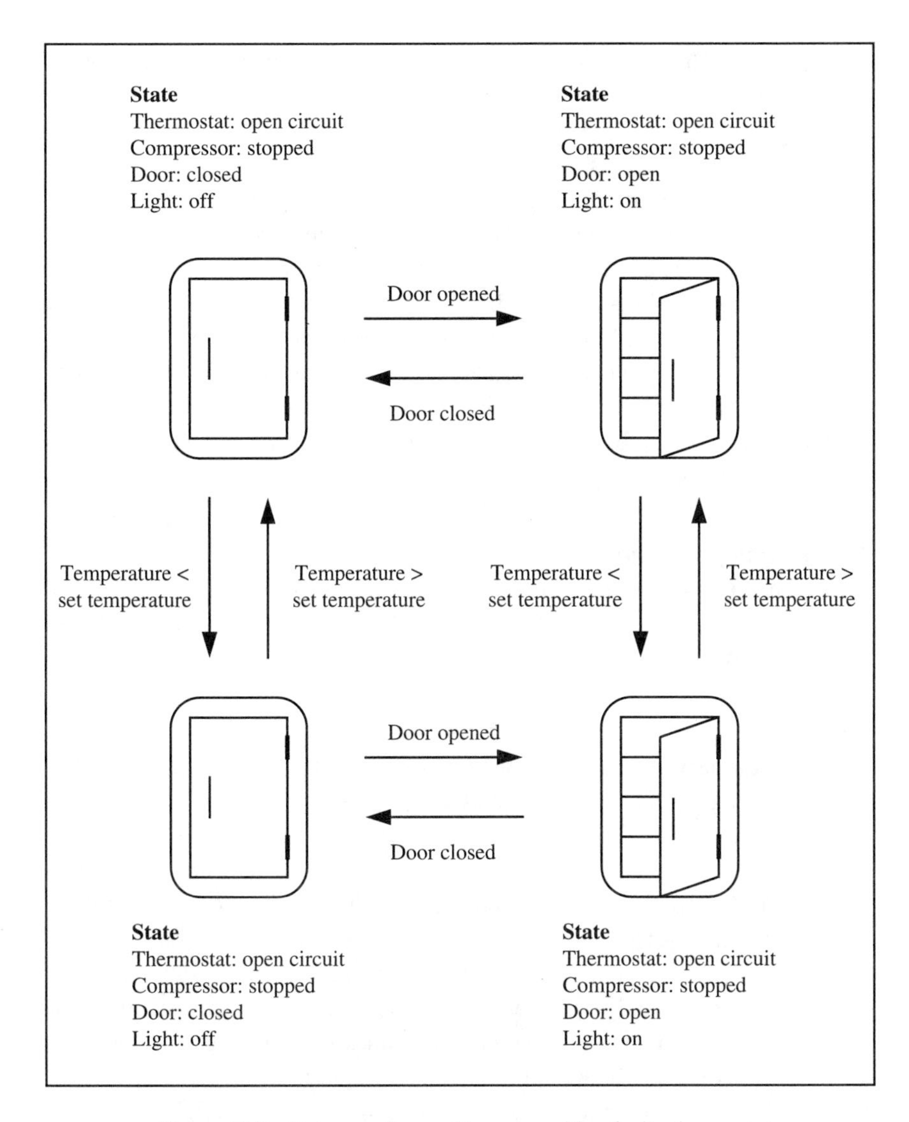

Figure 11.7 State map for a refrigerator with a faulty thermostat

temperature. If the thermostat changes its state, it would send a message to the compressor, which in turn would change its state.

A map of possible states can be drawn up, with links indicating the ways in which one state can be changed into another. A simple state map for a correctly functioning refrigerator is shown in Figure 11.6. A faulty refrigerator will have a different state map. Figure 11.7 shows the case of a thermostat that is stuck in the "open circuit" position.

Price and Hunt [28] have modeled various mechanical devices using object-oriented programming techniques. They created an instance of the class `device_state` for every state of a given device. In their system, each `device_state` instance is made up of instance variables (e.g., recording external factors such as temperature) and links to components and functional groups that make up the device. This is sufficient information to completely restore a state. Price and Hunt found advantages in the ability to treat a device state as an object in its own right. In particular, the process of manipulating a state was simplified and kept separate from the other objects in the system. They were, thus, able to construct state maps similar to those shown in Figures 11.6 and 11.7, where each state was an instance of `device_state`. Instance links were used to represent transitions (such as opening the door) that could take one state to another.

11.4.3 Using the model

The details of how a model can assist the diagnostic task vary according to the specific device and the method of modeling it. In general, three potential uses can be identified:

- monitoring the device to check for malfunctions;
- finding a suspect component, thereby forming a tentative diagnosis; and
- confirming or refuting the tentative diagnosis by simulation.

The diagnostic task is to determine which nonstandard component behavior in the model could make the output values of the model match those of the physical system. An overall strategy is shown in Figure 11.8. A modification of this strategy is to place a weighting on the forward links between evidence and hypotheses. As more evidence is gathered, the weightings can be updated using the techniques described in Chapter 3 for handling uncertainty. The hypothesis with the highest weighting is tried first, and if it fails the next highest is considered, and so on. The weighting may be based solely on perceived likelihood, or it may include factors such as the cost or difficulty of fixing the fault. It is often worth trying a quick and cheap repair before resorting to a more expensive solution.

When a malfunction has been detected, the *single point of failure* assumption is often made. This is the assumption that the malfunction has only one root cause. Such an approach is justified by Fulton and Pepe [20] on the basis that no two failures are truly simultaneous. They argue that one failure will always follow the other either independently or as a direct result.

A model can assist a diagnostic system that is confronted with a problem that lies outside its expertise. Since the function of a component is contained

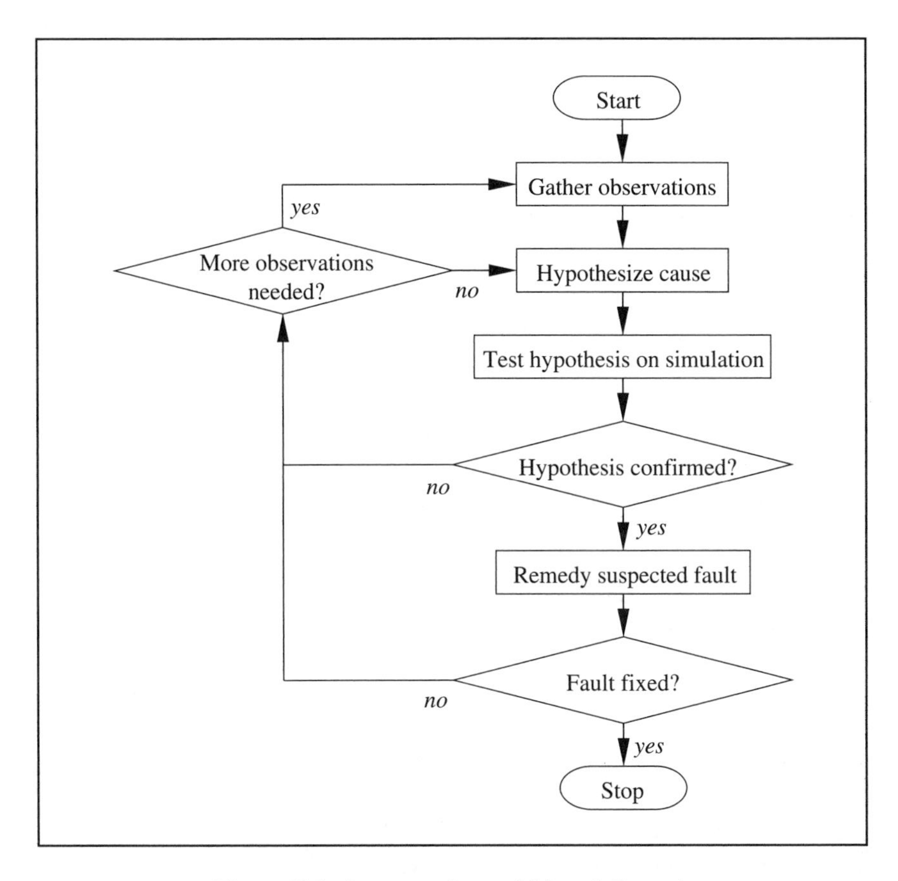

Figure 11.8 A strategy for model-based diagnosis

within the class definition, its behavior in novel circumstances may be predicted. If details of a specific type of component are lacking, comparisons can be drawn with sibling components in the class hierarchy.

11.4.4 Monitoring

A model can be used to simulate the behavior of a device. The output (e.g., data, a substance, or a signal) from one component forms the input to another component. If we alter one input to a component, the corresponding output may change, which may alter another input, and so on, resulting in a new set of sensor readings' being recorded. Comparison with real world sensors provides a monitoring facility.

The RESCU system [18] uses model-based reasoning for monitoring inaccessible plant parameters. In a chemical plant, for instance, a critical parameter to monitor might be the temperature within the reaction chamber. However, it may not be possible to measure the temperature directly, as no

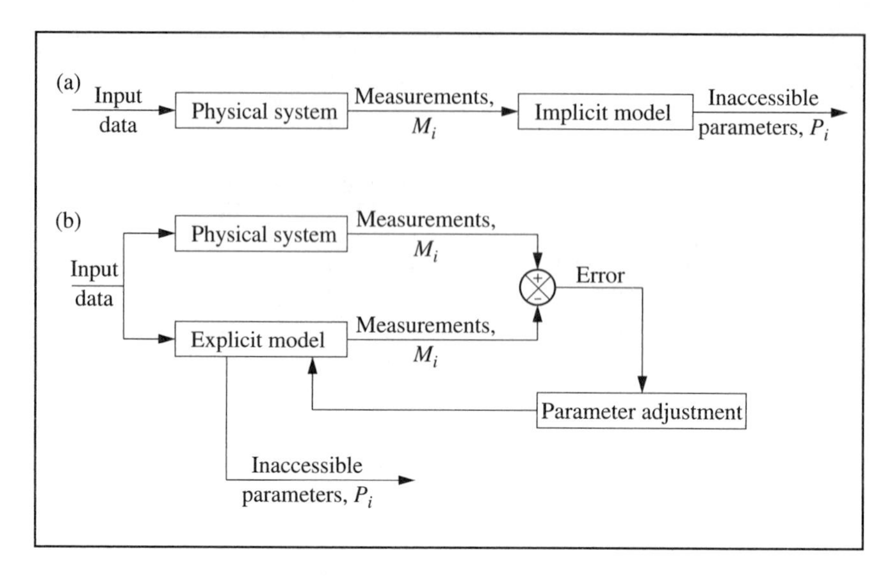

Figure 11.9 Monitoring inaccessible parameters by process modeling: (a) using an implicit model, (b) using an explicit model; (adapted from [18])

type of thermometer would be able to survive the highly corrosive environment. Therefore, the temperature has to be inferred from temperature measurements at the chamber walls, and from gas pressure and flow measurements.

Rules or algorithms that can translate the available measurements (M_i) into the inaccessible parameters (P_i) are described by Leitch et al. [18] as an *implicit model* (Figure 11.9(a)). Obtaining a reliable implicit model is often difficult, and *model-based reasoning* normally refers to the use of *explicit models* such as those described in Section 11.4.2 above. The real system and the explicit model are operated in parallel, the model generating values for the available measurements (M_i) and the inaccessible parameters (P_i). The model parameters (possibly including the parameters P_i) are adjusted to minimize the difference between the values of M_i generated by the model and the real system. This difference is called the *error*. By modifying the model in response to the error we have provided a feedback mechanism (Figure 11.9(b)), discussed further in Chapter 14. This mechanism ensures that the model accurately mimics the behavior of the physical system. If a critical parameter P_i in the model deviates from its expected value, it is assumed that the same has occurred in the physical system and an alarm is triggered.

Analogue devices (electrical or mechanical) can fail by varying degrees. When comparing a parameter in the physical system with the modeled value, we must decide how far apart the values have to be before we conclude that a discrepancy exists. Two ways of dealing with this problem are:

- to apply a tolerance to the expected value, so that an actual value lying beyond the tolerance limit is treated as a discrepancy;
- to give the value a degree of membership of fuzzy sets (e.g., `much too high`, `too high`, `just right`, and `too low`). The degree of membership of these fuzzy sets determines the extent of the response needed. This is the essence of fuzzy control (see Chapter 14).

It is sometimes possible to anticipate a malfunction before it actually happens. This is the approach adopted in EXPRES [29], a system for anticipating faults in the customer distribution part of a telephone network. The network is routinely subjected to electrical tests that measure characteristics such as resistance, capacitance, and inductance between particular points. A broken circuit, for example, would show up as a very high resistance. Failures can be anticipated and avoided by monitoring changes in the electrical characteristics with reference to the fault histories of the specific components under test and of similar components.

11.4.5 Tentative diagnosis

The strategy for diagnosis shown in Figure 11.8 requires the generation of a hypothesis, i.e., a tentative diagnosis. The method of forming a tentative diagnosis depends on the nature of the available data. In some circumstances, a complete set of symptoms and sensor measurements is immediately available, and the problem is one of interpreting them. More commonly in diagnosis, a few symptoms are immediately known and additional measurements must be taken as part of the diagnostic process. The tentative diagnosis is a best guess, or hypothesis, of the actual cause of the observed symptoms. We will now consider three ways of generating a tentative diagnosis.

The shotgun approach

Fulton and Pepe [20] advocate collecting a list of all objects that are upstream of the unexpected sensor reading. All of these objects are initially under suspicion. If several sensors have recorded unexpected measurements, only one sensor need be considered as it is assumed that there is only one cause of the problem. The process initially involves diagnosis at the functional grouping level. Having identified the faulty functional group, it may be sufficient to stop the diagnosis and simply replace the functional group in the real system. Alternatively the process may be repeated by examining individual components within the failed functional group to locate the failed component.

Structural isolation

KEATS [22] has been used for diagnosing faults in analogue electronic circuits, and makes use of the *binary chop* or *structural isolation* strategy of Milne [30]. Initially the electrical signal is sampled in the middle of the signal chain and compared with the value predicted by the model. If the two values correspond, then the fault must be downstream of the sampling point, and a measurement is then taken halfway downstream. Conversely, if there is a discrepancy between the expected and actual values, the measurement is taken halfway upstream. This process is repeated until the defective functional grouping has been isolated. If the circuit layout permits, the process can be repeated within the functional grouping in order to isolate a specific component. Motta et al. [22] found in their experiments that the structural isolation strategy closely resembles the approach adopted by human experts. They have also pointed out that the technique can be made more sophisticated by modifying the binary chop strategy to reflect heuristics concerning the most likely location of the fault.

The heuristic approach

Either of the above two strategies can be refined by the application of shallow (heuristic) knowledge. As noted above, IDM [19] has two knowledge bases, containing deep and shallow knowledge. If the shallow knowledge base has failed to find a quick solution, the deep knowledge base seeks a tentative solution using a set of five guiding heuristics:

(i) if an output from a functional unit is unknown, find out its value by testing or by interaction with a human operator;
(ii) if an output from a functional unit appears incorrect, check its input;
(iii) if an input to a functional unit appears incorrect, check the source of the input;
(iv) if the input to a functional unit appears correct but the output is not, assume that the functional unit is faulty;
(v) examine components that are *nodes* before those that are *conduits*, as the former are more likely to fail in service.

11.4.6 Fault simulation

Both the correct and the malfunctioning behavior of a device can be simulated using a model. The correct behavior is simulated during the monitoring of a device (Section 11.4.4). Simulation of a malfunction is used to confirm or refute a tentative diagnosis. A model allows the effects of changes in a device or in its input data to be tested. A diagnosis can be tested by changing the behavior of the suspected component within the model and checking that the model produces the symptoms that are observed in the real system. Such tests cannot be conclusive, as other faults might also be capable of producing the same symptoms, as noted in Section 11.2 above. Suppose that a real refrigerator is not working and makes no noise. If the thermostat on the refrigerator is suspected of being permanently open-circuit, this malfunction can be incorporated into the model and the effects noted. The model would show the same symptoms that are observed in the real system. The hypothesis is then confirmed as the most likely diagnosis.

Most device simulations proceed in a step-by-step manner, where the output of one component (component *A*) becomes the input of another (component *B*). Component *A* can be thought of as being "upstream" of component *B*. An input is initially supplied to the components that are furthest upstream. For instance, the thermostat is given a cabinet temperature, and the power cord is given a simulated supply of electricity. These components produce outputs, which become the inputs to other components, and so on. This sort of simulation can run into difficulties if the model includes a feedback loop. In these circumstances, a defective component not only produces an unexpected output, but also has an unexpected input. The output can be predicted only if the input is known, and the input can be predicted only if the output is known. One approach to this problem is to supply initially all of the components with their correct input values. If the defective behavior of the feedback component is modeled, its output and input values would be expected to converge on a failure value after a number of iterations. Fink and Lusth [19] found that convergence was achieved in all of their tests, but they acknowledge that there might be cases where this does not happen.

11.4.7 Fault repair

Once a fault has been diagnosed, the next step is normally to fix the fault. Most fault diagnosis systems offer some advice on how a fault should be fixed. In many cases this recommendation is trivial, given the diagnosis. For example, the diagnosis `worn bearings` might be accompanied by the recommendation `replace worn bearings`, while the diagnosis `leaking pipe` might lead to the recommendation `fix leak`. A successful repair provides definite confirmation of a diagnosis. If a repair fails to cure a problem, then the diagnostic process must recommence. A failed repair may not mean that a diagnosis was incorrect. It is possible that the fault which has now been fixed has caused a second fault that also needs to be diagnosed and repaired.

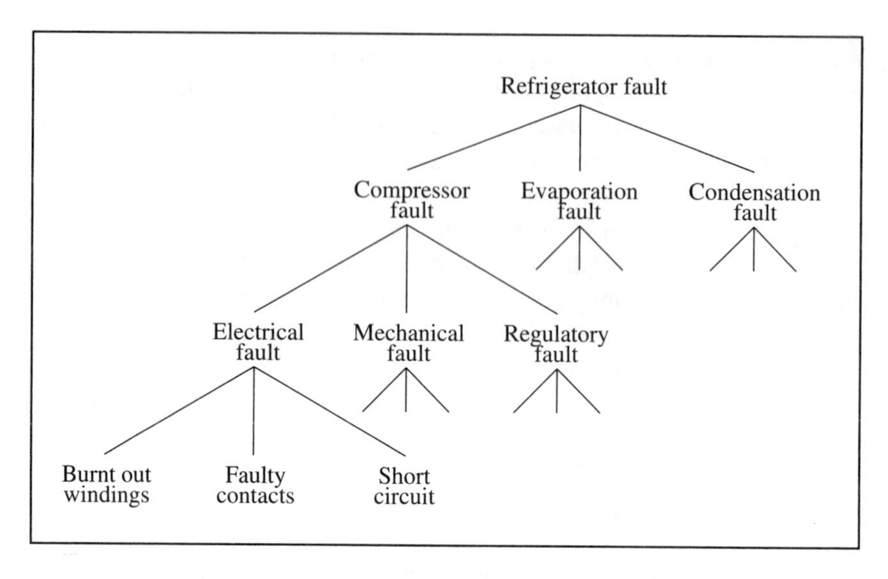

Figure 11.10 A problem tree for a refrigerator

11.4.8 Using problem trees

Some researchers [3, 31] favor the explicit modeling of faults, rather than inferring faults from a model of the physical system. Dash [3] builds a hierarchical tree of possible faults (a *problem tree*, see Figure 11.10), similar to those used for classifying case histories (Figure 6.3(a)). Unlike case-based reasoning, problem trees cover all anticipated faults, whether or not they have occurred previously. By applying deep or shallow rules to the observed symptoms, progress can be made from the root (a general problem description) to the leaves of the tree (a specific diagnosis). The tree is, therefore, used as a means of steering the search for a diagnosis. An advantage of this approach is that a partial solution is formed at each stage in the reasoning process. A partial solution is generated even if there is insufficient knowledge or data to produce a complete diagnosis.

11.4.9 Summary of model-based reasoning

Some of the advantages of model-based reasoning are listed below.

- A model is less cumbersome to maintain than a rule base. Real-world changes are easily reflected in changes in the model.
- The model need not waste effort looking for sensor verification. Sensors are treated identically to other components, and therefore a faulty sensor is as likely to be detected as any other fault.

- Unusual failures are just as easy to diagnose as common ones. This is not the case in a rule-based system, which is likely to be most comprehensive in the case of common faults.
- The separation of function, structure, and state may help a diagnostic system to reason about a problem that is outside its area of expertise.
- The model can simulate a physical system, for the purpose of monitoring or for verifying a hypothesis.

Model-based reasoning only works well in situations where there is a complete and accurate model. It is inappropriate for physical systems that are too complex to model properly, such as medical diagnosis or weather forecasting.

An ideal would be to build a model for monitoring and diagnosis directly from CAD data generated during the design stage. The model would then be available as soon as a device entered service.

11.5 Case study: a blackboard system for interpreting ultrasonic images

One of the aims of this book is to demonstrate the application of a variety of techniques, including knowledge-based systems, computational intelligence, and procedural computation. The more complicated problems have many facets, where each facet may be best suited to a different technique. This is true of the interpretation of ultrasonic images, which will be discussed as a case study in the remainder of this chapter. A *blackboard system* has been used to tackle this complex interpretation problem [17]. As we saw in Chapter 9, blackboard systems allow a problem to be divided into subtasks, each of which can be tackled using the most suitable technique.

An image interpretation system attempts to understand the processed image and describe the world represented by it. This requires symbolic reasoning (using rules, objects, relationships, list-processing, or other techniques) as well as numerical processing of signals [32]. ARBS (Algorithmic and Rule-based Blackboard System) has been designed to incorporate both processes. Numerically intensive signal processing, which may involve a large amount of raw data, is performed by conventional routines. Facts, causal relationships, and strategic knowledge are symbolically encoded in one or more knowledge bases. This explicit representation allows encoded domain knowledge to be modified or extended easily. Signal processing routines are used to transform the raw data into a description of its

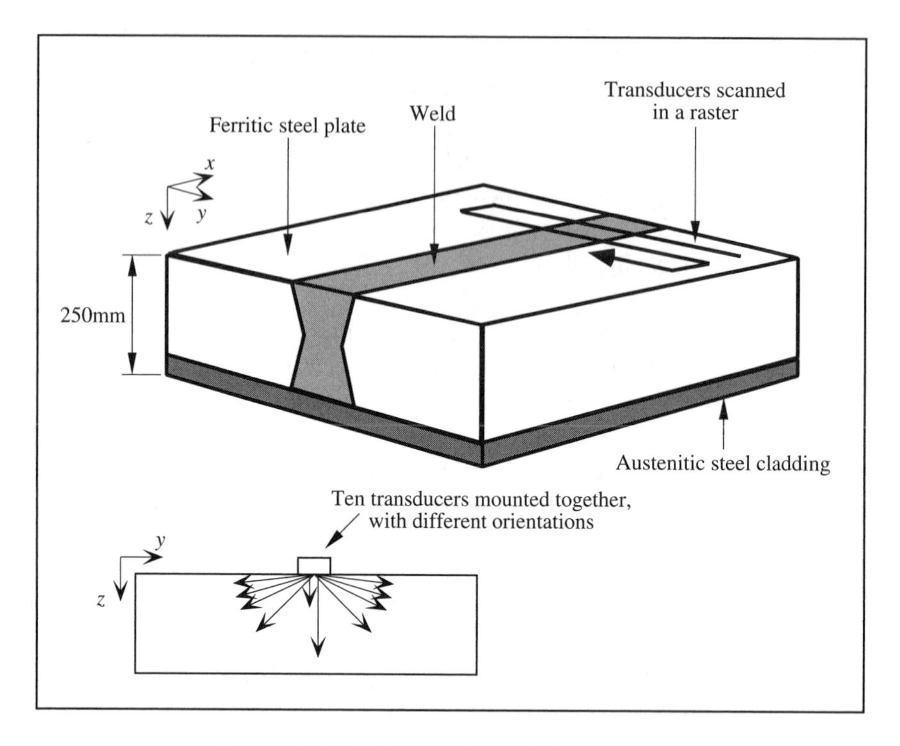

Figure 11.11 Gathering ultrasonic b-scan images

key features (i.e., into a symbolic image) and knowledge-based techniques are used to interpret this image.

The architecture allows signal interpretation to proceed under the control of knowledge in such a way that, at any one time, only the subset of the data that might contribute to a solution is considered. When a particular set of rules needs access to a chunk of raw data, it can have it. No attempt has been made to write rules that look at the whole of the raw data, as the preprocessing stage avoids the need for this. Although a user interface is provided to allow monitoring and intervention, ARBS is designed to run independently, producing a log of its actions and decisions.

11.5.1 Ultrasonic imaging

Ultrasonic imaging is widely used for the detection and characterization of features, particularly defects, in manufactured components. The technique belongs to a family of nondestructive testing methods, which are distinguished by the ability to examine components without causing any damage. Various arrangements can be used for producing ultrasonic images. Typically, a transmitter and receiver of ultrasound (i.e., high frequency sound at approximately 1–10MHz) are situated within a single probe that makes contact

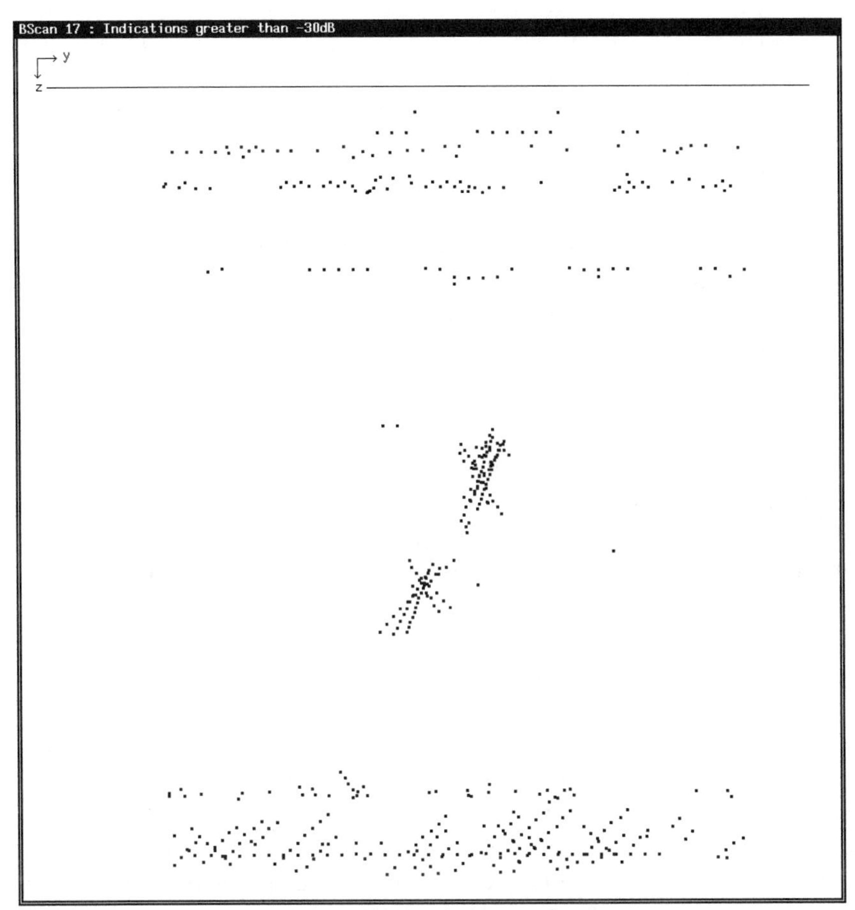

Figure 11.12 A typical b-scan image

with the surface of the component. The probe emits a short pulse of ultrasound and then detects waves returning as a result of interactions with the features within the specimen. If the detected waves are assumed to have been produced by single reflections, then the time of arrival can be readily converted to the depth (z) of these features. By moving the probe in one dimension (y), an image can be plotted of y versus z, with intensity often represented by color or grayscale. This image is called a *b-scan*, and it approximates to a cross section through a component. It is common to perform such scans with several probes pointing at different angles into the specimen, and to collect several b-scans by moving the probes in a raster (Figure 11.11).

A typical b-scan image is shown in Figure 11.12. A threshold has been applied, so that only received signals of intensity greater than –30db are displayed, and these appear as black dots on the image. Ten probes were used, pointing into the specimen at five different angles. Because ultrasonic beams are not collimated, a point defect is detected over a range of y values, giving rise to a characteristic arc of dots on the b-scan image. Arcs produced by a single probe generally lie parallel to each other and normal to the probe direction. The point of intersection of arcs produced by different probes is a good estimate of the location of the defect that caused them.

The problems of interpreting a b-scan are different from those of other forms of image interpretation. The key reason for this is that a b-scan is not a direct representation of the inside of a component, but rather it represents a set of wave interactions (reflection, refraction, diffraction, interference, and mode conversion) that can be used to *infer* some of the internal structure of the component.

11.5.2 Knowledge sources in ARBS

Each knowledge source (KS) in ARBS is contained in a record. Records are data structures consisting of different data types, and are provided in most modern computer languages. Unlike lists, the format of a record has to be defined before values can be assigned to any of its various parts (or *fields*). The record structure that is used to define an ARBS knowledge source is shown in Figure 11.13. There is a set of preconditions, in the `preconditions` field, which must be satisfied before the KS can be activated. The preconditions are expressed using the same syntax as the rules described below. ARBS has a control module which examines each KS in turn, testing the preconditions and activating the KS if the preconditions are satisfied. This is the simplest strategy, akin to forward-chaining within a rule-based system. More sophisticated strategies can be applied to the selection of knowledge sources, just as more complex inference engines can be applied to rules.

When a KS is activated, it applies its knowledge to the current state of the blackboard, adding to it or modifying it. The entry in the `KS type` field states whether the KS is procedural, rule-based, neural network-based, or genetic algorithm-based. If a KS is rule-based then it is essentially a rule-based system in its own right. The `rules` field contains the names of the rules to be used and the `inference mode` field contains the name of the inference engine. When the rules are exhausted, the KS is deactivated and any actions in the `actions` field of the KS are performed. These actions usually involve reports to the user or the addition of control information to the blackboard. If the KS is procedural or contains a neural network or genetic algorithm then the required code is simply included in the `actions` field and the fields relating to rules are ignored.

ARBS includes provisions for six different types of knowledge sources. One is procedural, one is for neural networks, one is for genetic algorithms, and the other three are all rule-based, but with different types of inference mechanisms. We have already met two of the inference mechanisms, namely,

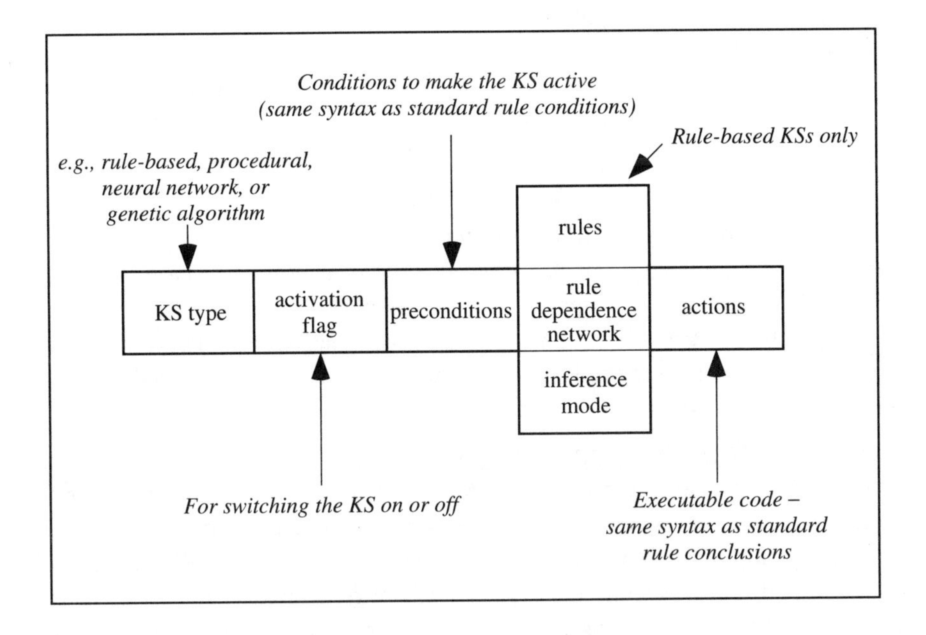

Figure 11.13 A knowledge source in ARBS is stored as a record

multiple and single instantiation of variables with a hybrid inference engine based on a rule dependence network (see Chapter 2). The third type of rule-based knowledge source is used for generating hypotheses. Such hypotheses can then be tested and thereby supported, confirmed, weakened, or refuted. ARBS uses this hypothesize-and-test approach (see Section 11.2 above) to handle the uncertainty that is inherent in problems of abduction, and the uncertainty that arises from sparse or noisy data.

The blackboard architecture is able to bring together the most appropriate tools for handling specific tasks. Procedural tasks are not entirely confined to procedural knowledge sources, since rules within a rule-based knowledge source can access procedural code from either their condition or conclusion parts. In the case of ultrasonic image interpretation, procedural KSs written in C++ are used for fast, numerically intensive data-processing, and rule-based KSs are used to represent specialist knowledge. In addition, neural network KSs may be used for judgmental tasks, involving the weighing of evidence from various sources. Judgmental ability is often difficult to express in rules, and neural networks were incorporated into ARBS as a response to this difficulty. The different types of knowledge sources used in this application of ARBS are summarized below.

Procedural KS

When activated, a procedural KS simply runs its associated procedure and is then deactivated. An example is the procedure that preprocesses the image, using the Hough Transform to detect lines of indications (or dots).

Rule-based KS with a hybrid inference engine

When activated, this type of KS behaves as a conventional-rule based system. A rule's conditions are examined and, if they are satisfied by a statement on the blackboard, the rule fires and the actions dictated by its conclusion are carried out. Single or multiple instantiation of variables can be selected.

Rule-based hypothesis-generating KS

This KS initiates a "hypothesize-and-test" cycle. The inference engine selects a rule and asserts its conclusion as a hypothesis. Its conditions are now referred to as *expectations* and placed in a list on the blackboard. Other KSs may then be selected to test whether the expectations are confirmed by the image data. The hypothesis-generating rules use a different syntax from other rules as they are used in a different way. The hypothesize-and-test method reduces the solution search space by focusing attention on those KSs relevant to the current hypothesis.

Neural network KS

This type of KS is used, for example, in small-scale classification tasks, where domain knowledge is difficult to express in rule form. The use of neural networks in this application is described in more detail in Section 11.5.6 below.

11.5.3 Rules in ARBS

Rules are used in ARBS in two contexts:

- to express domain knowledge within a rule-based knowledge source;
- to express the applicability of a knowledge source.

Just as knowledge sources are activated in response to information on the blackboard, so too are individual rules within a rule-based knowledge source. The main functions of ARBS rules are to look up information on the blackboard, to draw inferences from that information, and to post new information on the blackboard. Rules can access procedural code for performing such tasks as numerical calculations or database lookup.

In this section we will be concerned only with deductive rules, rather than the hypothesis-generating rules mentioned in Section 11.5.2 above. The rules are implemented as lists (see Chapter 10), delineated by square brackets. Each rule is a list comprising four elements: a number to identify the rule, a condition part, the word `implies`, and a conclusion part. A rule is defined as follows:

```
rule :: [number condition implies conclusion].
```

The double colon (::) means "takes the form of." The word `implies` is shown in bold type as it is recognized as a key word in the ARBS syntax. The words in italics have their own formal syntax description. The condition may comprise subconditions joined with Boolean operators `AND` and `OR`, while the conclusion can contain several action statements. There is no explicit limit on the number of subconditions or subconclusions that can be combined in this way.

In the syntax description shown in Box 11.1, many items have several alternative forms. Where square brackets are shown, these delimit lists. Notice that both `condition` and `conclusion` are defined recursively, thereby allowing any number of subconditions or subconclusions. It can also be seen that where multiple subconditions are used, they are always combined in pairs. This avoids any ambiguity over the order of precedence of the Boolean combinations.

Statements on the blackboard are recognized by pattern-matching, where the `?` symbol is used to make assignments to variables in rules, as shown in Section 2.6. Additionally, the `~` symbol has been introduced to indicate that the name of the associated variable is replaced by its assigned value when the rule is evaluated. Blackboard information that has been assigned to variables by pattern-matching can thereby be used elsewhere within the rule.

Rules are interpreted by sending them to a parser, i.e., a piece of software that breaks down a rule into its constituent parts and interprets them. This is achieved by pattern-matching between rules and templates that describe the rule syntax. The ARBS rule parser first extracts the condition statement and evaluates it, using recursion if there are embedded subconditions. If a rule is selected for firing and its overall condition is found to be true, all of the conclusion statements are interpreted and carried out.

Atomic conditions (i.e., conditions that contain no subconditions) can be evaluated in any of the following ways:

- test for the presence of information on blackboard, and look up the information if it is present;

```
rule::=         [rule_no condition implies conclusion]

rule_no::=      <constant>

condition::=    [condition and condition],
                [condition or condition],
                [present statement partition],
                [absent statement partition],
                [run [<procedure_name> [input_parameters]] output_parameter],
                [compare [operand operator operand] nil]

conclusion::=   [conclusion conclusion conclusion ...]
                [add statement partition],
                [remove statement partition],
                [report statement nil],
                [run [<procedure_name> [input_parameters]] output_parameter]

partition::=     <list_name>          /* a partition of the blackboard */

statement::=    ~<variable_name>,     /* uses the value of the variable */
                ?<variable_name>,     /* matches a value to the variable */
                ~~<list_name>,        /* as ~ but separates the list elements */
                ??<list_name>,        /* as ? but separates the list elements */
                <string>,
                <constant>
                ~[run [<procedure_name> [input_parameters]] output_parameter],
                <a list containing any of the above>

operand::=      ~[run [<procedure_name> [input_parameters]] output_parameter],
                ~<variable_name>,
                ?<variable_name>,
                <constant>

operator::=     eq,    /* equal */
                lt,    /* less than */
                gt,    /* greater than */
                le,    /* less than or equal */
                ge,    /* greater than or equal */
                ne     /* not equal */
```

Box 11.1 ARBS rule syntax

ARBS key words are shown in bold type. For clarity, some details of the syntax have been omitted or altered. A different syntax is used for hypothesis generation

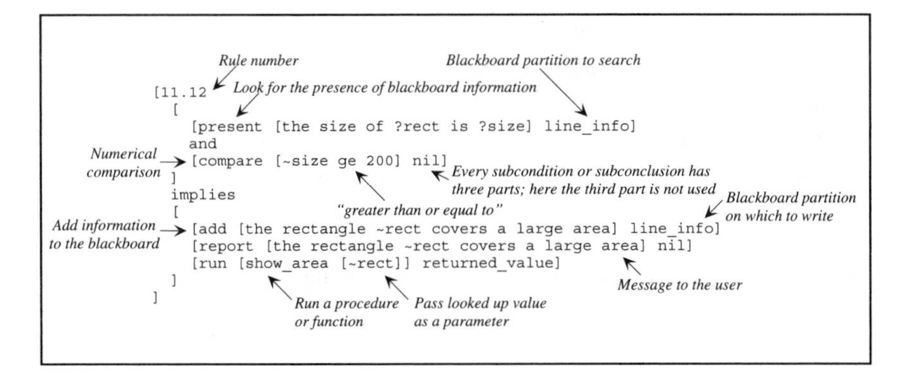

Figure 11.14 A rule in ARBS

- call algorithms or external procedures that return Boolean or numerical results;
- numerically compare variables, constants, or algorithm results.

The conclusions, or subconclusions, can comprise any of the following:

- add or remove information to or from the blackboard;
- call an algorithm or external procedure, and optionally add the results to the blackboard;
- report actions to the operator.

In ARBS, the blackboard, like the rules, is made up from lists. The retrieval of information from the blackboard by pattern-matching is best demonstrated by considering a rule, as shown in Figure 11.14. This rule examines the partition of the blackboard called `line_info`. Like all the blackboard partitions, `line_info` is a list. Supposing that `line_info` contains the sublist:

```
[the size of rectangle_B is 243]
```

then the first subcondition of the rule is true, and the local variables `rect` and `size` would become bound to `rectangle_B` and `243` respectively. This simple subcondition has thus checked for the presence of information on the blackboard and retrieved information. The second subcondition checks to see whether the value of `size` is ≥200. Notice the use of the ~ symbol, which instructs the ARBS parser to replace the word `size` with its current value. As both subconditions are true, the actions described in the conclusions part of the rule are carried out. The first subconclusion involves adding information to the blackboard. The other two conclusions are for the purposes of logging the system's actions. A message reporting the deduction that `rectangle_B` covers a large area is sent to the user, and the rectangle is displayed on the processed image.

Procedures can be directly accessed from within the condition or conclusion parts of any rules. This is achieved by use of the ARBS key word `run`. The rule parser knows that the word immediately following `run` is the procedure name, and that its parameters, if any, are in the accompanying list. The following subconclusion is taken form a KS in the ultrasonic interpretation system:

```
[add [~[run [group_intersections [~coord_list]] result]
      are the groups of points of intersection] line_info]
```

When the rule is fired, the function `group_intersections` is called with the value of `coord_list` as its parameter. The value returned is added to the blackboard as part of the list:

```
... are the groups of points of intersection]
```

Although alterations to the syntax are possible, the syntax shown in Box 11.1 has shown the flexibility necessary to cope with a variety of applications [33, 34].

11.5.4 Inference engines in ARBS

The strategy for applying rules is a key decision in the design of a system. In many types of rule-based system, this decision is irrevocable, committing the rule-writer to either a forward- or backward-chaining system. However, the blackboard architecture allows much greater flexibility, as each rule-based knowledge source can use whichever inference mechanism is most appropriate. ARBS makes use of the hybrid inference mechanism described in Chapter 2, which it can use with either single or multiple instantiation of variables (see Section 2.7.1).

The hybrid mechanism requires the construction of a network representing the dependencies between the rules. A separate dependence network is built for each rule-based knowledge source by a specialized ARBS module, prior to running the system. The networks are saved and only need to be regenerated if the rules are altered. The code to generate the networks is simplified by the fact that the only interaction between rules is via the blackboard. For rule *A* to enable rule *B* to fire, rule *A* must either add something to the blackboard that rule *B* needs to find or remove something that rule *B* requires to be absent. When a rule-based knowledge source is activated, the rules within the

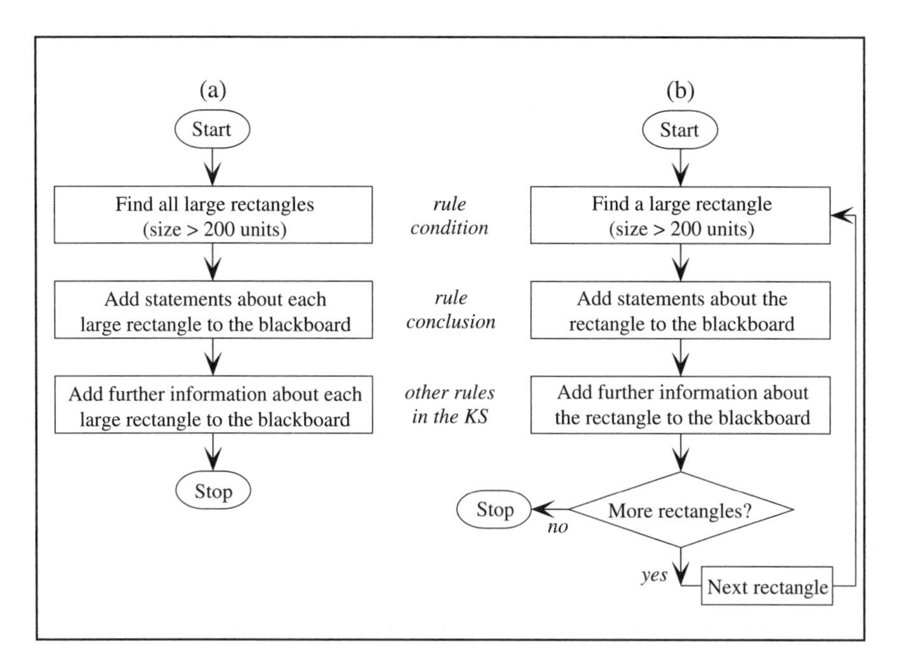

Figure 11.15 Firing Rule 11.12 using:
(a) multiple instantiation of variables, and (b) single instantiation of variables

knowledge source are selected for examination in the order dictated by the dependence network.

Several rules in the ultrasonic interpretation rule base need to be fired for each occurrence of a particular feature in the image. As an example, Rule 11.12 in Figure 11.14 is used to look at rectangular areas that have been identified as areas of interest on an image, and to pick out those that cover a large area. The condition clause is:

```
[
  [present [the size of ?rect is ?size] line_info]
  and
  [compare [~size ge 200] nil]
]
```

A list matching `[the size of ?rect is ?size]` is sought in the portion of the blackboard called `line_info`, and appropriate assignments are made to the variables `rect` and `size`. However, rather than finding only one match to the list template, we actually require all matches. In other words, we wish to find all rectangles and check the size of them all. This can be achieved by either single or multiple instantiation. The resultant order in which the image areas are processed in shown in Figure 11.15. In this example, there is no

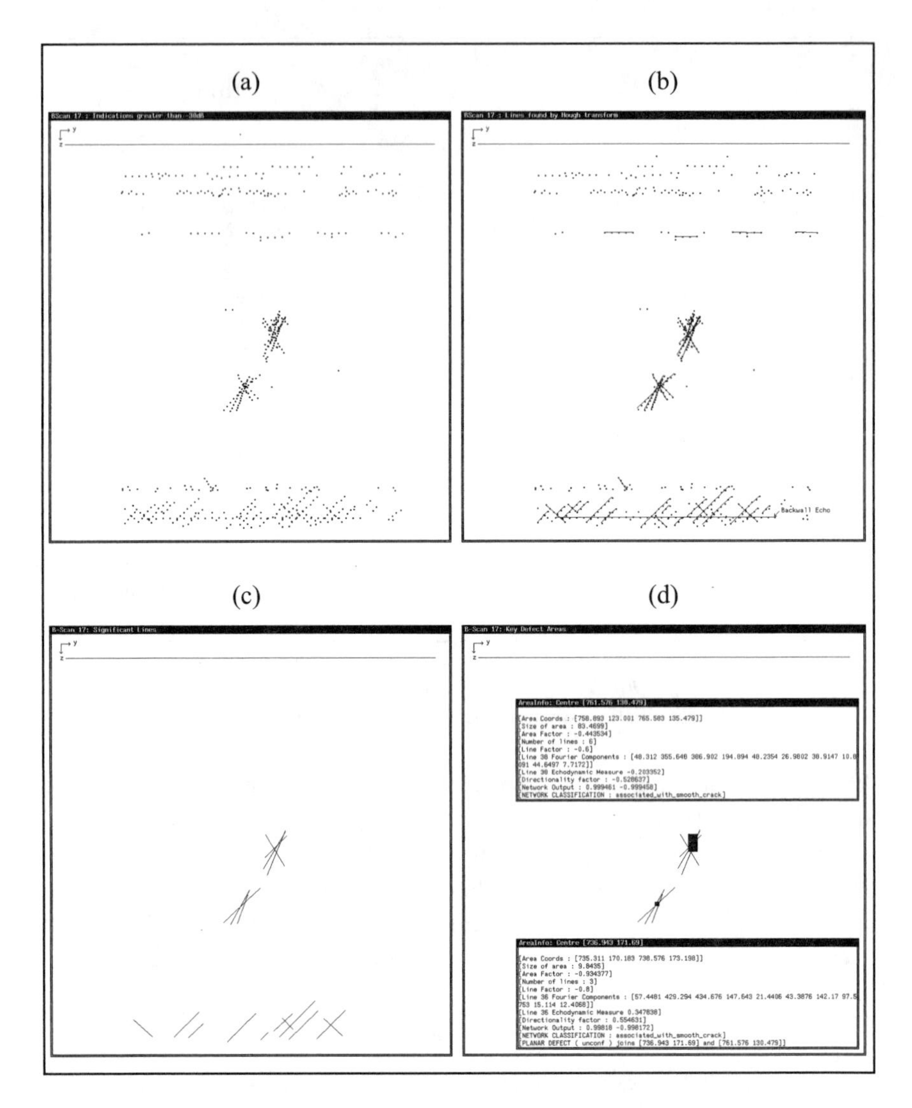

Figure 11.16 The stages of interpretation of a b-scan image:
(a) before interpretation
(b) with lines found by a Hough transform
(c) significant lines
(d) conclusion: a crack runs between the two marked areas

strong reason to prefer one strategy over the other, although multiple instantiation in ARBS is more efficient. In spite of the reduced efficiency, it will be shown in Chapter 14 that single instantiation may be preferable in problems where a solution must be found within a limited time frame.

11.5.5 The stages of image interpretation

The problem of ultrasonic image interpretation can be divided into three distinct stages: arc detection, gathering information about the regions of intersecting arcs, and classifying defects on the basis of the gathered information. These stages are now described in more detail.

Arc detection using the Hough transform

The first step toward defect characterization is to place on the blackboard the important features of the image. This is achieved by a procedural knowledge source that fits arcs to the data points (Figure 11.16(b)). In order to produce these arcs, a Hough transform [35] was used to determine the groupings of points. The transform was modified so that isolated points some distance from the others would not be included. The actual positions of the arcs were determined by least squares fitting.

This preprocessing phase is desirable in order to reduce the volume of data and to convert it into a form suitable for knowledge-based interpretation. Thus, knowledge-based processing begins on data concerning approximately 30 linear arcs rather than on data concerning 400–500 data points. No information is lost permanently. If, in the course of its operations, ARBS judges that more data concerning a particular line would help the interpretation, it retrieves from file the information about the individual points that compose that line and represents these point data on the blackboard. It is natural, moreover, to work with lines rather than points in the initial stages of interpretation. Arcs of indications are produced by all defect types, and much of the knowledge used to identify flaws is readily couched in terms of the properties of lines and the relations between them.

Gathering the evidence

Once a description of the lines has been recorded on the blackboard, a rule-based KS picks out those lines that are considered significant according to criteria such as intensity, number of points, and length. Rules are also used to recognize the back wall echo and lines that are due to probe reverberation. Both are considered "insignificant" for the time being. Key areas in the image are generated by finding points of intersection between significant lines and then grouping them together. Figure 11.16(d) shows the key areas found by applying this method to the lines shown in Figure 11.16(c). For large smooth cracks, each of the crack tips is associated with a distinct area. Other defects are entirely contained in their respective areas.

Rule-based KSs are used to gather evidence about each of the key areas. The evidence includes:

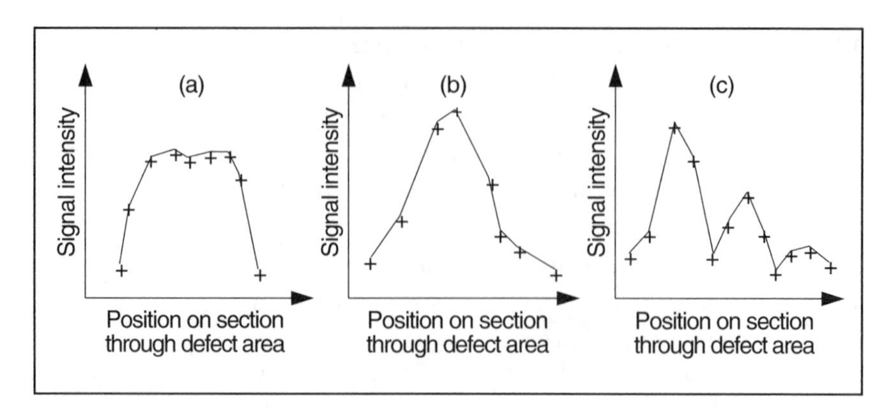

Figure 11.17 Echodynamics across (a) a crack face, (b) a crack tip, pore, or inclusion, (c) an area of porosity

- the size of the area;
- the number of arcs passing through it;
- the shape of the echodynamic (defined below);
- the intensity of indications;
- the sensitivity of the intensity to the angle of the probe.

The *echodynamic* associated with a defect is the profile of the signal intensity along one of the "significant" lines. This is of considerable importance in defect classification as different profiles are associated with different types of defect. In particular, the echodynamics for a smooth crack face, a spheroidal defect (e.g., a single pore or an inclusion) or crack tip, and a series of gas pores are expected to be similar to those in Figure 11.17 [36]. In ARBS, pattern classification of the echodynamic is performed using a fast Fourier transform and a set of rules to analyze the features of the transformed signal. A neural network has also been used for the same task (see Section 11.5.6 below).

The sensitivity of the defect to the angle of the probe is another critical indicator of the nature of the defect in question. Roughly spherical flaws, such as individual gas pores, have much the same appearance when viewed from different angles. Smooth cracks on the other hand have a much more markedly directional character — a small difference in the direction of the probe may result in a considerable reduction (or increase) in the intensity of indications. The directional sensitivity of an area of intersection is a measure of this directionality and is represented in ARBS as a number between –1 and 1.

Defect classification

Quantitative evidence about key areas in an image is derived by rule-based knowledge sources, as described above. Each piece of evidence provides clues as to the nature of the defect associated with the key area. For instance, the indications from smooth cracks tend to be sensitive to the angle of the probe, and the echodynamic tends to be plateau-shaped. In contrast, indications from a small round defect (e.g., an inclusion) tend to be insensitive to probe direction and have a cusp-shaped echodynamic. There are several factors like these that need to be taken into account when producing a classification, and each must be weighted appropriately.

Two techniques for classifying defects based upon the evidence have been tried using ARBS: a rule-based hypothesize-and-test approach and a neural network. In the former approach, hypotheses concerning defects are added to the blackboard. These hypotheses relate to smooth or rough cracks, porosity, or inclusions. They are tested by deriving from them *expectations* (or *predictions*) relating to other features of the image. On the basis of the correspondence between the expectations and the image, ARBS arrives at a conclusion about the nature of a defect, or, where this is not possible with any degree of certainty, it alerts the user to a particular problem case.

Writing rules to verify the hypothesized defect classifications is a difficult task, and in practice the rules needed continual refinement and adjustment in the light of experience. The use of neural networks to combine the evidence and produce a classification provides a means of circumventing this difficulty since they need only a representative training set of examples, instead of the formulation of explicit rules.

11.5.6 The use of neural networks

Neural networks have been used in ARBS for two quite distinct tasks, described below.

Defect classification using a neural network

Neural networks can perform defect classification, provided there are sufficient training examples and the evidence can be presented in numerical form. In this study [17], there were insufficient data to train a neural network to perform a four-way classification of defect types, as done under the hypothesize-and-test method. Instead, a backpropagation network was trained to classify defects as either critical or noncritical on the basis of four local factors: the size of the area, the number of arcs passing through it, the shape of the echodynamic, and the sensitivity of the intensity to the angle of the probe. Each of these local factors was expressed as a number between −1 and 1.

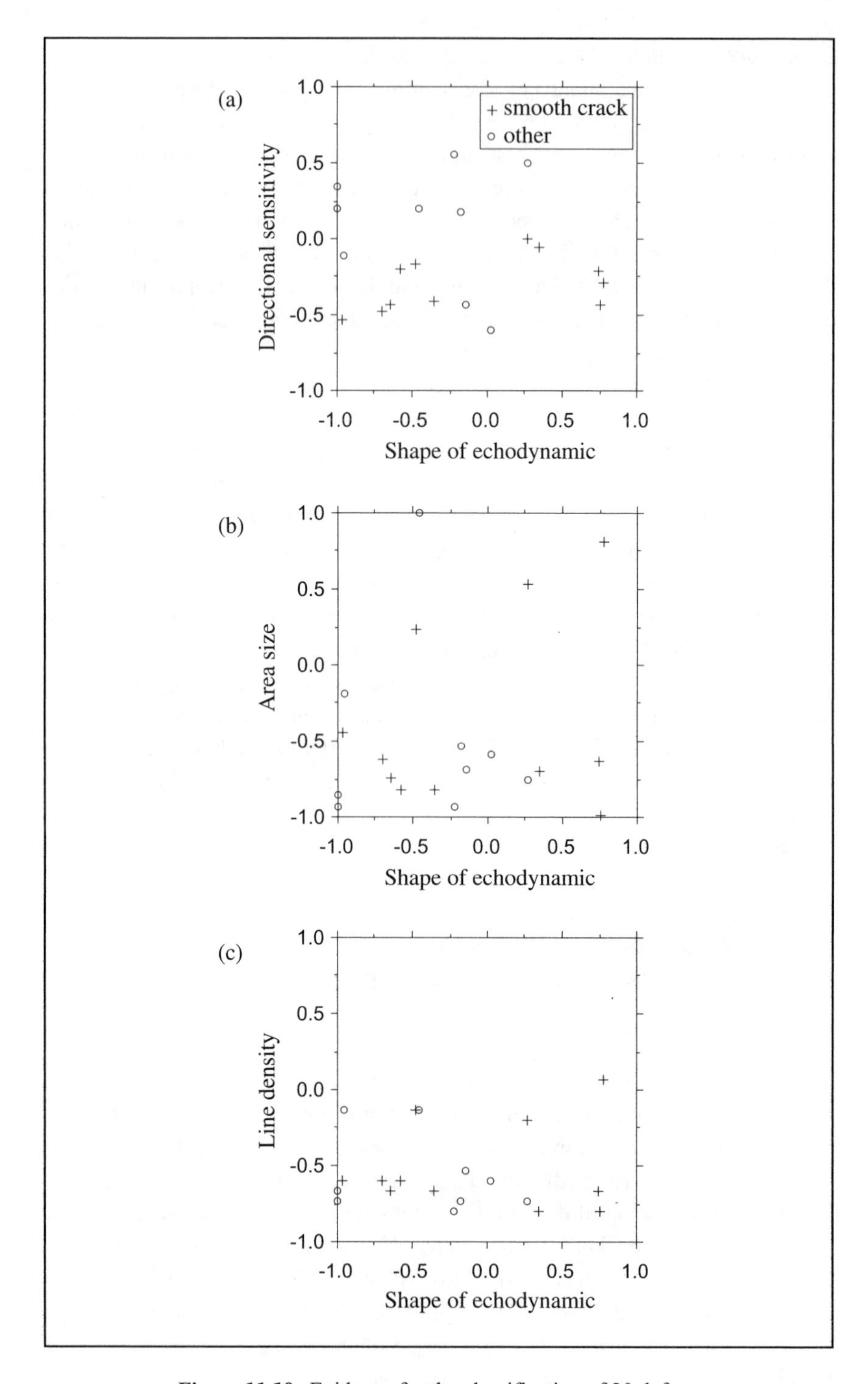

Figure 11.18 Evidence for the classification of 20 defects

Figure 11.18 shows plots of evidence for the 20 defects that were used to train and test a neural network. The data are, in fact, points in four-dimensional space, where each dimension represents one of the four factors considered. Notice that the clusters of critical and noncritical samples might not be linearly separable. This means that traditional numerical techniques for finding linear discriminators [37] are not powerful enough to produce good classification. However, a multilayer perceptron (MLP, see Chapter 8) is able to discriminate between the two cases, since it is able to find the three-dimensional surface required to separate them. Using the leave-one-out technique (Section 8.4.5), an MLP with two hidden layers correctly classified 16 out of 20 images from defective components [17].

Echodynamic classification using a neural network

One of the inputs to the classification network requires a value between –1 and 1 to represent the shape of the echodynamic. This value can be obtained by using a rule-based knowledge source that examines the Fourier components of the echodynamic and uses heuristics to provide a numerical value for the shape. An alternative approach is to use another neural network to generate this number.

An echodynamic is a signal intensity profile across a defect area and can be classified as a cusp, plateau, or wiggle. Ideally a neural network would make a three-way classification, given an input vector derived from the amplitude components of the first n Fourier coefficients, where $2n$ is the echodynamic sample rate. However, cusps and plateaux are difficult to distinguish since they have similar Fourier components, so a two-way classification is more practical, with cusps and plateaux grouped together. A multilayer perceptron has been used for this purpose.

Combining the two applications of neural networks

The use of two separate neural networks in distinct KSs for the classification of echodynamics and of the potential defect areas might seem unnecessary. Because the output of the former feeds, via the blackboard, into the input layer of the latter, this arrangement is equivalent to a hierarchical MLP (Section 8.4.4). In principle, the two networks could have been combined into one large neural network, thereby removing the need for a preclassified set of echodynamics for training. However, such an approach would lead to a loss of modularity and explanation facilities. Furthermore, it may be easier to train several small neural networks separately on subtasks of the whole classification problem than to attempt the whole problem at once with a single large network. These are important considerations when there are many subtasks amenable to connectionist treatment.

11.5.7 Rules for verifying neural networks

Defect classification, whether performed by the hypothesize-and-test method or by neural networks, has so far been discussed purely in terms of evidence gathered from the region of the image that is under scrutiny. However, there are other features in the image that can be brought to bear on the problem. Knowledge of these features can be expressed easily in rule form and can be used to *verify* the classification. The concept of the use of rules to verify the outputs from neural networks was introduced in Chapter 9. In this case, an easily identifiable feature of a b-scan image is the line of indications due to the back wall echo. A defect in the sample, particularly a smooth crack, will tend

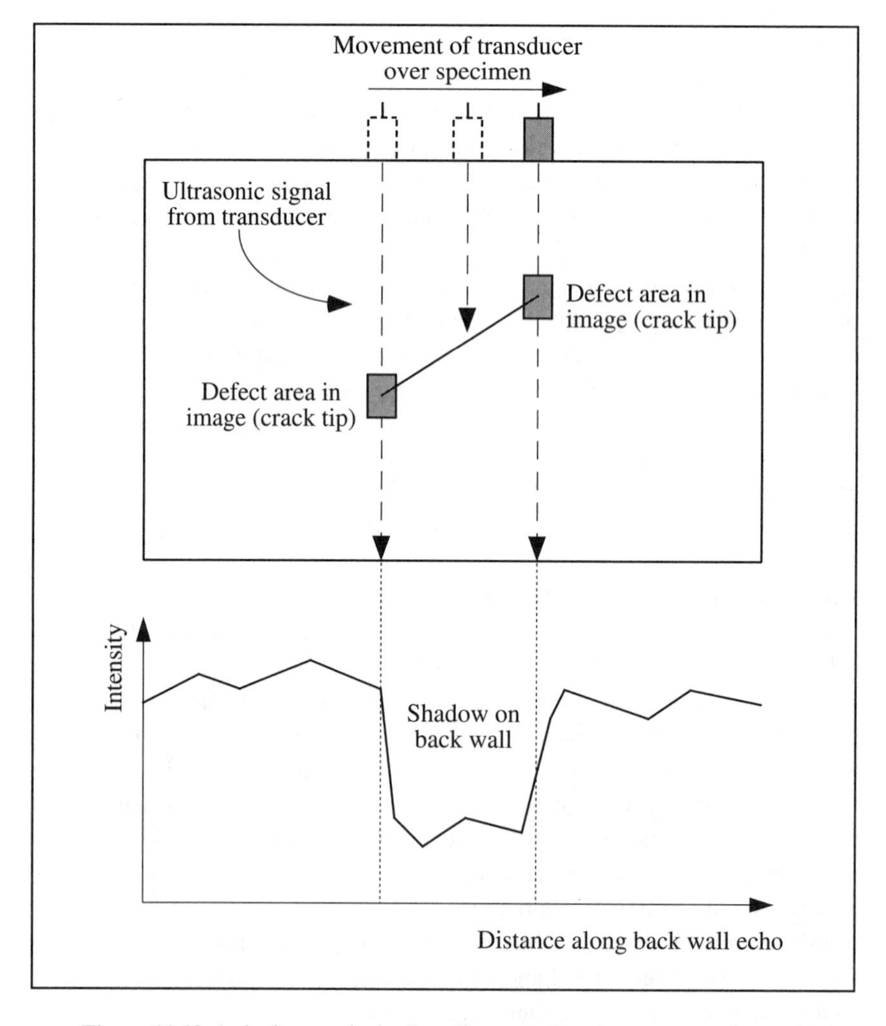

Figure 11.19 A shadow on the back wall can confirm the presence of a crack

to cast a "shadow" on the back wall directly beneath it (Figure 11.19). The presence of a shadow in the expected position can be used to verify the location and classification of a defect. In this application, the absence of this additional evidence is not considered a strong enough reason to reject a defect classification. Instead, the classification in such cases is marked for the attention of a human operator, as there are grounds for doubt over its accuracy.

11.6 Summary

This chapter has introduced some of the techniques that can be used to tackle the problems of automated interpretation and diagnosis. Diagnosis is considered to be a specialized case of the more general problem of interpretation. It has been shown that a key part of diagnosis and interpretation is abduction, the process of determining a cause, given a set of observations. There is uncertainty associated with abduction, since causes other than the selected one might give rise to the same set of observations. Two possible approaches to dealing with this uncertainty are to explicitly represent the uncertainty using the techniques described in Chapter 3, or to *hypothesize-and-test*. As the name implies, the latter technique involves generating a hypothesis (or best guess), and either confirming or refuting the hypothesis depending on whether it is found to be consistent with the observations.

Several different forms of knowledge can contribute to a solution. We have paid specific attention to rules, case histories, and physical models. We have also shown that neural networks and conventional "number-crunching" can play important roles when included as part of a blackboard system. Rules can be used to represent both shallow (heuristic) and deep knowledge. They can also be used for the generation and verification of hypotheses. Case-based reasoning, introduced in Chapter 6, involves comparison of a given scenario with previous examples and their solutions. Model-based reasoning relies on the existence of a model of the physical system which can be used for monitoring, generation of hypotheses, and verification of hypotheses by simulation.

Blackboard systems have been introduced as part of a case study into the interpretation of ultrasonic images. These systems allow various forms of knowledge representation to come together in one system. They are, therefore, well suited to problems that can be broken down into subtasks, where the most suitable form of knowledge representation for different subtasks is not necessarily the same. Each module of knowledge within a blackboard system is called a knowledge source (KS).

A neural network KS was shown to be effective for combining evidence generated by other KSs and for categorizing the shape of an echodynamic. This approach can be contrasted with the use of a neural network alone for interpreting images. The blackboard architecture avoids the need to abandon rules in favor of neural networks or vice versa, since the advantages of each can be incorporated into a single system. Rules can represent knowledge explicitly, whereas neural networks can be used where explicit knowledge is hard to obtain. Although neural networks can be rather impenetrable to the user and are unable to explain their reasoning, these deficiencies can be reduced by using them for small-scale localized tasks with reports generated in between.

References

1. Arroyo-Figueroa, G., Alvarez, Y., and Sucar, L. E., "SEDRET — an intelligent system for the diagnosis and prediction of events in power plants," *Expert Systems with Applications*, vol. 18, pp. 75–86, 2000.
2. Dague, P., Jehl, O., Deves, P., Luciani, P., and Taillibert, P., "When oscillators stop oscillating," International Joint Conference on Artificial Intelligence (IJCAI'91), Sydney, vol. 2, pp. 1109–1115, 1991.
3. Dash, E., "Diagnosing furnace problems with an expert system," *SPIE-Applications of Artificial Intelligence VIII*, vol. 1293, pp. 966–971, 1990.
4. Huang, J. K., Ho, M. T., and Ash, R. L., "Expert systems for automated maintenance of a Mars oxygen production system," *Journal of Spacecraft and Rockets*, vol. 29, pp. 425–431, 1992.
5. Bykat, A., "Nicbes-2, a nickel-cadmium battery expert system," *Applied Artificial Intelligence*, vol. 4, pp. 133–141, 1990.
6. Kang, C. W. and Golay, M. W., "A Bayesian belief network-based advisory system for operational availability focused diagnosis of complex nuclear power systems," *Expert Systems with Applications*, vol. 17, pp. 21–32, 1999.
7. Lu, Y., Chen, T. Q., and Hamilton, B., "A fuzzy diagnostic model and its application in automotive engineering diagnosis," *Applied Intelligence*, vol. 9, pp. 231–243, 1998.
8. Yamashita, Y., Komori, H., Aoki, E., and Hashimoto, K., "Computer aided monitoring of pump efficiency by using ART2 neural networks," *Kagaku Kogaku Ronbunshu*, vol. 26, pp. 457–461, 2000.
9. Huang, J. T. and Liao, Y. S., "A wire-EDM maintenance and fault-diagnosis expert system integrated with an artificial neural network,"

International Journal of Production Research, vol. 38, pp. 1071–1082, 2000.

10. Balakrishnan, A. and Semmelbauer, T., "Circuit diagnosis support system for electronics assembly operations," *Decision Support Systems*, vol. 25, pp. 251–269, 1999.
11. Netten, B. D., "Representation of failure context for diagnosis of technical applications," *Advances in Case-Based Reasoning*, vol. 1488, pp. 239–250, 1998.
12. Hunt, J., "Case-based diagnosis and repair of software faults," *Expert Systems*, vol. 14, pp. 15–23, 1997.
13. Prang, J., Huemmer, H. D., and Geisselhardt, W., "A system for simulation, monitoring and diagnosis of controlled continuous processes with slow dynamics," *Knowledge-Based Systems*, vol. 9, pp. 525–530, 1996.
14. Maderlechner, G., Egeli, E., and Klein, F., "Model guided interpretation based on structurally related image primitives," in *Knowledge-based expert systems in industry*, Kriz, J. (Ed.), pp. 91–97, Ellis Horwood, 1987.
15. Zhang, Z. and Simaan, M., "A rule-based interpretation system for segmentation of seismic images," *Pattern Recognition*, vol. 20, pp. 45–53, 1987.
16. Ampratwum, C. S., Picton, P. D., and Hopgood, A. A., "A rule-based system for optical emission spectral analysis," Symposium on Industrial Applications of Prolog (INAP'97), Kobe, Japan, pp. 99–102, 1997.
17. Hopgood, A. A., Woodcock, N., Hallam, N. J., and Picton, P. D., "Interpreting ultrasonic images using rules, algorithms and neural networks," *European Journal of Nondestructive Testing*, pp. 135–149, 1993.
18. Leitch, R., Kraft, R., and Luntz, R., "RESCU: a real-time knowledge based system for process control," *IEE Proceedings-D*, vol. 138, pp. 217–227, 1991.
19. Fink, P. K. and Lusth, J. C., "Expert systems and diagnostic expertise in the mechanical and electrical domains," *IEEE Transactions on Systems, Man, and Cybernetics*, vol. 17, pp. 340–349, 1987.
20. Fulton, S. L. and Pepe, C. O., "An introduction to model-based reasoning," *AI Expert*, pp. 48–55, January 1990.
21. Scarl, E. A., Jamieson, J. R., and Delaune, C. I., "Diagnosis and sensor validation through knowledge of structure and function," *IEEE*

Transactions on Systems, Man, and Cybernetics, vol. 17, pp. 360–68, 1987.

22. Motta, E., Eisenstadt, M., Pitman, K., and West, M., "Support for knowledge acquisition in the Knowledge Engineer's Assistant (KEATS)," *Expert Systems*, vol. 5, pp. 6–28, 1988.
23. Dague, P., Deves, P., Zein, Z., and Adam, J. P., "DEDALE: an expert system in VM/Prolog," in *Knowledge-based expert systems in industry*, Kriz, J. (Ed.), pp. 57–68, Ellis Horwood, 1987.
24. Price, C. J., "Function-directed electrical design analysis," *Artificial Intelligence in Engineering*, vol. 12, pp. 445–456, 1998.
25. Cunningham, P., "A case study on the use of model-based systems for electronic fault diagnosis," *Artificial Intelligence in Engineering*, vol. 12, pp. 283–295, 1998.
26. Dekleer, J. and Williams, B. C., "Diagnosing multiple faults," *Artificial Intelligence*, vol. 32, pp. 97–130, 1987.
27. Harel, D., "On visual formalisms," *Communications of the ACM*, vol. 31, pp. 514–530, 1988.
28. Price, C. J. and Hunt, J., "Simulating mechanical devices," in *Pop-11 Comes of Age: the advancement of an AI programming language*, Anderson, J. A. D. W. (Ed.), pp. 217–237, Ellis Horwood, 1989.
29. Jennings, A. J., "Artificial intelligence: a tool for productivity," Institution of Engineers (Australia) National Conference, Perth, Australia, 1989.
30. Milne, R., "Strategies for diagnosis," *IEEE Transactions on Systems, Man, and Cybernetics*, vol. 17, pp. 333–339, 1987.
31. Steels, L., "Diagnosis with a function-fault model," *Applied Artificial Intelligence*, vol. 3, pp. 129–153, 1989.
32. Walker, N. and Fox, J., "Knowledge-based interpretation of images: a biomedical perspective," *Knowledge Engineering Review*, vol. 2, pp. 249–264, 1987.
33. Hopgood, A. A., "Rule-based control of a telecommunications network using the blackboard model," *Artificial Intelligence in Engineering*, vol. 9, pp. 29–38, 1994.
34. Hopgood, A. A., Phillips, H. J., Picton, P. D., and Braithwaite, N. S. J., "Fuzzy logic in a blackboard system for controlling plasma deposition processes," *Artificial Intelligence in Engineering*, vol. 12, pp. 253–260, 1998.

35. Duda, R. O. and Hart, P. E., "Use of the Hough transform to detect lines and curves in pictures," *Communications of the ACM*, vol. 15, pp. 11–15, 1972.
36. Halmshaw, R., *Non-Destructive Testing*, Edward Arnold, 1987.
37. Duda, R. O. and Hart, P. E., *Pattern Classification and Scene Analysis*, Wiley, 1973.

Further reading

- Hamscher, W., Luca Console, L., and De Kleer, J. (Eds.), *Readings in Model-Based Diagnosis*, Morgan Kaufmann, 1992.
- Price, C. J., *Computer-Based Diagnostic Systems*, Springer Verlag, 2000.

Chapter twelve

Systems for design and selection

12.1 The design process

Before discussing how knowledge-based systems can be applied to design, it is important to understand what we mean by the word *design*. Traditionally, design has been broken down into engineering design and industrial design:

> *Engineering design is the use of scientific principles, technical information, and imagination in the definition of a mechanical structure, machine, or system to perform specified functions with the maximum economy and efficiency.* [1]

> *Industrial design seeks to rectify the omissions of engineering, a conscious attempt to bring form and visual order to engineering hardware where technology does not of itself provide these features.* [1]

We will take a more catholic view of design, in which no distinction is drawn between the technical needs of engineering design and the aesthetic approach of industrial design. Our working definition of design will be the one used by Sriram et al.:

> *[Design is] the process of specifying a description of an artifact that satisfies constraints arising from a number of sources by using diverse sources of knowledge.* [2]

Some of the constraints must be predetermined, and these constitute the *product design specification* (PDS). Other constraints may evolve as a result of decisions made during the design process. The PDS is an expression of the *requirements* of a product, rather than a specification of the product itself. The latter, which emerges during the design process, is the design. The design can

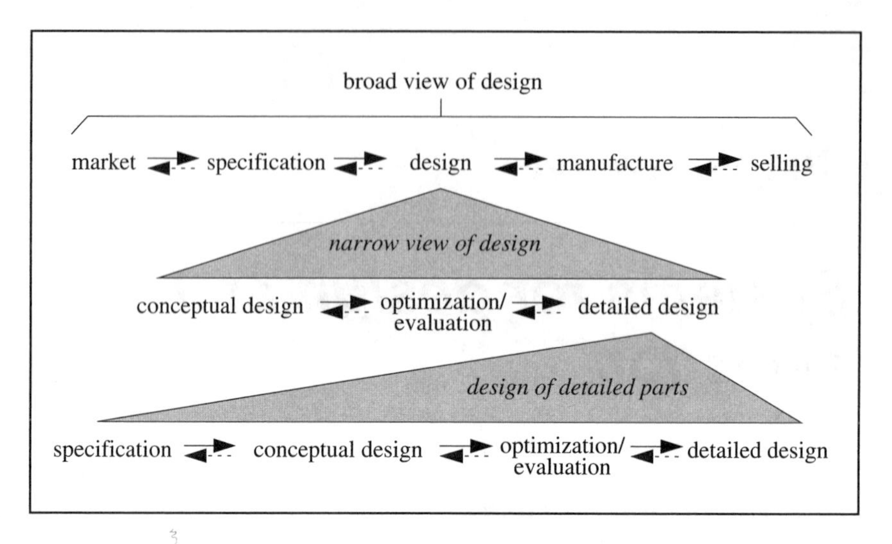

Figure 12.1 The principal phases of design

be interpreted for manufacture or construction, and it allows predictions about the performance of the product to be drawn.

Different authors have chosen to analyze the design process in different ways. An approximate consensus is that the broadest view of the design process comprises the following phases (Figure 12.1):

market — specification — design (narrow view) — manufacture — selling

The narrow view of design leads from a product design specification to the manufacturing stage. It can be subdivided as follows:

conceptual design — optimization/evaluation — detailed design

The purpose of each phase in the broad design process is as follows:

(i) *Market*: This phase is concerned with determining the need for a product. A problem is identified, resources allocated, and end-users targeted.
(ii) *Specification*: A product design specification (PDS) is drawn up that describes the requirements and performance specifications of the product. The PDS for a motorcar might specify a product that can transport up to four people in comfort, traveling on roads at speeds up to the legal limit.
(iii) *Conceptual design*: Preliminary design decisions are made at this stage, with the aim of satisfying a few key constraints. Several alternatives would normally be considered. Decisions taken at the conceptual design

stage determine the general form of the product, and so have enormous implications for the remainder of the design process. The conceptual design for a motorcar has altered little since the Model T Ford was unveiled in 1908. It describes a steel body with doors and windows, a wheel at each corner, two seats at the front (one of which has access to the controls), two seats at the back, and so on.

(iv) *Optimization/evaluation*: The conceptual design is refined, for instance by placing values on numerical attributes such as length and thickness. The performance of the conceptual design is tested for its response to external effects and its consistency with the product design specification. The optimization and evaluation stage for a motorcar might include an assessment of the relationship between the shape of the body and its drag coefficient. If the conceptual design cannot be made to meet the requirements, a new one is needed.

(v) *Detailed design*: The design of the product and its components are refined so that all constraints are satisfied. Decisions taken at this stage might include the layout of a car's transmission system, the position of the ashtray, the covering for the seats, and the total design of a door latch. The latter example illustrates that the complete design process for a component may be embedded in the detailed design phase of the whole assembly (Figure 12.1).

(vi) *Manufacture*: A product should not be designed without consideration of how it is to be manufactured, as it is all too easy to design a product that is uneconomical or impossible to produce. For a product that is to be mass-produced, the manufacturing plant needs to be designed just as rigorously as the product itself. Different constraints apply to a one-off product, as this can be individually crafted but mass-production techniques such as injection molding are not feasible.

(vii) *Selling*: The chief constraint for most products is that they should be sold at a profit. The broad view of design, therefore, takes into account not only how a product can be made, but also how it is to be sold.

Although the design process has been portrayed as a chronological series of events, in fact, there is considerable interaction between the phases — both forwards and backwards — as constraints become modified by the design decisions that are made. For instance, a decision to manufacture one component from polyethylene rather than steel has ramifications for the design of other components and implications for the manufacturing process. It may also alter the PDS, as the polymer component may offer a product that is cheaper but less structurally rigid. Similarly, sales of a product can affect the market, thus linking the last design phase with the first.

In our description of conceptual and detailed design, we have made reference to the choice of materials from which to manufacture the product. Materials selection is one of the key aspects of the design process, and one where considerable effort has been placed in the application of intelligent systems. The process of materials selection is discussed in detail in Section 12.8. Selection is also the key to other aspects of the design process, as attempts are made to select the most appropriate solution to the problem.

The description of the design process that has been proposed is largely independent of the nature of the product. The product may be a single component (such as a turbine blade) or a complex assembly (such as a jet engine); it may be a one-off product or one that will be produced in large numbers. Many designs do not involve manufacture at all in the conventional sense. An example that is introduced in Section 12.4 is the design of a communications network. This is a high-level design, which is not concerned with the layout of wires or optical fibers, but rather with the overall configuration of the network. The product is a *service* rather than a physical thing. Although selection is again one of the key tasks, materials selection is not applicable in this case.

In summary, we can categorize products according to whether they are:

- service-based or physical products;
- single component products or assemblies of many components;
- one-off products or mass-produced products.

Products in each category will have different requirements, leading to a different PDS. However, these differences do not necessarily alter the design process.

Three case studies are introduced in this chapter. The specification of a communications network is used to illustrate the importance and potential complexity of the product design specification. The processes of conceptual design, optimization and evaluation, and detailed design are illustrated with reference to the floor of a passenger aircraft. This case study will introduce some aspects of the materials selection problem, and these are further illustrated by the third case study, which concerns the design of a kettle.

12.2 Design as a search problem

Design can be viewed as a search problem, as it involves searching for an optimum or adequate design solution. Alternative solutions may be known in advance (these are *derivation* problems), or they may be generated

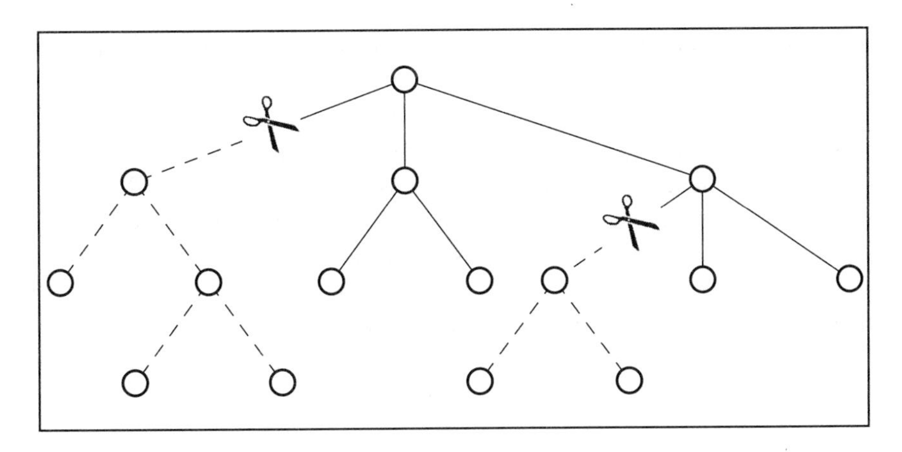

Figure 12.2 Pruning the search tree by eliminating classes of design that are unfeasible

automatically (these are *formulation* problems). Designs may be tested as they are found in order to check whether they are feasible and meet the design requirements. This is the *generate and test* method. In application areas such as diagnosis (see Chapter 11) it may be sufficient to terminate the search as soon as a solution is found. In design there are likely to be many solutions, and we would like to find "the best." The search may, therefore, continue in order to find many feasible designs from which a selection can be made.

Search becomes impractical when large numbers of unreasonable designs are included. Consider, for example, the design of a house. In order to generate solutions automatically, we might write a computer program that generates every conceivable combination of shapes and sizes of rooms, walls, roofs, and foundations. Of this massive number of alternatives, only a small proportion would be feasible designs. In order to make the search problem manageable, some means of eliminating the unfeasible designs is needed. Better still would be a means of eliminating whole families of ill-conceived designs before the individual variants have been produced. The design-generator could be modified by heuristics so that it produced only designs with the roof above the walls and with the walls above the foundations. This would have the effect of pruning the search space (Figure 12.2). The search space can also be reduced by decomposing the design problem into subproblems of designing the rooms, roof, and foundations separately, each with its own smaller search tree.

The search problem is similar to the proposition that a monkey playing random notes on a grand piano will eventually play a Beethoven symphony. The fault in this proposition is that the search space of compositions is so immense that the monkey would not stumble across the symphony within a practical time-frame. Only a composer with knowledge of suitable musical arrangements could hope to generate the symphony, as he or she is able to prune the search space of compositions.

Even if we succeed in pruning the search space so that only feasible designs are considered, we will still be left with the problem of selecting between alternatives. The selection problem is discussed in Section 12.8 below with particular reference to materials selection for design. The same techniques can be applied to selection between design alternatives.

Although heuristic rules can limit the search space, they do not offer unique solutions. This is because abductive rather than deductive rules are required (Chapter 1), as with diagnosis (Chapter 11). Consider this simple deductive rule:

```
/* Rule 12.1 */
IF ?x is a room with a bath and a toilet THEN ?x is a bathroom
```

If a room fits the description provided by the condition part of the rule, we could use the rule to classify that room as a bathroom. The abductive interpretation of this rule is:

```
/* Rule 12.2 */
IF a room is to be a bathroom
THEN it must have a bath and a toilet
```

The abductive rule poses two problems. First, we have made the closed-world assumption (see Chapters 1 and 2), and so the rule will never produce bathroom designs that have a shower and toilet but no bath. Second, the rule leads only to a partial design. It tells us that the bathroom will have a toilet and bath, but fails to tell us where these items should be placed or whether we need to add other items such as a basin.

12.3 Computer aided design

The expression *computer aided design*, or CAD, is used to describe computing tools that can assist in the design process. Most early CAD systems were intended primarily to assist in drawing a design, rather than directly supporting the decision-making process. CAD systems of this type carry out computer aided *drafting* rather than computer aided *design*. Typically, these drafting systems allow the designer to draw on a computer screen using a mouse, graphics tablet, or similar device. All dimensions are automatically calculated, and the design can be easily reshaped, resized, or otherwise modified. Such systems have had an enormous impact on the design process since they remove much of the tedium and facilitate alterations. Furthermore, CAD has altered the designers' working environment, as the traditional large flat drafting boards have been replaced by computer workstations.

Early CAD systems of this type do not make decisions and have little built-in intelligence. They do, however, frequently make use of object-oriented programming techniques. Each line, box, circle, etc., that is created can be represented as an object instance. Rather than describing such systems in more detail, this chapter will concentrate on the use of intelligent systems that can help designers make design decisions.

12.4 The product design specification (PDS): a telecommunications case study

12.4.1 Background

The product design specification (PDS) is a statement of the requirements of the product. In this section we will consider a case study concerning the problems of creating a PDS that can be accessed by a knowledge-based design system. In this case study, the "product" is not a material product but a service, namely, the provision of a communications network. The model used to represent the PDS is called the *common traffic model* (CTM) [3], because it is common to a variety of forms of communication traffic (e.g., analog voice, packetized data, or synchronous data).

The common traffic model allows different views of a communications network to be represented simultaneously. The simplest view is a set of requirements defined in terms of links between sites and the applications (e.g., fax or database access) to be used on these links. The more specialized views contain implementation details, including the associated costs. The model allows nontechnical users to specify a set of communications requirements, from which a knowledge-based system can design and cost a network, thereby creating a specialized view from a nonspecialized one. The model consists of object class definitions, and a PDS is represented as a set of instances of these classes.

12.4.2 Alternative views of a network

Suppose that a small retailer has a central headquarters, a warehouse, and a retail store. The retailer may require various communications applications, including customer order by fax, customer order by phone, and stock reorder (where replacement stock is ordered from suppliers). The retailer views the network in terms of the sites and the telecommunications applications that are carried between them. This is the simplest viewpoint, which defines the PDS. From a more technical viewpoint, the network can be broken down into voice and data components. For the voice section, each site has a fixed number of lines connecting it to the network via a private switching system, while the data section connects the head office to the other sites. The most detailed view of the network (the service-provider's viewpoint) includes a definition of the equipment and services used to implement the network. The detailed description is based on one of several possible implementations, while the less specialized views are valid regardless of the implementation.

There are several possible views of the network, all of which are valid and can be represented by the common traffic model. It is the translation from the customer's view (defined in terms of the applications being used) to the service-provider's view (defined in terms of the equipment and services supplied) that determines the cost and efficiency of the communications network. This translation is the design task.

12.4.3 Implementation

The requirements of the network are represented as a set of object instances. For example, if the customer of the telecommunications company has an office in New York, that office is represented as an object with a name and position, and is an instance of the object class `Customer_site`.

The common traffic model was originally designed using Coad and Yourdon's object-oriented analysis (OOA) [4], but is redrawn in Figure 12.3 using the Unified Modeling Language (UML) introduced in Chapter 4. The model is implemented as a set of object classes that act as templates for the object instances that are created when the system is used to represent a PDS. Various interclass relationships are employed. For example, a `Dispersion_link` is represented as a specialization of a `Link`. Similarly, an aggregation relationship is used to show that a `Network` comprises several instances of `Link`. Associations are used to represent physical connections, such as the connection between a `Link` and the instances of `Site` at its two ends.

The fact that instance connections are defined at the class level can be confusing. The common traffic model is defined entirely in terms of object classes, these being the templates for the instances that represent the user's communication needs. Although the common traffic model is only defined in terms of classes, it specifies the relationships that exist between instances when they are created.

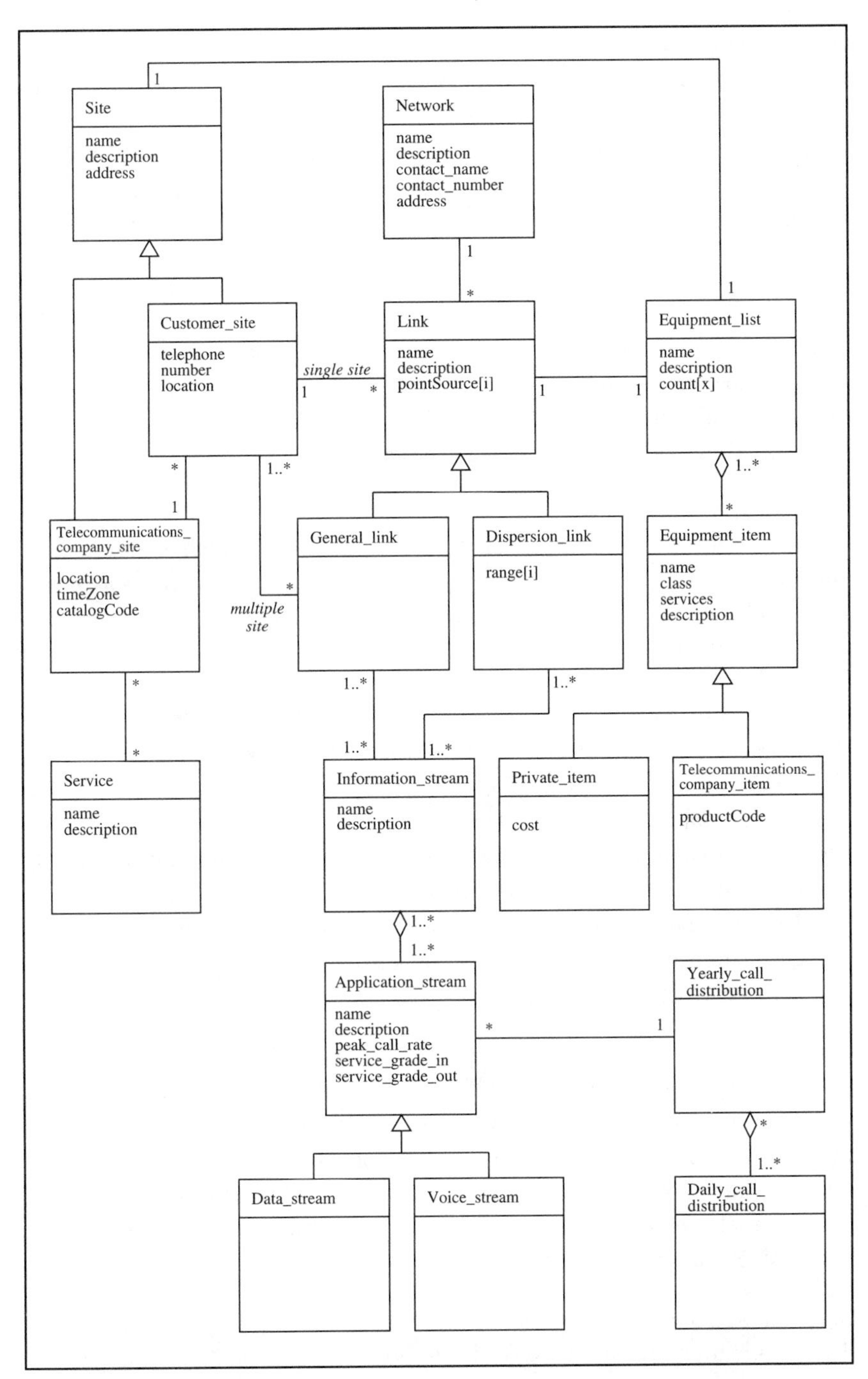

Figure 12.3 The main object classes and attributes in the common traffic model (adapted from [3])

12.4.4 The classes

The classes that make up the common traffic model and the relationships between them are shown in Figure 12.3. A detailed understanding of Figure 12.3 is not necessary for this case study. Instead it is hoped that the figure conveys the general idea of using object-oriented analysis for generating both the PDS and a detailed network description. The main classes of the common traffic model are briefly described below.

Network

The `Network` object contains the general information relating to the network, but is independent of the network requirements. It includes information such as contact people and their addresses. The specification of the network is constructed from a set of `Link` objects (described below).

Link

A `Link` identifies the path between customer sites along which an information stream (described below) occurs. Instance connections are used to associate links with appropriate customer sites, information streams, and equipment. Conceptually three types of link are defined:

- *Multipoint links*, where information is exchanged between a single nominated site and a number of other sites. The links are instances of the class `General_link`, where an attribute (`pointsource`) indicates whether calls are initiated by the single site or by the multiple sites.
- *Point-to-point links*, which are treated as multipoint links, but where only one of the multiple sites is specified.
- *Dispersion links*, which carry application traffic that does not have a fixed destination site. This type of link applies to customers who want access to a public switched network.

Site

Two classifications of sites are defined, namely, the customer's sites and the telecommunications company's sites. The latter specify sites that are part of the supplier's network, such as telephone exchanges. For most telecommunications services, the design and costing of a network is dependent on its spatial layout. For this reason, the common traffic model has access to a geographical database.

Information stream

The information streams specify the traffic on a link in terms of a set of application streams. Two subclasses of `Application_stream` are defined, `Data_stream` and `Voice_stream`. The first specifies digital applications, while the second specifies analog applications. Each application stream has a peak call rate and associated yearly and daily traffic profiles. Application streams can be broken down further into individual calls.

Equipment

An equipment list specifies a set of items that are present at a site or on a link. Two subclasses of equipment item are defined: those that are owned by the telecommunications company and those that are privately owned.

12.4.5 Summary of PDS case study

The common traffic model illustrates a formalized approach to creating a product design specification, showing that the PDS and its implementation need to be carefully thought out before a knowledge-based design system can be employed. The common traffic model has proved an effective tool for representing a set of communication requirements in a way that satisfies more than one viewpoint. Nontechnical users can specify the PDS in terms of the types of use that they have in mind for the network. The common traffic model can also be used to represent the detailed network design, which may be one of many that are technically possible.

12.5 Conceptual design

It has already been noted in Section 12.1 that conceptual design is the stage where broad decisions about the overall form of a product are made. A distinction can be drawn between cases where the designer is free to innovate and more routine cases where the designer is working within tightly bound constraints. An example of the former case would be the design of a can opener. Many designs have appeared in the past and the designer may call upon his or her experience of these. However, he or she is not bound by those earlier design decisions. In contrast, a designer might be tasked with arranging the layout of an electronic circuit on a VLSI (very large scale integration) chip. While this is undoubtedly a complex task, the conceptual design has already been carried out, and the designer's task is one that can be treated as a problem of mathematical optimization. We will call this *routine design*.

Brown and Chandrasekaran [5] subdivide the innovative design category between *inventions* (such as the first helicopter) and more modest *innovations* (such as the first twin-rotor helicopter). Both are characterized by the lack of any prescribed strategy for design, and rely on a spark of inspiration. The invention category makes use of new knowledge, whereas the innovation category involves the reworking of existing knowledge or existing designs. The three categories of design can be summarized as follows:

- invention;
- innovative use of existing knowledge or designs;
- routine design.

Researchers have different opinions of how designers work, and it is not surprising that markedly different software architectures have been produced. For instance, Sriram et al. [2] claim to have based their CYCLOPS system on the following set of observations about innovative design:

(i) designers use multiple objectives and constraints to guide their decisions, but are not necessarily bound by them;

(ii) as new design criteria emerge they are fed back into the PDS;

(iii) designers try to find an optimum solution rather than settling on a satisfactory one;

(iv) extensive use is made of past examples.

Demaid and Zucker [6] have no quarrel with observations (i), (ii), and (iv). However, in contrast to observation (iii), they emphasize the importance of choosing *adequate* materials for a product rather than trying to find an *optimum* choice.

The CYCLOPS [2] and FORLOG [7] systems assume that innovative design can be obtained by generating a variety of alternatives and choosing between them. CYCLOPS makes use of previous design histories and attempts to adapt them to new domains. The success of this approach depends upon the ability to find diverse novel alternatives. In order to increase the number of past designs that might be considered, the design constraints are relaxed. Relaxation of constraints is discussed in Section 12.8.5 as part of an overall discussion of techniques for selecting between alternatives. CYCLOPS also has provision for modification of the constraints in the light of past experience.

As well as selecting a preexisting design for use in a novel way, CYCLOPS allows adaptation of the design to the new circumstances. This is achieved through having a stored explanation of the precedent designs. The example cited by Sriram et al. [2] relates to houses in Thailand. Thai villagers put their houses on stilts to avoid flooding, and this forms a precedent design. The underlying explanation for the design, which is stored with it, is that stilts raise the structure. The association with flooding may not be stored at all, as this is not fundamental to the role of the stilts. CYCLOPS might then use this precedent to raise one end of a house that is being designed for construction on a slope.

Most work in knowledge-based systems for design relies on the application of a predetermined strategy. Dyer et al. [8] see this as a limitation on innovation and have incorporated the idea of *brainstorming* into EDISON, a system for designing simple mechanical devices. Some of the key features of EDISON are:

- brainstorming by $\left\{\begin{array}{l}\text{mutation}\\ \text{generalization}\\ \text{analogy}\end{array}\right.$
- problem-solving heuristics;
- class hierarchies of mechanical parts;
- heuristics describing relationships between mechanical parts.

EDISON makes use of metarules (see Chapter 2) to steer the design process between the various strategies that are provided. Brainstorming and problem solving often work in tandem, as brainstorming tends to generate new problems. Brainstorming involves retrieving a previous design from memory and applying *mutation*, *generalization*, and *analogical reasoning* until a new functioning device is "invented." *Mutation* is achieved through a set of heuristics describing general modifications that can be applied to a variety of products. For example, slicing a door creates two slabs, each covering half a door frame. This operation results in a problem: the second slab is not connected to the frame. Two typical problem-solving heuristics might be:

```
Hinged joints allow rotation about pin
Hinged joints prohibit motion in any other planes
```

These rules provide information about the properties of hinges. Application of similar *problem-solving* rules might result in the free slab's being connected either to the hinged slab or to the opposite side of the frame. In one case we have invented the swinging saloon door; in the other case the accordion door (Figure 12.4).

Generalization is the process of forming a generic description from a specific item. For instance, a door might be considered a subclass of the general class of entrances (Figure 12.5). Analogies can then be drawn (*analogical reasoning*) with another class of entrance, namely, a cat flap, leading to the invention of a door that hangs from hinges mounted at the top. Generalization achieves the same goal as the deep explanations used in the adaptation mode of CYCLOPS, described above.

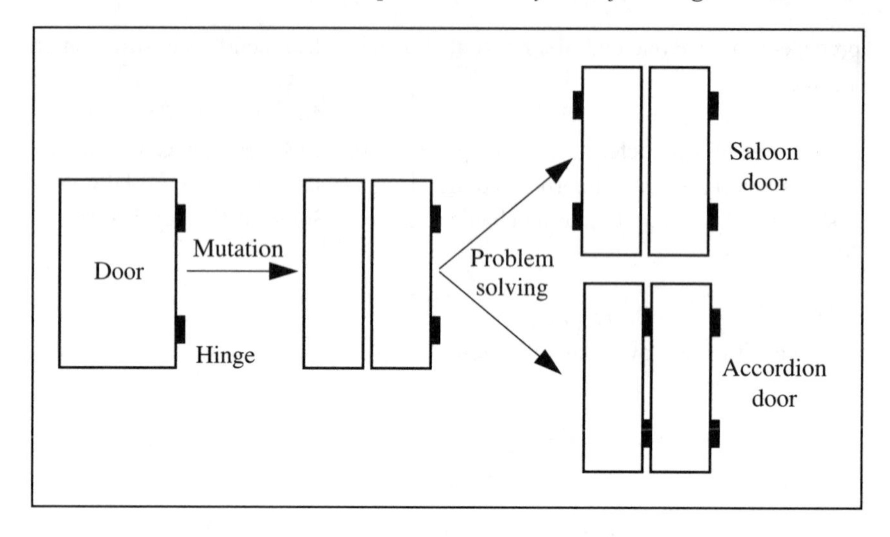

Figure 12.4 Inventing new types of doors by mutation and problem solving (after Dyer et al. [8])

Murthy and Addanki [9] have built a system called PROMPT in which innovative structural designs are generated by reasoning from first principles, i.e., using the fundamental laws of physics. Fundamental laws can lead to unconventional designs that heuristics based on conventional wisdom might have failed to generate. Other authors [10, 11] have proposed a *systematic* approach to innovation which generates only feasible solutions, rather than large numbers of solutions from which the feasible ones must be extracted. In this approach, the goals are first determined and then the steps needed to satisfy these goals are found. These steps have their own subgoals, and so the processes proceeds recursively.

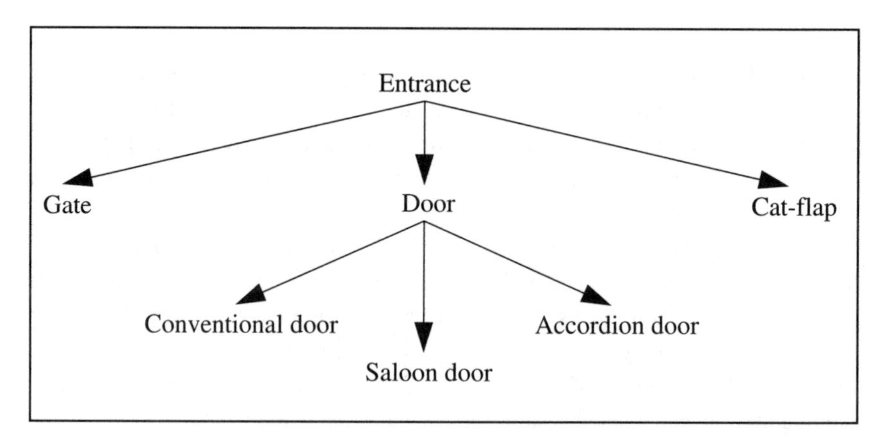

Figure 12.5 Hierarchical classification of types of entrances

12.6 Constraint propagation and truth maintenance

The terms *constraint propagation* and *truth maintenance* are commonly used in the field of artificial intelligence to convey two separate but related ideas. They have particular relevance to design, as will be illustrated by means of some simple examples. Constraints are limitations or requirements that must be met when producing a solution to a problem (such as finding a viable design). Imagine that we are designing a product, and that we have already made some conceptual design decisions. Propagation of constraints refers to the problem of ensuring that new constraints arising from the decisions made so far are taken into account in any subsequent decisions. For instance, a decision to manufacture a car from steel rather than (say) fiberglass introduces a constraint on the design of the suspension, namely, that it must be capable of supporting the mass of the steel body.

Suppose that we wish to investigate two candidate solutions to a problem, such as a steel-bodied car and a fiberglass car. Truth maintenance refers to the problem of ensuring that more detailed investigations, carried out subsequently, are associated with the correct premise. For example, steps must be taken to ensure that a lightweight suspension design is associated only with the lightweight (fiberglass) design of car to which it is suited.

In order to illustrate these ideas in more detail, we have adapted the example provided by Dietterich and Ullman [7]. The problem is to place two batteries into a battery holder. There are four possible ways in which the batteries can be inserted, as shown in Figure 12.6. This situation is described by the following Prolog clauses (Chapter 10 includes an overview of the syntax and workings of Prolog):

```
terminal(X):- X=positive;X=negative.
    % battery terminal may be positive or negative

layout(T,B):- terminal(T),terminal(B).
    % layout defined by identifying top and bottom terminals
```

We can now query our Prolog system so that it will return all valid arrangements of the batteries:

```
?- layout(Top,Bottom).
Top = Bottom = positive;
Top = positive,  Bottom = negative;
Top = negative,  Bottom = positive;
Top = Bottom = negative
```

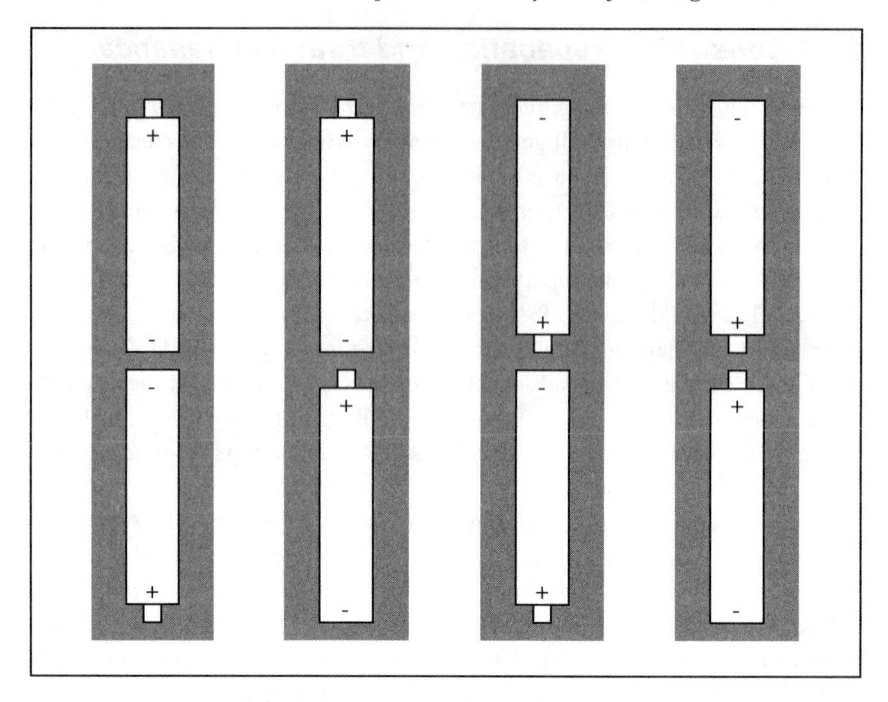

Figure 12.6 Four possible ways of inserting batteries into a holder

Now let us introduce the constraint that the batteries must be arranged in series. This is achieved by adding a clause to specify that terminals at the top and bottom of the battery holder must be of opposite sign:

```
terminal(X):- X=positive;X=negative.

layout(T,B):- terminal(T),terminal(B),
not(T=B).              % top terminal not equal to bottom terminal
```

We can now query our Prolog system again:

```
?- layout(Top,Bottom).
Top = positive,  Bottom = negative;
Top = negative,  Bottom = positive;
no
```

We will now introduce another constraint, namely, that a positive terminal must appear at the top of the battery holder:

```
terminal(X):- X=positive;X=negative.
```

```
layout(T,B):- terminal(T),terminal(B),
not(T=B),
T=positive.% positive terminal at top of holder
```

There is now only one arrangement of the batteries that meets the constraints:

```
?- layout(Top,Bottom).
Top = positive,  Bottom = negative;
no
```

This is an example of constraint propagation, because it shows how a constraint affecting one part of the design (i.e., the orientation of the battery at the top of the holder) is propagated to determine some other part of the design (i.e., the orientation of the other battery). In this particular example, constraint propagation has been handled by the standard facilities of the Prolog language. Many researchers, including Dietterich and Ullman [7], have found the need to devise their own means of constraint propagation in large design systems.

Truth maintenance becomes an important problem if we wish to consider more than one solution to a problem at a time, or to make use of nonmonotonic reasoning (see Chapter 11). For instance, we might wish to develop several alternative designs, or to assume that a particular design is feasible until it is shown to be otherwise. In order to illustrate the concept of truth maintenance, we will stay with our example of arranging batteries in a holder. However, we will veer away from a Prolog representation of the problem, as standard Prolog can consider only one solution to a problem at a time.

Let us return to the case where we had specified that the two batteries must be in series, but we had not specified an orientation for either. There were, therefore, two possible arrangements:

```
[Top = positive,  Bottom = neg]
```

or:

```
[Top = negative,  Bottom = positive]
```

It is not sufficient to simply store these four assertions together in memory:

```
Top = positive
Bottom = negative
Top = negative
Bottom = positive
```

For these statements to exist concurrently, it would be concluded that the two terminals of a battery are identical (i.e., negative = positive). This is clearly not the intended meaning. A frequently used solution to this difficulty is to label each fact, rule, or assertion, such that those bearing the same label are recognized as interdependent and, therefore, "belonging together." This is the basis of deKleer's assumption-based truth maintenance system (ATMS) [12, 13, 14]. If we choose to label our two solutions as `design1` and `design2`, then our four assertions might be stored as:

```
Top = positive          {design1}
Bottom = negative       {design1}
Top = negative          {design2}
Bottom = positive       {design2}
```

Let us now make explicit the rule that the two terminals of a battery are different:

```
not (positive = negative)    {global}
```

The English translation for these labels would be "if you believe the `global` assumptions, then you must believe `not(positive = negative)`." Similarly for `design1`, "if you believe `design1`, then you must also believe `Top = negative` and `Bottom = positive`." Any deductions made by the inference engine should be appropriately labeled. For instance the deduction:

```
negative = positive            {design1, design2}
```

is compatible with the sets of beliefs defined by `design1` and `design2`. However, this deduction is incompatible with our global rule, and so a warning of the form:

```
INCOMPATIBLE                   {design1, design2, global}
```

should be produced. This tells us that we cannot believe `design1`, `design2`, and `global` simultaneously. It is, however, all right to believe (`design1` and `global`) or (`design2` and `global`). This is the behavior we want, as there are two separate designs, and the inference engine has simply discovered that the two designs cannot be combined together.

12.7 Case study: the design of a lightweight beam

12.7.1 Conceptual design

To illustrate some of the ideas behind the application of knowledge-based systems to conceptual design, we will consider the design of a lightweight beam. The beam is intended to support a passenger seat in a commercial aircraft. The whole aircraft will have been designed, and we are concerned with the design of one component of the whole assembly. The total design process for the beam is part of the detailed design process for the aircraft. The intended loading of the beam tends to cause it to bend, as shown in Figure 12.7. The objectives are for the beam to be:

- stiff enough that the deflection (*D*) is kept small;
- strong enough to support the load without fracture;
- as light as possible, so as to maximize the ratio of cargo weight to fuel consumption.

Together, these three objectives form the basis of the product design specification (PDS). The PDS can be made more specific by placing limits on the acceptable deflection (*D*) under the maximum design load (*F*). A limit could also be placed on the mass of the beam. However, a suitable mass limit is difficult to judge, as it presupposes the form of the beam (i.e., its conceptual design) and the materials used. For this reason, we will simply state that the beam is required to be as light as possible within the constraints of fulfilling the other two requirements. In practice, a number of additional constraints will apply, such as materials costs, manufacturing costs, and flammability.

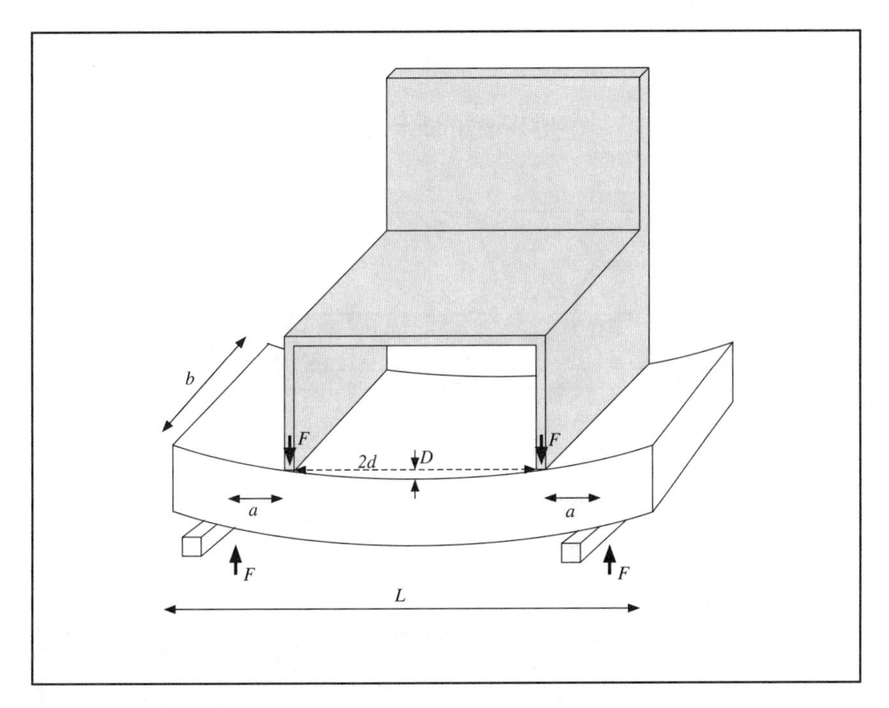

Figure 12.7 Four-point loading of a beam supporting a chair

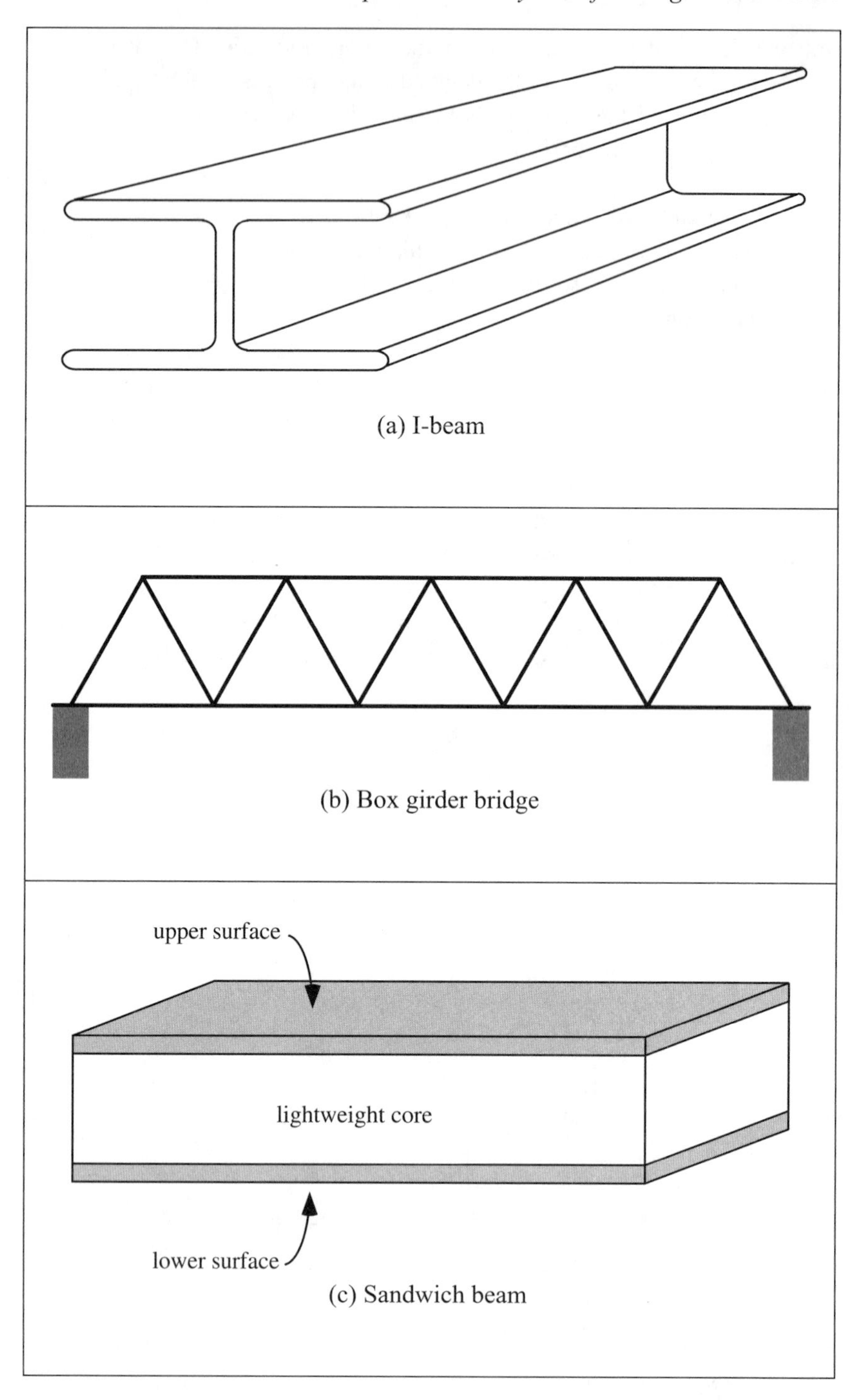

Figure 12.8 Some alternative conceptual designs for load-bearing beams

Kim and Suh [15] propose that the design process in general can be based upon two axioms, which can be implemented as metarules (see Chapter 2):

```
axiom 1: maintain independence of functional requirements
axiom 2: minimize information content
```

Our statement of the PDS fulfills these two axioms, because we have identified three concise and independent requirements.

Many knowledge-based systems for conceptual design attempt to make use of past designs (e.g., CYCLOPS, mentioned above), as indeed do human designers. Some past designs that are relevant to designing the beam are shown in Figure 12.8. These are:

- I-beams used in the construction of buildings;
- box girder bridges;
- sandwich structures used in aircraft wings.

All three structures have been designed to resist bending when loaded. For this knowledge to be useful, it must be accompanied by an explanation of the underlying principles of these designs, as well as their function. The principle underlying all three designs is that strength and stiffness are mainly provided by the top and bottom surfaces, while the remaining material keeps the two surfaces apart. The heaviest parts of the beam are, therefore, concentrated at the surfaces, where they are most effective. This explanation could be expressed as a rule, or perhaps by hierarchical classification of the structural objects that share this property (Figure 12.9). A set of conceptual design rules might seize upon the `beams` class as being appropriate for the current application because they maximize both (*stiffness/mass*) and (*strength/mass*) in bending.

At this stage in the design procedure, three markedly different conceptual designs have been found that fulfill the requirements as determined so far. A key difference between the alternatives is shape. So a knowledge-based system for conceptual design might seek information about the shape requirements of the beam. If the beam needs both to support the seats and to act as a floor that passengers can walk on, it should be flat and able to fulfill the design requirements over a large area. Adding this criterion leaves only one suitable conceptual design, namely the sandwich beam. If the human who is interacting with the system is happy with this decision, the new application can be added to the `applications` attribute of the `Sandwich_beam` class so that this experience will be available in future designs (Figure 12.9).

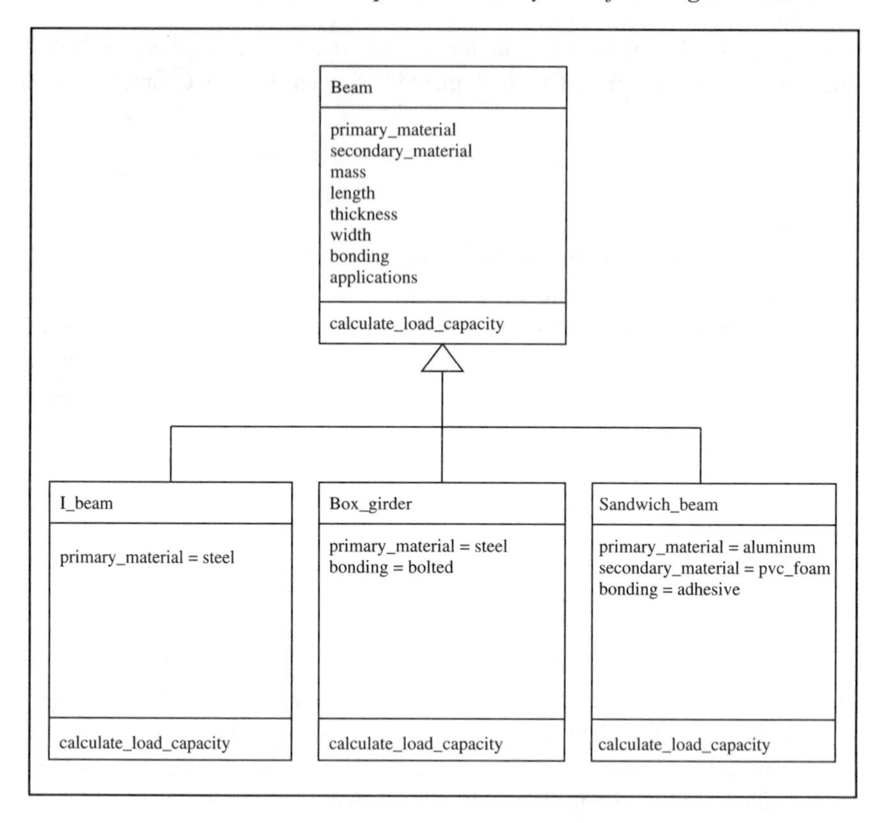

Figure 12.9 Hierarchical classification of beams

12.7.2 Optimization and evaluation

The optimization and evaluation stage of the design process involves performing calculations to optimize performance and to check whether specifications are being met. As this phase is primarily concerned with numerical problems, the tasks are mainly handled using procedural programming techniques. However, the numerical processes can be made more effective and efficient by the application of rules to steer the analysis. For instance, a design system may have access to a library of optimization procedures, the most appropriate for a specific task being chosen by a rule-based selection module.

Three important forms of numerical analysis are:

- mathematical optimization;
- finite-element analysis;
- specialized modeling.

Several techniques for mathematical optimization were described in Chapter 7, including hill-climbing, simulated annealing, and genetic algorithms. *Finite-element analysis* is a general technique for modeling complex shapes. In order to analyze the performance of a three-dimensional physical product, a technique has to be devised for representing the product numerically within the computer. For regular geometric shapes, such as a cube or sphere, this poses no great problem. But the shape of real products, such as a saucepan handle or a gas turbine blade, can be considerably more complex. Since the shape of an object is defined by its surfaces, or *boundaries*, the analysis of performance (e.g., the flow of air over a turbine blade) falls into the class of *boundary-value problems*. Finite-element analysis provides a powerful technique for obtaining approximate solutions to such problems. The technique is based on the concept of breaking up an arbitrarily complex surface or volume into a network of simple interlocking shapes. The performance of the whole product is then taken to be the sum of each constituent part's performance. There are many published texts that give a full treatment of finite-element analysis (e.g., [16, 17]).

Mathematical optimization or finite-element analysis might be used in their own right or as subtasks within a customized model. If equations can be derived that describe the performance of some aspects of the product under design, then it is obviously sensible to make use of them. The rest of this section will, therefore, concentrate on the modeling of a physical system, with particular reference to the design of a sandwich beam.

In the case of the sandwich beam, expressions can be derived that relate the minimum mass of a beam that meets the stiffness and strength requirements to dimensions and material properties. Mass, stiffness, and strength are examples of *performance variables*, as they quantify the performance of the final product. The thicknesses of the layers of the sandwich beam are *decision variables*, as the designer must choose values for them in order to achieve the required performance. Considering first the stiffness requirement, it can be shown [18, 19] that the mass of a beam that just meets the stiffness requirement is given by:

$$M \approx bL\left(\frac{2\rho_s fFad^2}{DE_s bt_c^2} + \rho_c t_c\right) \qquad (12.1)$$

where:
M = mass of beam
b, L, a, d = dimensions defined in Figure 12.7
F = applied load
f = safety factor (f = 1.0 for no margin of safety)

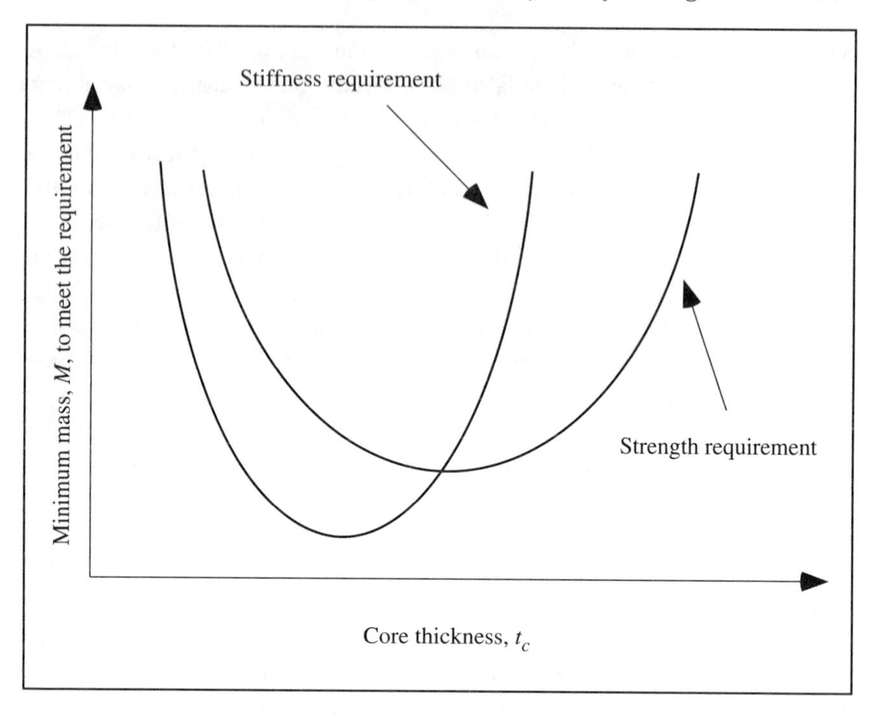

Figure 12.10 Mass of a sandwich beam that just meets stiffness and strength requirements

t_S, ρ_S, E_S = thickness, density and Young's modulus of surface material
t_C, ρ_C, E_C = thickness, density and Young's modulus of core material

Equation 12.1 is written in terms of the core thickness t_C. For each value of core thickness, there is a corresponding surface thickness t_S that is required in order fulfill the stiffness requirement:

$$t_s \approx \frac{fFad^2}{DE_s bt_c^2} \tag{12.2}$$

Thus, given a choice of materials, the plan view dimensions (b, L, a, and d), and the maximum deflection D under load F, there is a unique pair of values of t_C and t_S that correspond to the minimum mass beam that meets the requirement. If this were the only requirement, the analysis would be complete. However, as well as being sufficiently stiff, the beam must be sufficiently strong, i.e., it must not break under the design load. A new pair of equations can be derived that describe the strength requirement:

$$M \approx bL\left(\frac{2\rho_s fFa}{\sigma_f bt_c} + \rho_c t_c\right) \tag{12.3}$$

$$t_s \approx \frac{fFa}{\sigma_f bt_c} \tag{12.4}$$

where σ_f is the failure stress of the surface material.

Assuming a choice of core and surface materials and given the plan dimensions and loading conditions, Equations 12.1 and 12.3 can be plotted to show mass as a function of core thickness, as shown in Figure 12.10. The position of the two curves in relation to each other depends upon the materials chosen. It should be noted that the minimum mass to fulfill the stiffness requirement may be insufficient to fulfill the strength requirement, or vice versa.

There are still two other complications to consider before the analysis of the beam is complete. First, the core material must not fail in shear. In order to achieve this, the following condition must be satisfied:

$$t_c \geq \frac{3fF}{2b\tau_c} \tag{12.5}$$

where τ_c is the critical shear stress for failure of the core material.

Finally, the upper surface, which is in compression, must not buckle. This condition is described by the following equation:

$$t_s \geq \frac{2fFa}{bt_c(E_s E_c G_c)^{1/3}} \tag{12.6}$$

where G_c is the shear modulus of the core material.

Armed with these numerical models, reasonable choices of layer thicknesses can be made. Without such models, a sensible choice would be fortuitous.

12.7.3 Detailed design

The detailed design phase allows the general view provided by the conceptual design phase to be refined. The optimization and evaluation phase provides the information needed to make these detailed design decisions. The decisions taken at this stage are unlikely to be innovative, as the design is constrained by decisions made during the conceptual design phase. In the case of the sandwich beam, the following decisions need to be made:

- choice of core material;
- choice of upper surface material;
- choice of lower surface material;
- choice of core thickness;
- choice of upper surface thickness;
- choice of lower surface thickness;
- method of joining the surfaces to the core.

There is clearly a strong interaction among these decisions. There is also an interaction with the optimization and evaluation process, as Equations 12.1–12.6 need to be reevaluated for each combination of materials considered. The decisions also need to take account of any assumptions or approximations that might be implicit in the analysis. For instance, Equations 12.1–12.4 were derived under the assumption that the top and bottom surfaces were made from identical materials and each had the same thickness.

12.8 Design as a selection exercise

12.8.1 Overview

It should be noted that the crux of both conceptual and detailed design is the problem of selection. Some of the techniques available for making selection decisions are described in the following sections. In the case of a sandwich beam, the selection of the materials and glue involves making a choice from a very large but finite number of alternatives. Thickness, on the other hand, is a continuous variable, and it is tempting to think that the "right" choice is yielded directly by the analysis phase. However, this is rarely the case. The requirements on, say, core thickness will be different depending on whether we are considering stiffness, surface strength, or core shear strength. The actual chosen thickness has to be a compromise. Furthermore, although thickness is a continuous variable, the designer may be constrained by the particular set of thicknesses that a supplier is willing to provide.

This section will focus on the use of scoring techniques for materials selection, although neural network approaches can offer a viable alternative [20]. The scoring techniques are based on awarding candidate materials a score for their performances with respect to the requirements, and then selecting the highest-scoring materials. We will start by showing a naive attempt at combining materials properties to reach an overall decision, before considering a more successful algorithm called AIM [21]. AIM will be illustrated by considering the selection of a polymer for the manufacture of a kettle.

For the purposes of this discussion, selection will be restricted to polymer materials. The full range of materials available to designers covers metals, composites, ceramics, and polymers. Each of these categories is vast, and restricting the selection to polymers still leaves us with a very complex design decision.

12.8.2 Merit indices

The analysis of the sandwich beam yielded expressions for the mass of a beam that just meets the requirements. These expressions contained geometrical measurements and physical properties of the materials. Examination of Equation 12.1 shows that the lightest beam that meets the stiffness requirement will have a low density core (ρ_C) and surfaces(ρ_S), and the surface material will have a high Young's modulus (E_S). However, this observation would not be sufficient to enable a choice between two materials where the first had a high value of E_S and ρ_S, and the second had a low value for each. Merit indices can help such decisions by enabling materials to be ranked according to *combinations* of properties. For instance, a merit index for the surface material of a sandwich beam would be E_S/ρ_S. This is because Equation 12.1 reveals that the important combination of properties for the surface material is the ratio

Minimum weight for given:	Merit index for surface material	Merit index for core material
stiffness	$\frac{E_s}{\rho_s}$	$\frac{1}{\rho_c}$
strength	$\frac{\sigma_f}{\rho_s}$	$\frac{\tau_c}{\rho_c}$
buckling resistance	$\frac{E_s^{1/3}}{\rho_s}$	$\frac{(E_c G_c)^{1/3}}{\rho_c}$

Table 12.1 Merit indices for a sandwich beam

Mode of loading	Minimize mass for given: stiffness	Minimize mass for given: ductile strength
Tie F, l specified r free 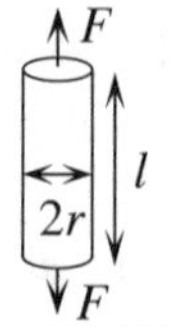	$\frac{E}{\rho}$	$\frac{\sigma_y}{\rho}$
Torsion bar T, l specified r free 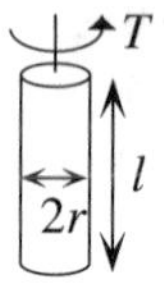	$\frac{G}{\rho}$	$\frac{\sigma_y}{\rho}$
Torsion tube T, l, r specified t free 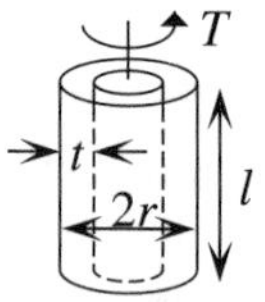	$\frac{G}{\rho}$	$\frac{\sigma_y}{\rho}$
Bending of rods and tubes F, l specified r or t free 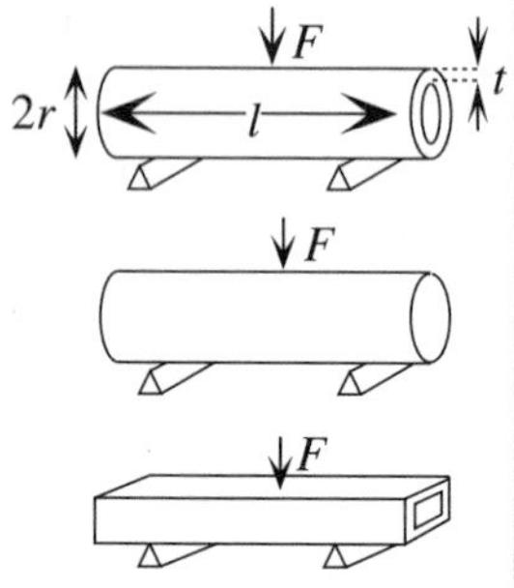	$\frac{E^{1/2}}{\rho}$	$\frac{\sigma_y^{2/3}}{\rho}$

Figure 12.11 Merit indices for minimum mass design (after Ashby [22]). E = Young's modulus; G = shear modulus; ρ = density; σ_y = yield stress. Reprinted from *Acta Metall.*, 37, M. F. Ashby, "On the engineering properties of materials," Copyright (1989), pp. 1273–1293, with permission from Elsevier Science

ρ_S/E_S. As the latter ratio is to be minimized, while merit indices are normally taken to be a quantity that is to be maximized, the merit index is the reciprocal of this ratio. By considering Equations 12.1–12.6, we can derive the merit indices shown in Table 12.1.

Merit indices can be calculated for each candidate material. Given these, tables can be drawn up for each merit index, showing the ranking order of the materials. Thus, merit indices go some way toward the problem of materials selection based on a combination of properties. However, if more than one merit index needs to be considered (as with the sandwich beam), the problem is not completely solved. Materials that perform well with respect to one merit index may not perform so well with another. The designer then faces the problem of finding the materials that offer the best compromise. The scoring techniques described in Sections 12.8.6 and 12.8.7 address this problem. Merit indices for minimum mass design of a range of mechanical structures are shown in Figure 12.11.

12.8.3 The polymer selection example

With the huge number of polymers available, a human designer is unlikely to have sufficient knowledge to make the most appropriate choice of polymer for a specific application. Published data are often unreliable and are generally produced by polymer manufacturers, who have a vested interest in promoting their own products. Even when adequate data are available, the problem of applying them to the product design is likely to remain intractable unless the designer is an expert in polymer technology or has on-line assistance. The selection system described here is intended to help the designer by making the best use of available polymer data. The quality of the recommendations made will be limited by the accuracy and completeness of these data. Use of a computerized materials selection system has the spin-off advantage of encouraging designers to consider and analyze their requirements of a material.

12.8.4 Two-stage selection

The selection system in this example is based on the idea of ranking a shortlist of polymers by comparing their relative performance against a set of materials properties. The length of the shortlist can be reduced by the prior application of numerical specifications, such as a minimum acceptable impact strength. The selection process then comprises two stages, as shown in Figure 12.12. First, any polymers that fail to meet the user's numerical specifications are eliminated. These specifications are *constraints* on the materials, and can be used to limit the number of candidate polymers. Constraints of this sort are sometimes described as *primary constraints*, indicating that they are nonnegotiable. A facility to alter the specifications helps the user of a selection system to assess the sensitivity of the system to changes in the constraints.

Second, the selection process requires the system to weigh the user's *objectives* to arrive at some balanced compromise solutions. The objectives are properties that are to be maximized or minimized as far as possible while satisfying constraints and other objectives. For instance, it may be desirable to maximize impact strength while minimizing cost. Cost is treated as a polymer

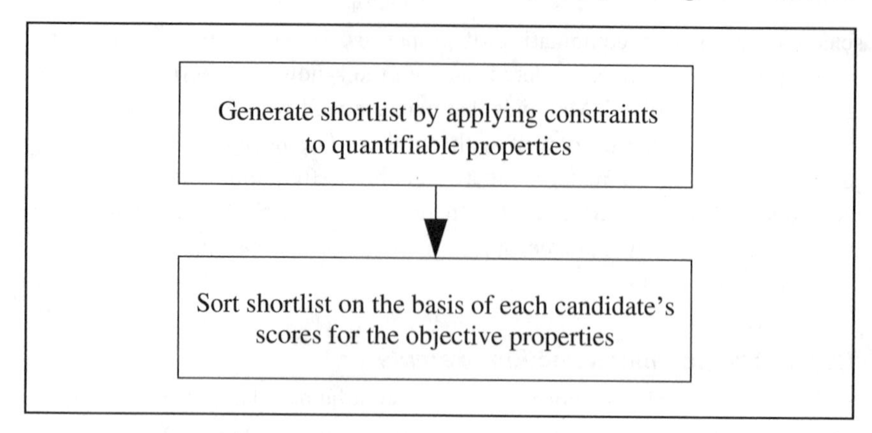

Figure 12.12 Two-stage selection

property in the same way as the other physical properties. Each objective has an importance rating — supplied by the user — associated with it. In the unlikely event of a polymer's offering outstanding performance for each material objective, this polymer will appear at the top of the list of recommendations made by the selection system. More typically, the properties being optimized represent conflicting requirements for each polymer. For example, a polymer offering excellent impact resistance may not be easily injection molded. For such problems there is no single correct answer, but several answers offering different levels of suitability. Objectives may also be known as *preferences* or *secondary constraints*.

12.8.5 Constraint relaxation

Several authors (e.g., Demaid and Zucker [6], Navichandra and Marks [23], and Sriram et al. [2]) have stressed the dangers of applying numerical constraints too rigidly and so risking the elimination of candidates that would have been quite suitable. Hopgood [21] and Navichandra and Marks [23] overcome this problem by relaxing the constraints by some amount. In Hopgood's system, the amount of constraint relaxation is described as a tolerance, which is specified by the user. Relaxation overcomes the artificial precision that is built into a specification. It could be that it is difficult to provide an accurately specified constraint, the property itself may be ill-defined, or the property definition may only approximate what we are really after. Application and relaxation of constraints can be illustrated by representing each candidate as a point on a graph where one property is plotted against another. A boundary is drawn between those materials that meet the constraints and those that do not, and relaxation of the constraints corresponds to sliding this boundary (Figure 12.13).

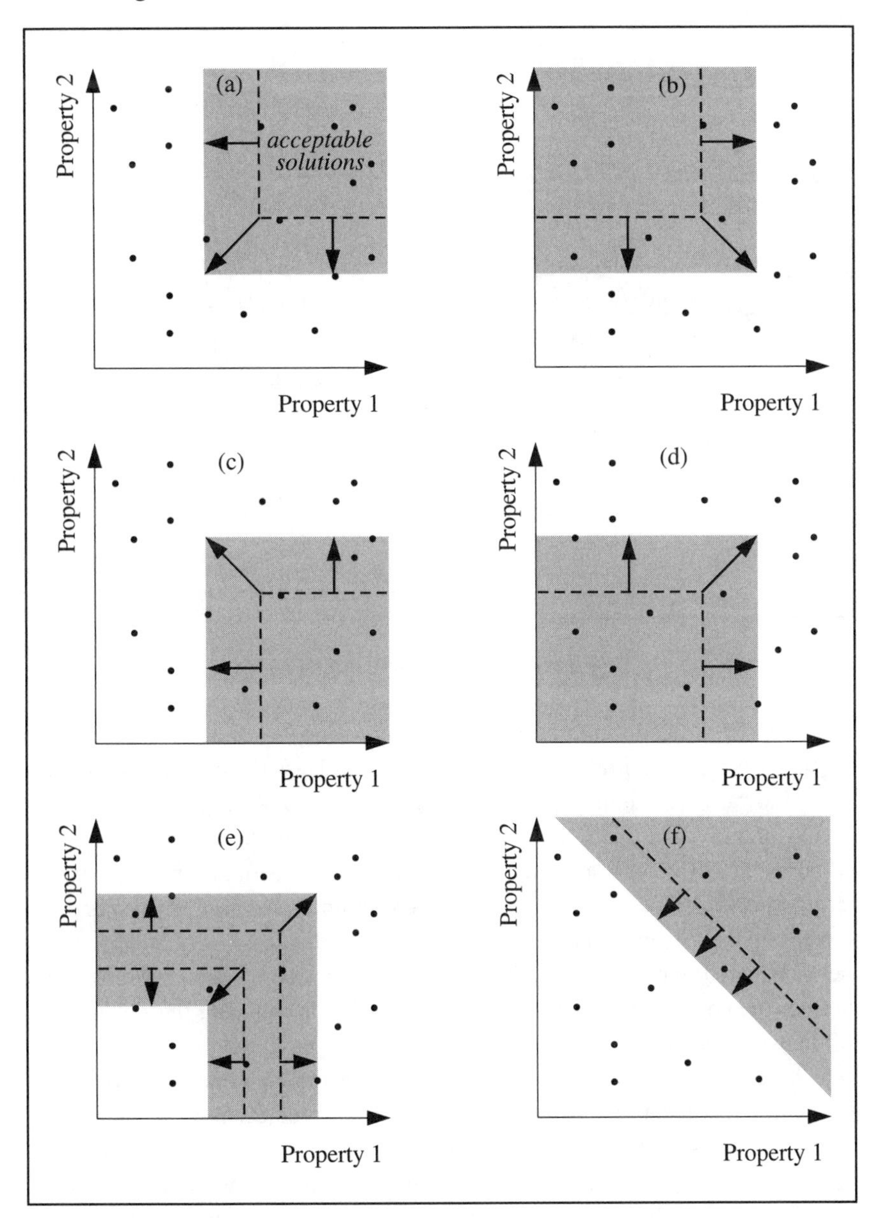

Figure 12.13 Relaxation of constraints:

(a) both constraints are minimum specifications
(b) property 1 has a maximum specification; property 2 has a minimum specification
(c) property 2 has a maximum specification; property 1 has a minimum specification
(d) both constraints are maximum specifications
(e) constraints are target values with associated tolerances
(f) constraint is a trade-off between interdependent properties

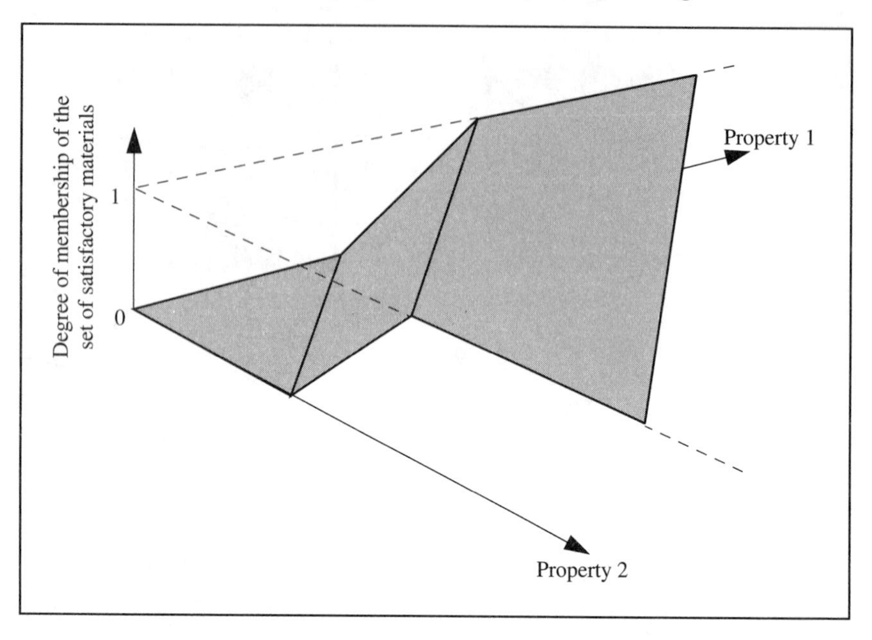

Figure 12.14 A fuzzy constraint

If our specification represents a minimum value that must be attained for a single property (e.g., impact resistance must be at least 1kJ/m), the boundary is moved toward the origin (Figure 12.13(a)). If one or both specifications are for a maximum value, then the boundary is moved in the opposite direction (Figures 12.13(b), (c), and (d)). Figure 12.13(e) illustrates the case where target specifications are provided, and constraint relaxation corresponds to increasing the tolerance on those specifications. Often the specifications cannot be considered independently, but instead some combination of properties defines the constraint boundary (Figure 12.13(f)). In this case there is a trade-off between the properties.

An alternative approach is to treat the category of satisfactory materials (i.e., those that meet the constraints) as a fuzzy set (see Chapter 3). Under such a scheme, those materials that possessed properties comfortably within the specification would be given a membership value of 1, while those that failed completely to reach the specification would be given a membership value of 0. Materials close to the constraint boundary would be assigned a degree of membership between 0 and 1 (Figure 12.14). The membership values for each material might then be taken into account in the next stage of the selection process, based on scoring each material.

Ashby [22] has plotted maps similar to those in Figure 12.13 using logarithmic scales. These "Ashby maps" are a particularly effective means of

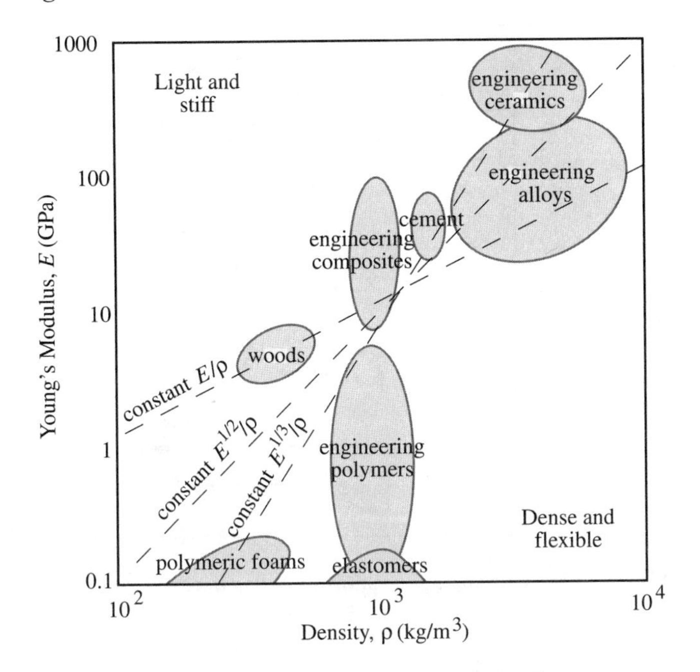

Figure 12.15 Ashby map for Young's modulus versus density [22]. Reprinted from *Acta Metall.*, 37, M. F. Ashby, "On the engineering properties of materials," Copyright (1989), pp. 1273–1293, with permission from Elsevier Science

representing a constraint on a merit index. Figure 12.15 shows the loci of points for which:

$$\frac{E}{\rho} = \text{constant}$$

$$\frac{E^{1/2}}{\rho} = \text{constant}$$

$$\frac{E^{1/3}}{\rho} = \text{constant}$$

E/ρ is a suitable merit index for the surface material of a stiff lightweight sandwich beam, $E^{1/2}/\rho$ is a suitable merit index for the material of a stiff lightweight tube, and $E^{1/3}/\rho$ is a suitable merit index for the material of a stiff lightweight plate. In Figure 12.15, the materials that meet the merit index specification most comfortably are those that are toward the top left side of the map.

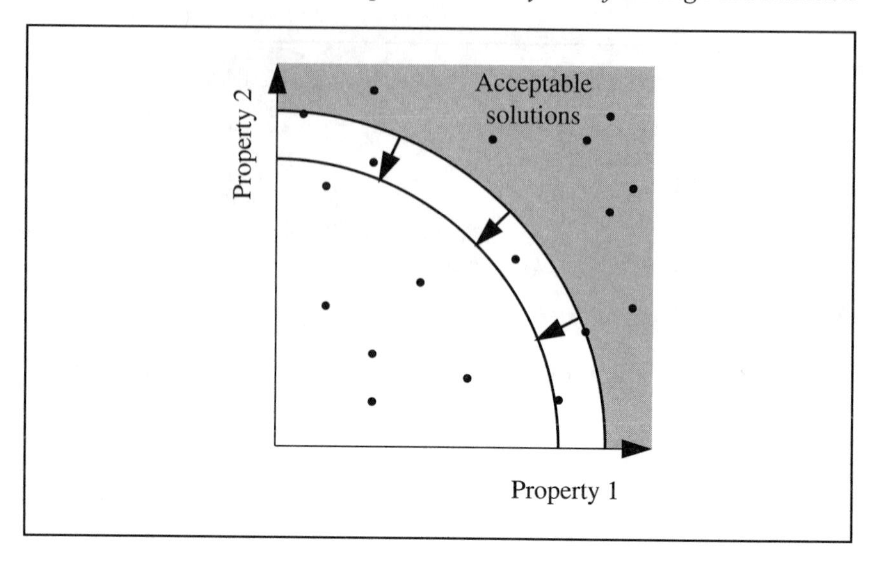

Figure 12.16 Constraint relaxation by sliding the Pareto boundary

When two desirable properties (such as strength and cheapness) are plotted against each other, the boundary of the population of acceptable materials may follow an arc, as shown in Figure 12.16, representing the trade-off between the properties. This boundary is known as the *Pareto boundary*. If more than two properties are considered, the boundary defines a surface in multidimensional space known as the *Pareto surface*. Materials that lie on the Pareto surface are said to be *Pareto optimal*, as an improvement in one property is always accompanied by deterioration in another if the acceptability criterion is maintained. Selection could be restricted to Pareto optimal candidates, but constraint relaxation allows materials *close* to the boundary to be considered as well (Figure 12.16). These same arguments apply to selection between design alternatives [2] as well as to selection between materials.

12.8.6 A naive approach to scoring

We shall now move on to the problem of sorting the shortlist into an order of preference. Let us assume the existence of a data file containing, for each polymer, an array of performance values (ranging from 0 to 9) for each of a number of different properties. The user can supply an importance weighting for each property of interest. A naive approach to determining a polymer's score is to multiply the two figures together for each property, and then to take the sum of the values obtained to be the overall score for that polymer. The polymers with the highest scores are recommended to the user. This scoring system is summarized below:

$$\text{Total score for polymer } i = \sum_j performance(i,j) \times weight(j) \qquad (12.7)$$

where:
performance(i, j) = performance value of polymer i for property j;
weight(j) = user-supplied weighting for property j.

An implication of the use of a summation of scores is that — even though a particular polymer may represent a totally inappropriate choice because of, for example, its poor impact resistance — it may still be highly placed in the ordered list of recommendations. An alternative to finding the arithmetic sum of all of the scores is to find their product:

$$\text{Product of scores for polymer } i = \prod_j performance(i,j) \times weight(j) \qquad (12.8)$$

When combining by multiplication, a poor score for a given property is less readily compensated by the polymer performance for other properties. A polymer that scores particularly badly on a given criterion tends to be filtered out from the final list of recommendations. Thus, using the multiplication approach, good "all-round performers" are preferred to polymers offering performance that varies between extremes. This distinction between the two approaches is illustrated by the following simple example:

	score 1	score 2	score 3	combination by addition	combination by multiplication
polymer *A*	1	2	3	6	6
polymer *B*	2	2	2	6	8

In this example polymer *B* offers a uniform mediocre rating across the three properties, while the rating of polymer *A* varies from poor (score 1) to good (score 3). Under an additive scheme the polymers are ranked equal, while under the multiplication scheme polymer *B* is favored.

A little reflection will show that both of these approaches offer an inadequate means of combining performance values with weightings. Where a property is considered important (i.e., has a high weighting) and a polymer performs well with respect to that property (i.e., has a high performance value), the contribution to the polymer score is large. However, where a property is considered less important (low weighting) and a polymer performs poorly with respect to that property (low performance value), this combination produces

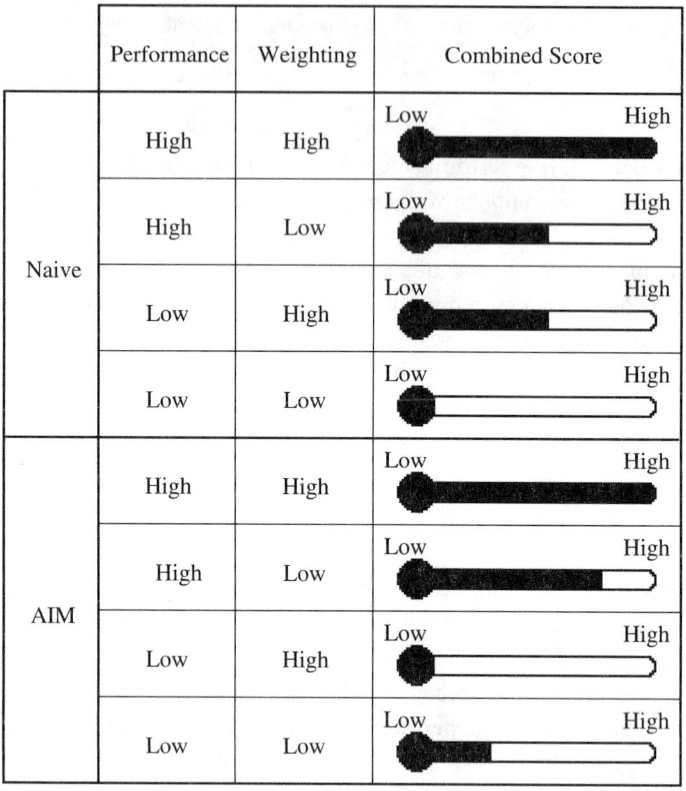

	Performance	Weighting	Combined Score
Naive	High	High	Low High
	High	Low	Low High
	Low	High	Low High
	Low	Low	Low High
AIM	High	High	Low High
	High	Low	Low High
	Low	High	Low High
	Low	Low	Low High

Figure 12.17 Comparison of naive and AIM scoring schemes

the smallest contribution to the polymer score. In fact, since the property in question has a low importance rating, the selection of the polymer should be still favored. The AIM algorithm (Section 12.8.7) was developed specifically to deal with this anomaly. The least appropriate polymer is actually one that has low performance values for properties with high importance weightings. Figure 12.17 compares the naive algorithms with AIM.

12.8.7 A better approach to scoring

The shortcomings of a naive approach to scoring have been noted above and used as a justification for the development of an improved algorithm, AIM [21]. Using AIM, the score for each polymer is given by:

$$\text{Total score for polymer } i = \prod_j \left((performance(i,j) - offset) \times weight(j) + scale\_shift\_term \right) \quad (12.9)$$

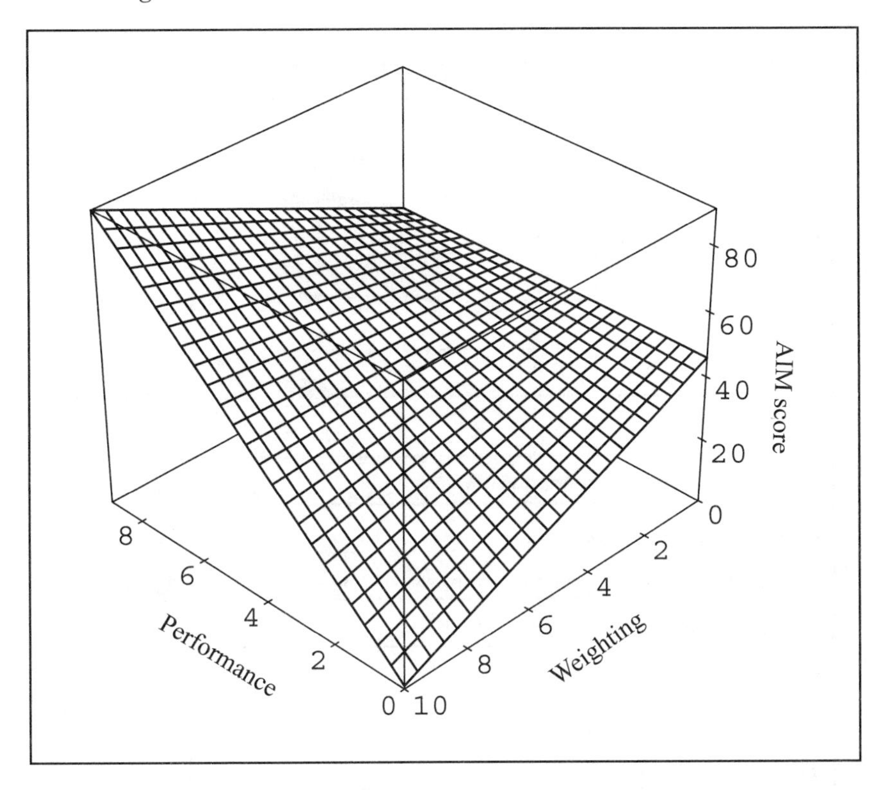

Figure 12.18 Combination of performance values with weightings for a single property, using AIM

where *scale_shift_term* is the smallest number that will ensure that the combined weight and performance rating is positive. In an implemented system [21], the following parameters were selected:

polymer performance rating range	0.0–9.0
weighting range	0.0–10.0
offset	4.5
scale_shift_term	46.0

The AIM equation for a single property, with these parameters inserted, is shown in Figure 12.18. Performance values lower than the offset value can be thought of as degrees of undesirability. On the weightings scale, zero means "I don't care."

Input:

Property	Constraint	Tolerance	Weighting
impact resistance			5
resistance to aqueous environments			8
maximum operating temperature	≥100°C	30°C	
glossiness			3
cheapness			6
ease of injection molding			9

Output:

Recommended polymers	Normalized score
Polypropylene copolymer	4.05
ABS (acrylonitrile-butadiene-styrene copolymer)	3.59
Polypropylene homopolymer	3.29
Fire-retardant polypropylene	2.53
30% glass-fibre coupled polypropylene	2.40
TPX (poly-4-methyl pent-1-ene)	2.06
(114 others meet the contraints)	

Figure 12.19 Use of the AIM polymer selection system during the design of a kettle

12.8.8 Case study: the design of a kettle

Figure 12.19 shows a possible set of inputs and outputs from a polymer selection system that uses AIM. After receiving a list of recommended polymers, the user may alter one or more previous inputs in order to test the effect on the system's recommendations. These "what if?" experiments are also useful for designers whose materials specifications were only vaguely formed when starting a consultation. In these circumstances, the system serves not only to make recommendations for the choice of polymer, but also to assist the designer in deciding upon the materials requirements. The interface contains gaps in places where an entry would be inappropriate. For instance, the user can indicate that "glossiness" is to be maximized, and supply a weighting. However, the user cannot supply a specification of the minimum acceptable glossiness, as only comparative data are available.

In the example shown in Figure 12.19, the designer is trying to determine a suitable polymer for the manufacture of a kettle. The designer has decided that the kettle must be capable of withstanding boiling water for intermittent periods. In addition, a high level of importance has been placed upon the need for the polymer to be injection moldable. The designer has also chosen material cheapness as a desired property, independent of manufacturing costs. Additionally the glossiness and impact resistance of the polymer are to be maximized, within the constraints of attempting to optimize as many of the chosen properties as possible.

As we have already noted, care must be taken when entering any numerical specifications. In this example it has been specified that a maximum operating temperature of at least 100°C is required. A tolerance of 30°C has been placed on this value to compensate for the fact that the polymer will only be intermittently subjected to this temperature. A polymer whose maximum operating temperature is 69°C would be eliminated from consideration, but one with a maximum operating temperature of 70°C would remain a candidate. In the current example, the temperature requirement is clearly defined, although the tolerance is more subjective. The tolerance is equivalent to constraint relaxation.

The recommendations shown in the example are reasonable. The preferred polymer (polypropylene) is sometimes used in kettle manufacture. The second choice (ABS, or acrylonitrile-butadiene-styrene copolymer) is used for the handles of some brands of kettles. The most commonly used polymer in kettle manufacture is an acetal copolymer, which was missing from the selection system's database. This illustrates the importance of having access to adequate data.

12.8.9 Reducing the search space by classification

The selection system described above relies on the ability to calculate a score for every polymer in the system database. In this example, only 150 polymers are considered, for which the data are complete (for a limited set of properties). However, even with the search constrained to polymer materials, there are in reality thousands of candidate polymers and grades of polymer. Countless more grades could be specified by slight variations in composition or processing. Clearly, a system that relies on a complete and consistent set of data for each material cannot cope with the full range of available materials. Even if the data were available, calculating scores for every single one is unnecessary, and bears no relationship with the approach adopted by a human expert, who would use knowledge about *families* of materials.

In general, chemically similar materials tend to have similar properties, as shown by the Ashby map in Figure 12.15. It would therefore be desirable to

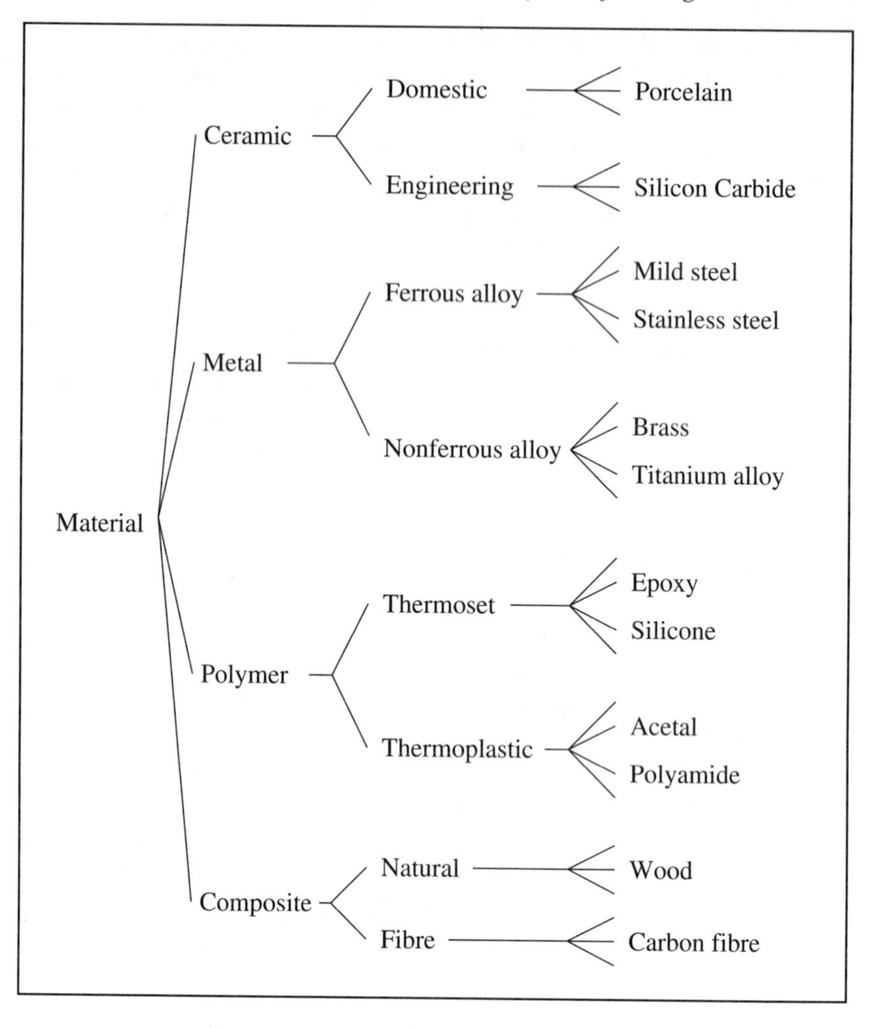

Figure 12.20 One of many possible ways to classify materials

restructure the database so that polymers are hierarchically classified, with polymers of a given type grouped together. Thus, given only a vague specification, many categories could be eliminated from consideration early on. Within the category of materials called *polymer*, several subcategories exist, such as acetal. The selection task is simplified enormously by using knowledge of the range of values for a given property that apply to a particular subcategory. The initial searches would then scan only polymer groups, based upon ranges of properties for polymers within that group.

Only when the search has settled on one or two such families is it necessary to consider individual grades of polymer within those groups. As

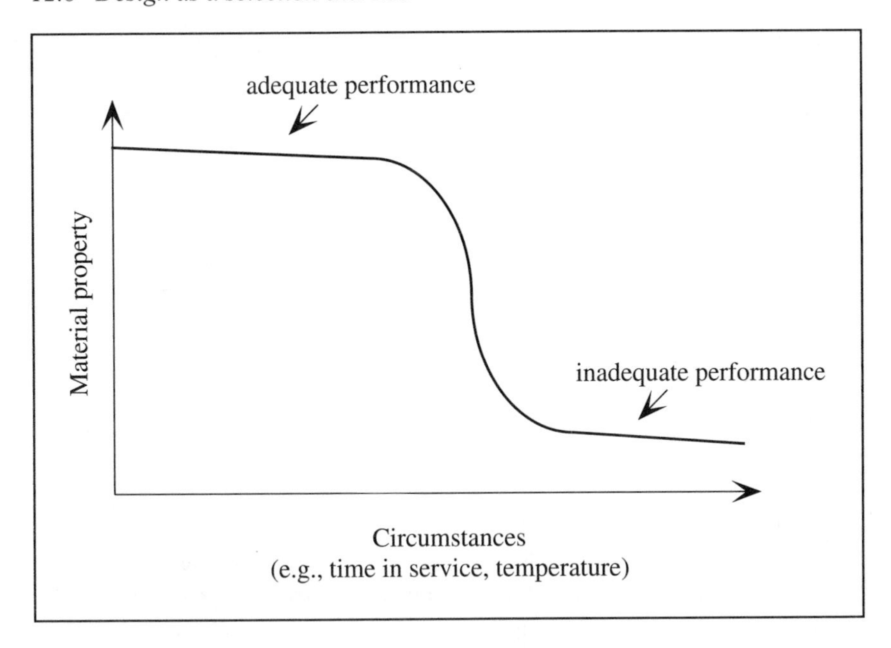

Figure 12.21 The "cliff edge" effect

such a classification of materials is hierarchical, it can be represented using object classes joined by specialization relationships (Chapter 4). One of many possible classification schemes is shown in Figure 12.20.

Demaid and Zucker [24] make use of their own specialized object-oriented system to allow a full and detailed description of real materials and also of the hypothetical "ideal" material for the job. They specifically aim to overcome the restrictions inherent in systems that rely on a single number to describe a complex property. The knowledge-based system incorporating AIM makes some attempt at this by using rules to modify data in certain circumstances [21]. However, the real problem is that a single value describing a material property, such as stiffness, can only be valid at one temperature, after a fixed duration, under a fixed load, and in a particular environment. So in order to choose a polymer that is sufficiently stiff to be suitable for a kettle body, we need more information than just its stiffness at room temperature. We also need to know its stiffness at 100°C and after (say) two years of daily use. To illustrate how acute the problems can be when dealing with polymers, Figure 12.21 shows how a property such as stiffness might vary with temperature or duration of exposure. The designer (or the intelligent selection system) needs to be aware that some polymers may have an adequate stiffness for many purposes at room temperature, but not necessarily after prolonged exposure to elevated temperatures.

12.9 Failure mode and effects analysis (FMEA)

An important aspect of design is the consideration of what happens when things go wrong. If any component of a product should fail, the designer will want to consider the impact of that failure on the following:

- *Safety*
  For example, would an explosion occur? Would a machine go out of control?

- *Indication of failure*
  Will the user of the product notice that something is amiss? For example, will a warning light illuminate or an alarm sound?

- *Graceful or graceless degradation*
  Will the product continue to function after a component has failed, albeit less efficiently? This capability is known as graceful degradation and has some advantages over designs in which the failure of a component is catastrophic. On the other hand, graceful degradation may require that the product contain more than the bare minimum of components, thereby increasing costs.

- *Secondary damage*
  Will the failure of one component lead to damage of other components? Are these other components more or less vital to the function of the product? Is the secondary damage more expensive to fix than the original damage?

The assessment of all possible effects from all possible failures is termed *failure mode and effects analysis* (FMEA). FMEA is not concerned with the *cause* of failures (this is a diagnosis problem — see Chapter 11) but the *effects* of failures. FMEA comprises the following key stages:

- identifying the possible failure modes;
- generating the changes to the product caused by the failure;
- identifying the consequences of those changes;
- evaluating the significance of the consequences.

The scoring technique discussed in Section 12.8.7 could feasibly be adapted for the fourth stage, i.e., evaluating the *significance* of failure mode effects. Price and Hunt's FLAME system [25, 26] uses product models in order to

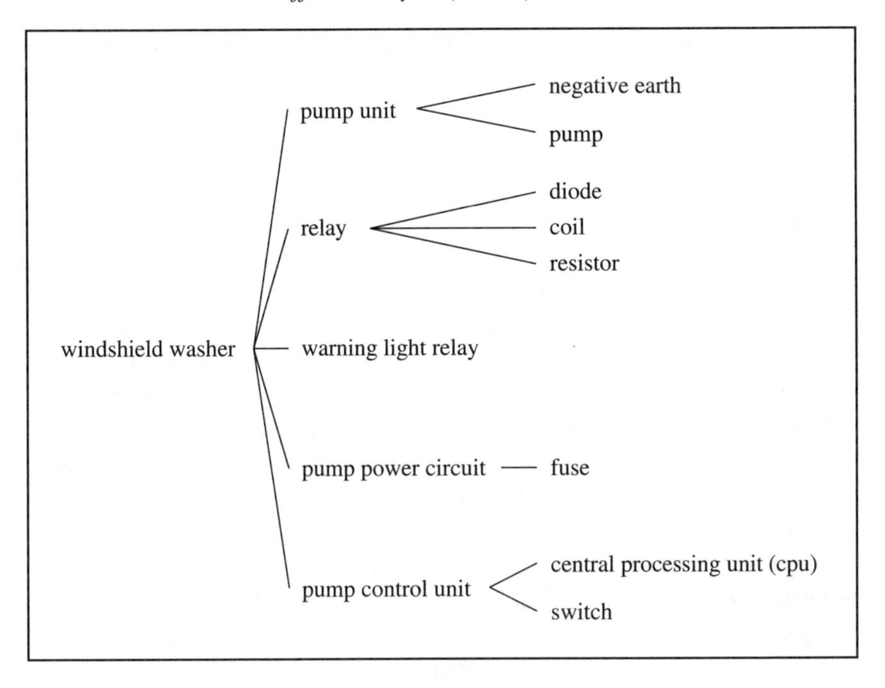

Figure 12.22 Functional decomposition of a windshield washer system (adapted from Price and Hunt [25])

automate the first three stages of FMEA. Two modeling approaches have been used — *functional* and *structural* modeling. Functional modeling involves the breakdown of a system into subsystems, where each subsystem fulfills a specific function. The subsystems may be further decomposed, leading to a hierarchical breakdown based on functionality. The encapsulated nature of each subsystem favors the use of object-oriented programming (see Chapter 4). In the case of the windshield washer system of a car (Figure 12.22), each subsystem is modeled by its response to one of three standard electrical inputs — positive voltage relative to earth, open circuit, or short-circuited to earth. The output from a subsystem then forms the input to another.

Price and Hunt argue that functional modeling is only adequate when the subsystems respond correctly to each of the modeled inputs. Under such circumstances, each subsystem can be relied upon to generate one of a few standard responses, which becomes the input to another subsystem. However, if the behavior of a subsystem is altered by a failure mode, a response may be generated that is not described in the functional model. If this response forms the input to another subsystem, the functional model can no longer cope. To model the functional response to *all* such inputs is impractical, as it would require a complete FMEA in advance. FLAME [25, 26] overcomes this problem by augmenting the functional model with a structural model, i.e., a simulation of the overall system, in order to analyze the inputs that are generated at each subsystem.

12.10 Summary

This chapter has addressed some of the issues in developing intelligent systems to support design decision making. Design can be viewed as a search problem in which alternatives must be found or generated and a selection made from among these. It is a particularly difficult task because it requires both creativity and a vast range of knowledge. Electrical and electronic engineering have been most amenable to the application of decision-support tools, as designs in these domains are often routine rather than innovative and can often be treated as optimization problems.

Selection between alternatives forms an integral part of the design problem. One important selection decision is the choice of materials, a problem that has been explored in some detail in this chapter. Similar techniques might be applied to other aspects of selection within design. Even within the apparently limited domain of materials selection, the range of relevant knowledge is so wide and the interactions so complex that current systems are rather inadequate.

We have seen by reference to the design of a telecommunication network that the design process can be applied to services as well as to manufactured products. This particular case study has also illustrated that producing a design specification can in itself be a complex task, and one that has to be formalized before computerized support tools can be considered. The concepts of constraint propagation and truth maintenance have been illustrated by considering the problem of arranging batteries in a battery holder. Conceptual design, optimization and evaluation, and detailed design have been illustrated by considering the design of an aircraft floor. This design exercise included both geometric design and materials selection. The final case study, concerning the design of a kettle, was used to illustrate some additional ideas for materials selection.

Computer aided design packages have been mentioned briefly. These are useful tools, but are often limited to drafting rather than decision making. The human designer remains at the center of the design process and a range of decision-support tools is being developed that will assist rather than replace the human designer. To this end, it is likely that the coming years will bring a greater degree of integration of CAD tools with intelligent systems for decision support.

References

1. Open University, *PT610: Manufacture, Materials, Design — Unit 7*, Open University Press, 1986.
2. Sriram, D., Stephanopoulos, G., Logcher, R., Gossard, D., Groleau, N., Serrano, D., and Navinchandra, D., "Knowledge-based system applications in engineering design: research at MIT," *AI Magazine*, pp. 79–96, Fall 1989.
3. Hopgood, A. A. and Hopson, A. J., "The common traffic model: a universal model for communications networks," Institution of Radio and Electronic Engineers Conference (IREECON'91), Sydney, pp. 61–64, 1991.
4. Coad, P. and Yourdon, E., *OOA: object-oriented analysis*, Prentice-Hall, 1990.
5. Brown, D. and Chandrasekaran, B., "Expert systems for a class of mechanical design activity," in *Knowledge Engineering in Computer-aided Design*, Gero, J. S. (Ed.), pp. 259–282, Elsevier, 1985.
6. Demaid, A. and Zucker, J., "A conceptual model for materials selection," *Metals and Materials*, pp. 291–297, May 1988.
7. Dietterich, T. G. and Ullman, D. G., "FORLOG: a logic-based architecture for design," in *Expert Systems in Computer-Aided Design*, Gero, J. S. (Ed.), pp. 1–17, Elsevier, 1987.
8. Dyer, M. G., Flowers, M., and Hodges, J., "Edison: an engineering design invention system operating naively," *Artificial Intelligence in Engineering*, vol. 1, pp. 36–44, 1986.
9. Murthy, S. S. and Addanki, S., "PROMPT: an innovative design tool," in *Expert Systems in Computer-Aided Design*, Gero, J. S. (Ed.), pp. 323–341, North-Holland, 1987.
10. Lirov, Y., "Systematic invention for knowledge engineering," *AI Expert*, pp. 28–33, July 1990.
11. Howe, A. E., Cohen, P. R., Dixon, J. R., and Simmons, M. K., "DOMINIC: a domain-independent program for mechanical engineering design," *Artificial Intelligence in Engineering*, vol. 1, pp. 23–28, 1986.
12. deKleer, J., "An assumption-based TMS," *Artificial Intelligence*, vol. 28, pp. 127–162, 1986.
13. deKleer, J., "Problem-solving with the ATMS," *Artificial Intelligence*, vol. 28, pp. 197–224, 1986.

14. deKleer, J., “Extending the ATMS,” *Artificial Intelligence*, vol. 28, pp. 163–196, 1986.
15. Kim, S. H. and Suh, N. P., “Formalizing decision rules for engineering design,” in *Knowledge-Based Systems in Manufacturing*, Kusiak, A. (Ed.), pp. 33–44, Taylor and Francis, 1989.
16. Cook, R., *Finite Element Modeling for Stress Analysis*, Wiley, 1995.
17. Hughes, T. J. R., *The Finite Element Method: linear static and dynamic finite element analysis*, Dover, 2000.
18. Reid, C. N. and Greenberg, J., “An exercise in materials selection,” *Metals and Materials*, pp. 385–387, July 1980.
19. Greenberg, J. and Reid, C. N., “A simple design task (with the aid of a microcomputer),” 2nd International Conference on Engineering Software, Southampton, UK, pp. 926–942, 1981.
20. Cherian, R. P., Smith, L. N., and Midha, P. S., “A neural network approach for selection of powder metallurgy materials and process parameters,” *Artificial Intelligence in Engineering*, vol. 14, pp. 39–44, 2000.
21. Hopgood, A. A., “An inference mechanism for selection, and its application to polymers,” *Artificial Intelligence in Engineering*, vol. 4, pp. 197–203, 1989.
22. Ashby, M. F., “On the engineering properties of materials,” *Acta Metallurgica et Materialia*, vol. 37, pp. 1273–1293, 1989.
23. Navichandra, D. and Marks, D. H., “Design exploration through constraint relaxation,” in *Expert Systems in Computer-Aided Design*, Gero, J. S. (Ed.), pp. 481–509, Elsevier, 1987.
24. Demaid, A. and Zucker, J., “Prototype-oriented representation of engineering design knowledge,” *Artificial Intelligence in Engineering*, vol. 7, pp. 47–61, 1992.
25. Price, C. J. and Hunt, J. E., “Automating FMEA through multiple models,” in *Research and development in expert systems VIII*, Graham, I. and Milne, R. (Eds.), pp. 25–39, Cambridge University Press, 1991.
26. Price, C. J., “Function-directed electrical design analysis,” *Artificial Intelligence in Engineering*, vol. 12, pp. 445–456, 1998.

Further reading

- Gero, J. S. (Ed.), *Proceedings of the International Conference Series on Artificial Intelligence in Design*, biannual.
- Gero, J. S. (Ed.), *Proceedings of the International Conference Series on Computational Models of Creative Design*, triennial.

Chapter thirteen

Systems for planning

13.1 Introduction

The concept of planning is one that is familiar to all of us, as we constantly make and adjust plans that affect our lives. We may make long-range plans, such as selecting a particular career path, or short-range plans, such as what to eat for lunch. A reasonable definition of planning is the process of producing a *plan*, where:

> *a plan is a description of a set of actions or operations, in a prescribed order, that are intended to reach a desired goal.*

Planning therefore concerns the analysis of actions that are to be performed in the future. It is a similar problem to designing (Chapter 12), except that it includes the notion of time. Design is concerned with the detailed description of an artifact or service, without consideration of when any specific part should be implemented. In contrast, the timing of a series of actions, or at least the order in which they are to be carried out, is an essential aspect of planning. Planning is sometimes described as "reasoning about actions," which suggests that a key aspect to planning is the consideration of questions of the form "what would happen if ...?"

Charniak and McDermott [1] have drawn an analogy between planning and programming, as the process of drawing up a plan can be thought of as programming oneself. However, they have also highlighted the differences, in particular:

- programs are intended to be repeated many times, whereas plans are frequently intended to be used once only;
- the environment for programs is predictable, whereas plans may need to be adjusted in the light of unanticipated changes in circumstances.

The need for computer systems that can plan, or assist in planning, is widespread. Potential applications include management decisions, factory configuration, organization of manufacturing processes, business planning and forecasting, and strategic military decision making. Some complex software systems may feature *internal planning*, i.e., planning of their own actions. For example, robots may need to plan their movements, while other systems may use internal planning to ensure that an adequate solution to a problem is obtained within an acceptable time scale.

According to our definition, planning involves choosing a set of operations and specifying either the timing of each or just their order. In many problems, such as planning a manufacturing process, the operations are known in advance. However, the tasks of allocating resources to operations and specifying the timing of operations can still be complicated and difficult. This specific aspect of planning is termed *scheduling* and is described in Sections 13.7 and 13.8.

Some of the principles and problems of planning systems are discussed in Section 13.2 and an early planning system (STRIPS) is described in Section 13.3. This forms the basis for the remainder of this chapter, where more sophisticated features are described.

13.2 Classical planning systems

In order to make automated planning a tractable problem, certain assumptions have to be made. All the systems discussed in this chapter (except for the reactive planners in Section 13.9) assume that the world can be represented by taking a "snapshot" at a particular time to create a world *state*. Reasoning on the basis of this assumption is termed *state-based reasoning*. Planning systems that make this assumption are termed *classical planning systems*, as the assumption has formed the basis of much of the past research. The aim of a classical planner is to move from an initial world state to a different world state, i.e., the goal state. Determining the means of achieving this is *means–ends analysis* (see Section 13.3.1).

The inputs to a classical system are well defined:

- a description of the initial world state;
- a set of actions or operators that might be performed on the world state;
- a description of the goal state.

The output from a classical planner consists of a description of the sequence of operators that, when applied to the current world state, will lead to the desired

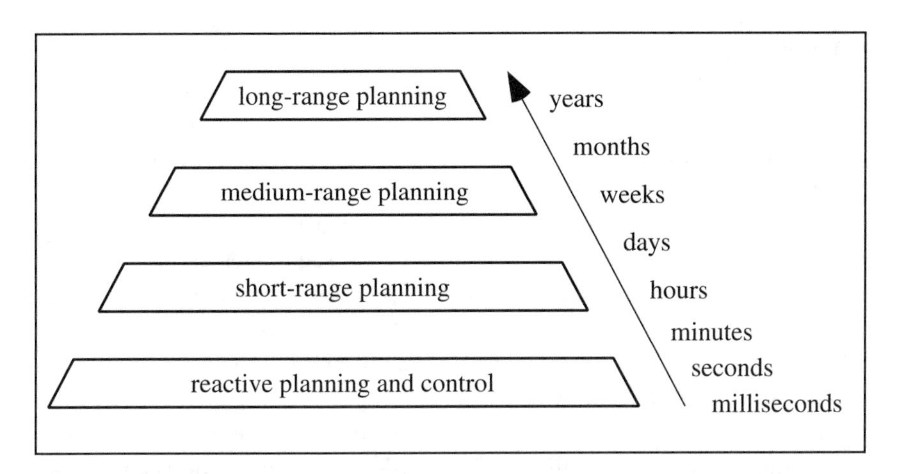

Figure 13.1 Classification by time scale of planning tasks (adapted from [2])

world state as described in the goal. Each operator in the sequence creates a new projected world state, upon which the next operator in the sequence can act, until the goal has been achieved.

The assumption that a plan can be based upon a snapshot of the world is only valid if the world does not change during the planning process. Classical planners may be inadequate for reasoning about a continuous process or dealing with unexpected catastrophes. If a system can react quickly enough to changes in the world to replan and to instigate the new plan, all in real time, the system is described as *reactive* (Section 13.9). As reactive planners must be continuously alert to changes in the world state, they are not classical systems.

Bel et al. [2] have categorized types of planning decision on the basis of the time scale to which they apply (Figure 13.1). Long-range planning decisions concern general strategy and investments; medium-range planning applies to decisions such as production planning, with a horizon of weeks or months; short-range planning or scheduling refers to the day to day management of jobs and resources. Control decisions have the shortest time horizon and are made in real time (see Chapter 14).

A significant problem in the construction of a planning system is deciding which aspects of the world state are altered by an operation (and by how much they are altered) and which are not affected at all. Furthermore, some operations may have a different effect depending on context, i.e., depending on some other aspect of the world state. Collectively, this is known as the *frame problem*, introduced in Chapter 1. As an example, consider the operation of walking from home to the shops. Two obvious changes in the world state are first that I am no longer at home, and second that I have arrived at the shops. However, there are a number of other changes, such as the amount of rubber left on the soles of my shoes. There are also many things that will *not* have changed as a result of my walk, such as the price of oil. Some changes in the world state will depend on context. For example, if the initial world state includes rainy weather then I will arrive at the shops wet, but otherwise I will arrive dry.

Describing every feasible variant of an action with every aspect of the world state is impractical, as the number of such interactions increases dramatically with the complexity of the model. One approach to dealing with the frame problem is the so-called *STRIPS assumption*, which is used in STRIPS (see below) and many other classical planners. This is the assumption that all aspects of the world will remain unchanged by an operation, apart from those that are changed explicitly by the modeled operation. The STRIPS assumption is, therefore, similar to the closed-world assumption (see Chapters 1 and 2).

The planning problem becomes even more complex if we introduce multiple agents (see Chapter 5). Imagine that we have programmed a robot as an autonomous agent, capable of planning and executing a set of operations in order to achieve some goal. Suppose now that our robot is joined by several other robots that also plan and execute various operations. It is quite probable that the plans of the robots will interfere, e.g., they may bump into each other or require the same tool at the same time. Stuart [3] has considered multiple robots, where each reasons independently about the beliefs and goals of others, so as to benefit from them. Other researchers such as Durfee et al. [4] have considered collaborative planning by multiple agents (or robots), where plan generation is shared so that each agent is allocated a specific task.

13.3 STRIPS

13.3.1 General description

STRIPS [5, 6] is one of the oldest AI planning systems, but it is nevertheless an informative one to consider. In its original form, STRIPS was developed in order to allow a robot to plan a route between rooms, moving boxes along the way. However, the principles are applicable to a number of planning domains, and we will consider the operational planning of a company that supplies components for aircraft engines.

Like all classical planning systems, STRIPS is based upon the creation of a world model. The world model comprises a number of objects as well as operators that can be applied to those objects so as to change the world model. So, for instance, given an object *alloy block*, the operator `turn` (i.e., on a lathe) can be applied so as to create a new object called *turbine disk*. The problem to be tackled is to determine a sequence of operators that will take the world model from an initial state to a goal state.

Operators cannot be applied under all circumstances, but instead each has a set of preconditions that must hold before it can be applied. Given knowledge of the preconditions and effects of each operator, STRIPS is able to apply a technique called *means–ends analysis* to solve planning problems. This technique involves looking for differences between the current state of the world and the goal state, and finding an operator that will reduce the difference. Many classical planning systems use means–ends analysis because it reduces the number of operators that need to be considered and, hence, the amount of search required. The selected operator has its own preconditions, the satisfaction of which becomes a new goal. STRIPS repeats the process recursively until the desired state of the world has been achieved.

13.3.2 An example problem

Consider the problem of responding to a customer's order for turbine disks. If we already have some suitable disks, then we can simply deliver these to the customer. If we do not have any disks manufactured, we can choose either to manufacture disks from alloy blocks (assumed to be the only raw material) or to subcontract the work. Assuming that we decide to manufacture the disks ourselves, we must ensure that we have an adequate supply of raw materials, that staff is available to carry out the work, and that our machinery is in good working order. In the STRIPS model we can identify a number of objects and operators, together with the preconditions and effects of the operators (Table 13.1).

Table 13.1 shows specific instantiations of objects associated with operators (e.g., `purchase` raw materials), but other instantiations are often possible (e.g. `purchase` anything). The table shows that each operator has preconditions that must be satisfied before the operator can be executed. Satisfying a precondition is a subproblem of the overall task, and so a developing plan usually has a hierarchical structure.

We will now consider how STRIPS might solve the problem of dealing with an order for a turbine disk. The desired (goal) state of the world is simply *customer has turbine disk*, with the other parameters about the goal state of the world left unspecified.

Let us assume for the moment that the initial world state is as follows:

- the customer has placed an order for a turbine disk;
- the customer does not have the turbine disk;
- staff is available;

- we have no raw materials;
- materials are available from a supplier;
- we can afford the raw materials;
- the machinery is working.

Means–ends analysis can be applied. The starting state is compared with the goal state and one difference is discovered, namely *customer has turbine disk.* STRIPS would now treat *customer has turbine disk* as its goal and would look for an operator that has this state in its list of effects. The operator that achieves the desired result is `deliver`, which is dependent on the conditions *turbine disk ready* and *order placed.* The second condition is satisfied already. The first condition becomes a subgoal, which can be solved in either of two ways: by subcontracting the work or by manufacturing the disk. We will assume for the moment that the disk is to be manufactured in-house. Three preconditions exist for the `manufacture` operator. Two of these are already satisfied (staff is available and the machinery is working), but the precondition that we have raw materials is not satisfied and becomes a new subproblem.

`Operator` and *object*	Precondition	Effect
`deliver` *product*	*product is ready and order has been placed by customer*	*customer has product*
`subcontract` (manufacture of) *product*	*subcontractor available and subcontractor cost is less than product price*	*product is ready*
`manufacture` *product*	*staff is available, we have the raw materials and the machinery is working*	*product is ready, our raw materials are reduced and the machinery is closer to its next maintenance period*
`purchase` *raw materials*	*we can afford the raw materials and the raw materials are available*	*we have the raw materials and money is subtracted from our account*
`borrow` *money*	*good relationship with our bank*	*money is added to our account*
`sell` *assets*	*assets exist and there is sufficient time to sell them*	*money is added to our account*
`repair` *machinery*	*parts are available*	*machinery is working and next maintenance period scheduled*

Table 13.1 Operators for supplying a product to customer. Objects are underlined

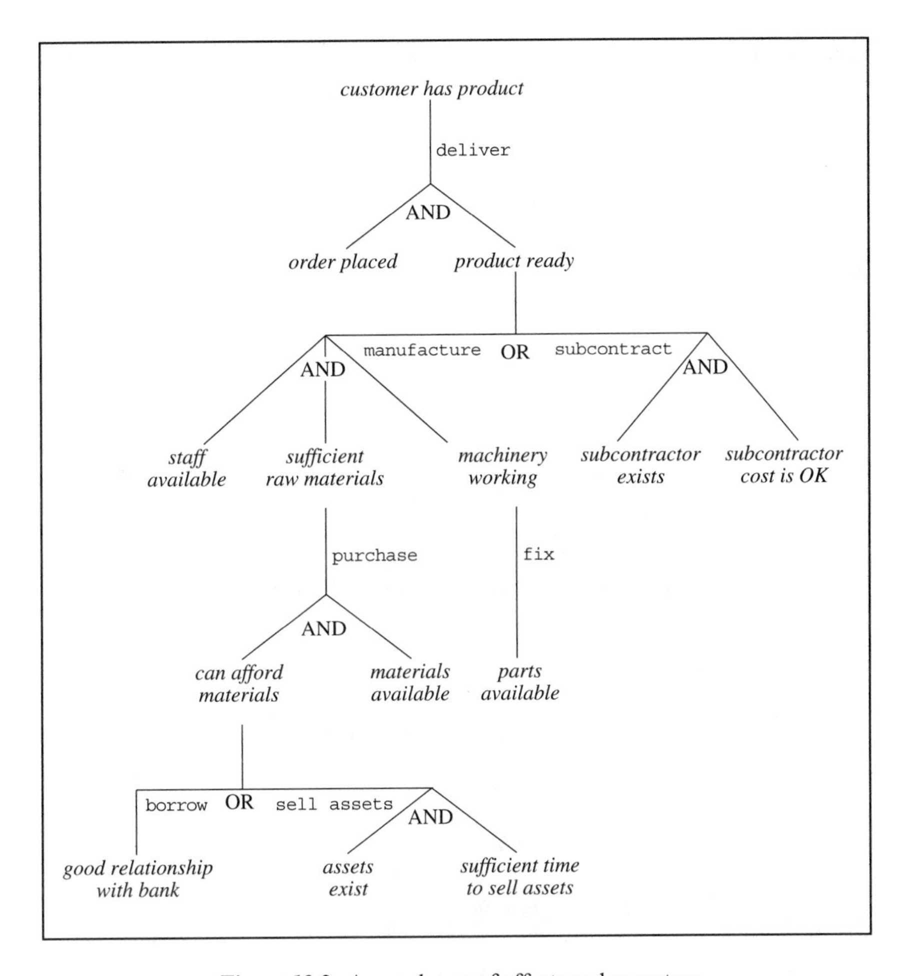

Figure 13.2 A search tree of effects and operators

This can be satisfied by the operator `purchase`, whose two preconditions (that we can afford the materials and that materials are available from a supplier) are already satisfied. Thus, STRIPS has succeeded in finding a plan, namely:

purchase raw materials — manufacture product — deliver product.

The full search tree showing all operators and their preconditions is shown in Figure 13.2, while Figure 13.3 shows the subset of the tree that was used in producing the plan derived above. Note that performing an operation changes the state of the world model. For instance, the operator `purchase`, applied to raw materials, raises our stock of materials to a sufficient quantity to fulfill the order. In so doing, it fulfills the preconditions of another operator, namely, `manufacture`. This operator is then also able to change the world state, as it results in the product's being ready for delivery.

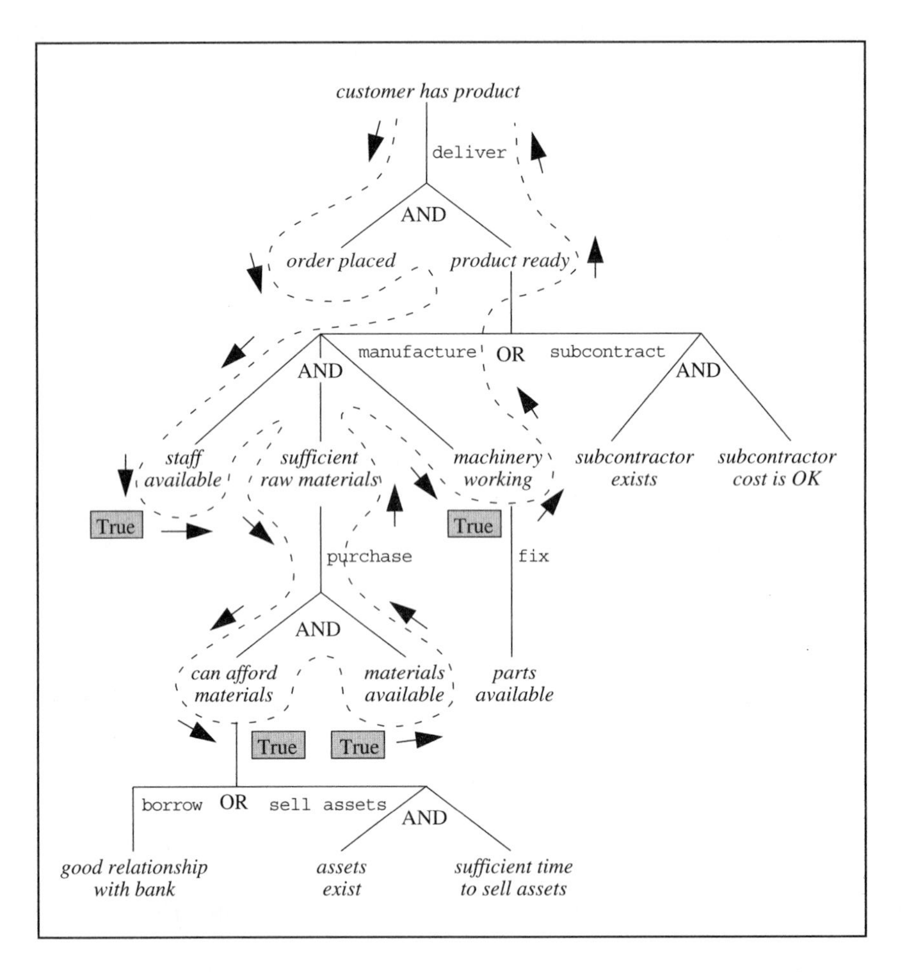

Figure 13.3 STRIPS searching for a plan to fulfill the goal *customer has product*, given the following initial world state:

- the customer has placed an order for a turbine disk
- the customer does not have the turbine disk
- staff are available
- we have no raw materials
- materials are available from a supplier
- we can afford the raw materials
- the machinery is working

13.3.3 A simple planning system in Prolog

STRIPS involves search, in a depth first fashion, through a tree of states linked by operators. STRIPS initially tries to establish a goal. If the goal has preconditions, these become new goals and so on until either the original goal is established or all possibilities have been exhausted. In other words, STRIPS performs backward-chaining in order to establish a goal. If it should find that a necessary condition cannot be satisfied for the branch that it is investigating, STRIPS will backtrack to the last decision point in the tree (i.e., the most recently traversed OR node). STRIPS is heavily reliant on backward-chaining and backtracking, features that are built into the Prolog language (Chapter 10). For this reason it should be fairly straightforward to program a STRIPS-like system in Prolog.*

There are a number of different ways in which we might build a system for our example of supply of a product to a customer. For the purposes of illustration, we will adopt the following scheme:

- The sequence of operators used to achieve a goal is stored as a list, representing a plan. For example, assuming that we have no money in our account and no raw materials, a plan for obtaining a turbine blade that is ready for delivery would be the list:

```
[borrow_money, purchase_materials, manufacture]
```

- Each effect is represented as a clause whose last argument is the plan for achieving the effect. For instance, the effect *our customer has turbine disk* is represented by the following clause:

```
has(our_customer, turbine_disk, Plan).
```

- If a particular effect does not require any actions, then the argument corresponding to its plan is an empty list. We may wish to set up certain effects as part of our initial world state, which do not require a plan of action. An initial world state in which a subcontractor exists would be represented as:

```
exists(subcontractor, []).
```

---

* The actual STRIPS program described by Fikes et al. [5, 6] was implemented in Lisp. It was considerably more sophisticated than the Prolog program presented here and it used a different representation of objects and operators.

```
has(Customer, Product, Newplan):-          % deliver product
  ready(Product, Plan1),
  order_placed(Product, Customer, Plan2),
  merge([Plan1, Plan2, deliver], Newplan).

ready(Product, Newplan):-    % subcontract the manufacturing
  exists(subcontractor, Plan1),
  subcontractor_price_OK(Plan2),
  merge([Plan1, Plan2, subcontract], Newplan).

subcontractor_price_OK([]):-
  % no action is required if the subcontractor's
  % price is OK, i.e. less than the disk price
  price(subcontractor,Price1,_),
  price(product,Price2,_),
  Price1 < Price2.

ready(Product, Newplan):-
                          % manufacture the product ourselves
  exists(staff,Plan1),
  sufficient_materials(Plan2),
  machine_working(Plan3),
  merge([Plan1, Plan2, Plan3, manufacture], Newplan).

sufficient_materials([]):-
  % no action required if we have sufficient stocks already
  current_balance(materials,Amount,_),
  Amount >= 1.

sufficient_materials(Newplan):-          % purchase materials
  can_afford_materials(Plan),
  merge([Plan, purchase_materials], Newplan).

can_afford_materials([]):-
  % no action is required if our bank balance is adequate
  current_balance(account,Amount,_),
  price(materials,Price,_),
  Amount >= Price.

can_afford_materials(Newplan):-               % borrow money
  bank_relationship(good, Plan),
  merge([Plan, borrow_money], Newplan).
```

Box 13.1 (a) A simple planner in Prolog (part I)

```
can_afford_materials(Newplan):-                  % sell assets
  exists(assets,Plan1),
  exists(time_to_sell,Plan2),
  merge([Plan1, Plan2, sell_assets], Newplan).

machine_working(Newplan):-                       % fix machinery
  exists(spare_parts,Plan2),
  merge([Plan1, Plan2, fix_machine], Newplan).

merge([],[]):-!.

merge([ [] | Hierarchical], Flat):-
  !,
merge(Hierarchical, Flat).

merge([X | Hierarchical], [X | Flat]):-
  atom(X), !,
  merge(Hierarchical, Flat).

merge([X | Hierarchical], [A | Flat]):-
  X = [A | Rest], !,
  merge([Rest | Hierarchical], Flat).

%set up the initial world state
price(product,1000,[]).
price(materials,200,[]).
price(subcontractor,500,[]).
current_balance(account,0,[]).
current_balance(materials,0,[]).
bank_relationship(good,[]).
order_placed(turbine,acme,[]).
exists(subcontractor,[]).
exists(spare_parts,[]).
exists(staff,[]).

%  The following are assumed false (closed world
%  assumption):
%     machine_working([]),
%     exists(assets,[]),
%     exists(time_to_sell,[]).
```

Box 13.1 (b) A simple planner in Prolog (part II)

- The preconditions for achieving an effect are represented as Prolog rules. For instance, one way in which a product can become ready is by subcontracting its manufacture. Assuming this decision, there are two preconditions, namely, that a subcontractor exists and that the cost of subcontracting is acceptable. The Prolog rule is as follows:

```
ready(Product, Newplan):-
  exists(subcontractor, Plan1),
  subcontractor_price_OK(Plan2),
  merge([Plan1, Plan2, subcontract], Newplan).
```

The final condition of the rule shown above is the `merge` relation which is used for merging subplans into a single sequence of actions. This is not a standard Prolog facility, so we will have to create it for ourselves. The first argument to `merge` is a list that may contain sublists, while the second argument is a list containing no sublists. The purpose of `merge` is to "flatten" a hierarchical list (the first argument) and to assign the result to the second argument. We can define `merge` by four separate rules, corresponding to different structures that the first argument might have. We can ensure that the four rules are considered mutually exclusive by using the cut facility (see Chapter 10).

The complete Prolog program is shown in Box 13.1. The program includes one particular initial world state, but of course this can be altered. The world state shown in Box 13.1 is as follows:

```
price(product,1000,[]).           % Turbine disk price is $1000
price(materials,200,[]).          % Raw material price is $200 per disk
price(subcontractor,500,[]).      % Subcontractor price is $500 per disk
current_balance(account,0,[]).    % No money in our account
current_balance(materials,0,[]).  % No raw materials
bank_relationship(good,[]).       % Good relationship with the bank
order_placed(turbine,acme,[]).    % Order has been placed by ACME, Inc.
exists(subcontractor,[]).         % A subcontractor is available
exists(spare_parts,[]).           % Spare parts are available
exists(staff,[]).                 % Staff is available
```

Because it is not specified that we have assets or time to sell them, or that the machinery is working, these are all considered false under the closed-world assumption.

We can now ask our Prolog system for suitable plans to provide a customer (ACME, Inc.) with a product (a turbine disk), as follows:

```
?- has(acme, turbine, Plan).
```

Prolog offers the following plans in response to our query:

```
Plan = [subcontract,deliver];
Plan = [borrow_money,purchase_materials,fix_machine,manufacture,
        deliver];
no
```

We can also ask for plans to achieve any other effect that is represented in the model. For instance, we could ask for a plan to give us sufficient raw materials, as follows:

```
?- sufficient_materials(Plan).
Plan = [borrow_money,purchase_materials];
no
```

Having discussed a simple planning system, the remainder of this chapter will concentrate on more sophisticated features that can be incorporated.

13.4 Considering the side effects of actions

13.4.1 Maintaining a world model

Means–ends analysis (see Section 13.3.1 above) relies upon the maintenance of a world model, as it involves choosing operators that reduce the difference between a given state and a goal state. Our simple Prolog implementation of a planning system does not explicitly update its world model, and this leads to a deficiency in comparison with the real STRIPS implementation. When STRIPS has selected an operator, it applies that operator to the current world model, so that the model changes to a projected state. This is important because an operator may have many effects, only one of which may be the goal that is being pursued. The new world model therefore reflects both the intended effects and the side effects of applying an operator, provided that they are both explicit in the representation of the operator. All other attributes of the world state are assumed to be unchanged by the application of an operator — this is the STRIPS assumption (see Section 13.2).

In the example considered in Section 13.3.3, the Prolog system produced a sequence of operators for achieving a goal, namely, to supply a product to a customer. What the system fails to tell us is whether there are any implications of the plan, other than achievement of the goal. For instance, we might like to be given details of our projected cash flow, of our new stocks of materials, or of the updated maintenance schedule for our machinery. Because these data are not necessary for achieving the goal — although they are affected by the planned operators — they are ignored by a purely backward-chaining mechanism. (See Chapter 2 for a discussion of forward- and backward-chaining). Table 13.1 indicates that purchasing raw materials has the effect of reducing our bank balance, and manufacturing reduces the time that can elapse before the machinery is due for servicing. Neither effect was considered in our Prolog system because these effects were not necessary for achieving the goal.

13.4.2 Deductive rules

SIPE [7, 8, 9] is a planning system that can deduce effects additional to those explicitly included in the operator representation. This is a powerful capability, as the same operator may have different effects in different situations, i.e., it may be context-sensitive. Without this capability, context-sensitivity can only be modeled by having different operators to represent the same action taking place in different contexts.

SIPE makes use of two types of deductive rules, *causal* rules and *state* rules. Causal rules detail the auxiliary changes in the world state that are associated with the application of an operator. For example, the operator `purchase` is intended to change the world state from *we have no raw materials* to *we have raw materials*. This change has at least one side-effect, i.e., that our bank account balance is diminished. This side-effect can be modeled as a causal rule.

State rules are concerned with maintaining the consistency of a world model, rather than explicitly bringing about changes in the model. Thus if the assertion *machinery is working* is true in the current world state, then a state rule could be used to ensure that the assertion *machinery is broken* is made false.

Causal rules react to changes between states, whereas state rules enforce constraints within a state. Example causal and state rules are shown in Box 13.2, using syntax similar to that in SIPE. Note that parameters are passed to the rules in place of the named arguments. The rules are, therefore, more general than they would be without the arguments. A rule is considered for firing if its *trigger* matches the world state *after* an operator has been applied. In the case of the causal rule `update_bank_balance`, the trigger is the world state *we have sufficient supplies*, which is brought about by the operator `purchase`. Because causal rules apply to a change in world state, they also contain a *precondition*, describing the world state *before* the operator was applied (e.g., `NOT(sufficient raw materials)`). A causal rule will only fire if its trigger is matched after an operator has been applied and its precondition had been matched immediately before the operator was applied. State rules are not directly concerned with the application of an operator, and so do not have a precondition. There is, however, provision for naming further conditions (additional to the trigger) that must be satisfied.

causal-rule:	update_bank_balance
arguments:	cost,old_balance,new_balance
trigger:	sufficient supplies of something
precondition:	NOT(sufficient supplies of something)
effects:	new_bank_balance = old_bank_balance - cost
state-rule:	Deduce_fixed
arguments:	machine1
trigger:	machine1 is working
other conditions:	<none>
effects:	Not(machine1 is broken)

Box 13.2 Causal and state rules in SIPE

In SIPE, when an operator is added to the current plan, causal rules are examined first in order to introduce any changes to the world model, and then state rules are applied to maintain consistency with constraints on the model. Other than the order of applicability, there is no enforced difference between causal and state rules. According to the syntax, both can have preconditions and a trigger, although there appears to be no justification for applying a precondition to a state rule.

13.5 Hierarchical planning

13.5.1 Description

Virtually all plans are hierarchical by nature, as exemplified by Figure 13.2, although they are not represented as such by all planning systems. STRIPS (a nonhierarchical planner) may produce the following plan for satisfying a customer's order:

> borrow money — purchase materials — fix machinery — manufacture — deliver.

Some of the actions in this plan are major steps (e.g., manufacture), whereas others are comparatively minor details (e.g., purchase materials). A *hierarchical planner* would first plan the major steps, for example:

> be ready to deliver — deliver.

The details of a step such as *be ready to deliver* might then be elaborated:

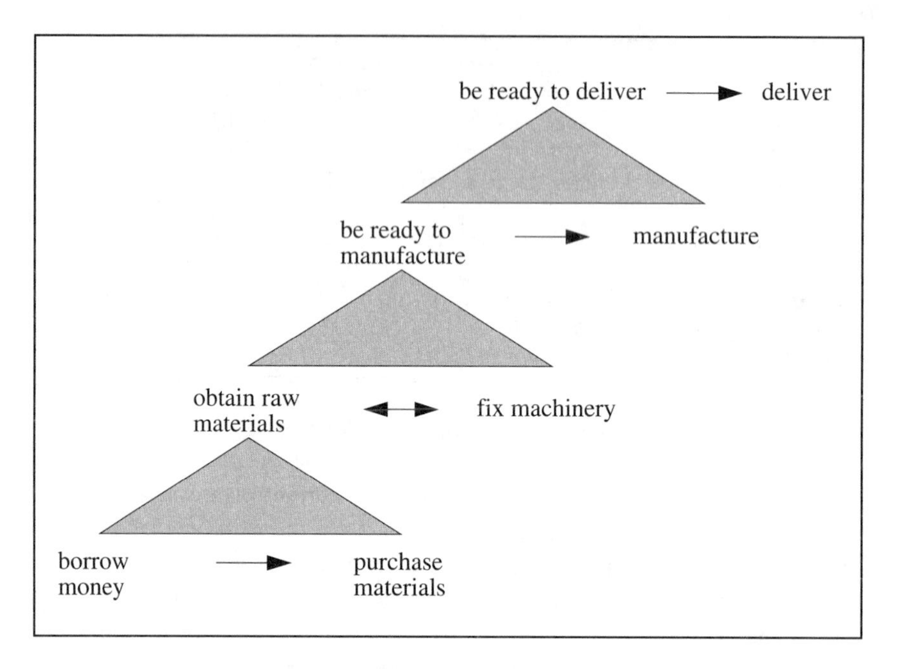

Figure 13.4 A hierarchical plan

be ready to manufacture — manufacture — deliver.

The step *be ready to manufacture* might be broken down into the following actions:

fix machinery — obtain raw materials.

The action *obtain raw materials* can then be elaborated further:

borrow money — purchase materials.

This hierarchical plan is depicted in Figure 13.4. An action that needs no further refinement is a *primitive* action. In some applications *purchase materials* may be considered a primitive action, but in other applications it may be necessary to elaborate this further (e.g., pick up phone — dial number — speak to supplier — and so on).

The distinction between a nonhierarchical planner (such as STRIPS) and a hierarchical planner (such as ABSTRIPS [10]), is that hierarchical planners explicitly represent the hierarchical nature of the plan. At the top of the hierarchy is a simplification or abstraction of the plan, while the lower levels contain the detailed requirements (Figure 13.4). A subplan is built for achieving each action in the main plan. While STRIPS does recognize that the achievement of some goals is dependent on subgoals, no distinction is drawn between goals that are major steps and those that are merely details. Furthermore, as the STRIPS hierarchy is not explicitly represented, it cannot be modified either by the user or by the system itself.

The method of hierarchical planning can be summarized as follows:

- sketch a plan that is complete but too vague to be useful;
- refine the plan into more detailed subplans until a complete sequence of problem-solving operators has been specified.

Since the plan is complete (though perhaps not useful in its own right) at each level of abstraction, the term *length-first* search is sometimes used to describe this technique for selecting appropriate operators that constitute a plan.

13.5.2 Benefits of hierarchical planning

Although means–ends analysis is an effective way of restricting the number of operators that apply to a problem, there may still be several operators to choose from, with no particular reason for preferring one over another. In other words, there may be several alternative branches of the search tree. Furthermore, there is no way of knowing whether the selected branch might lead to a dead end, i.e., one of its preconditions might fail.

Consider the example of satisfying a customer's order for a product. Suppose that STRIPS has chosen to apply the `manufacture` operator (i.e., the left-hand branch of the tree shown in Figure 13.2 has been selected). STRIPS would now verify that staff is available, plan to purchase raw materials (which in turn requires money to be borrowed), and then consider the state of the manufacturing equipment. Suppose that at this point it found that the machinery was broken, and spare parts were not available. The plan would have failed, and the planner would have to backtrack to the point where it chose to manufacture rather than subcontract. All the intermediate processing would have been in vain, since STRIPS cannot plan to manufacture the product if the machinery is inoperable. The search path followed is shown in Figure 13.5.

Part of the expense of backtracking in this example arises from planning several operations that are minor details compared with the more important issue of whether equipment for manufacturing is available. This is a relatively important question that one would expect to have been established earlier in the plan, before considering the details of how to obtain the money to buy the raw materials. The more natural approach to planning is to plan out the important steps first, and then fill in the details (i.e., to plan hierarchically).

Hierarchical planning is one way of postponing commitment to a particular action until more information about the appropriateness of the action is available. This philosophy (sometimes called the *principle of least commitment*) occurs in different guises and is discussed further in Section 13.6.

Hierarchical planning requires the use of levels of abstraction in the planning process and in the description of the domain, where an abstraction level is distinguished by the granularity (or level of detail) of its description. It is unfortunate that the term "hierarchical planning" is sometimes used with different meanings. For instance, the term is sometimes used to describe levels of metaplanning, i.e., planning the process of creating a plan.

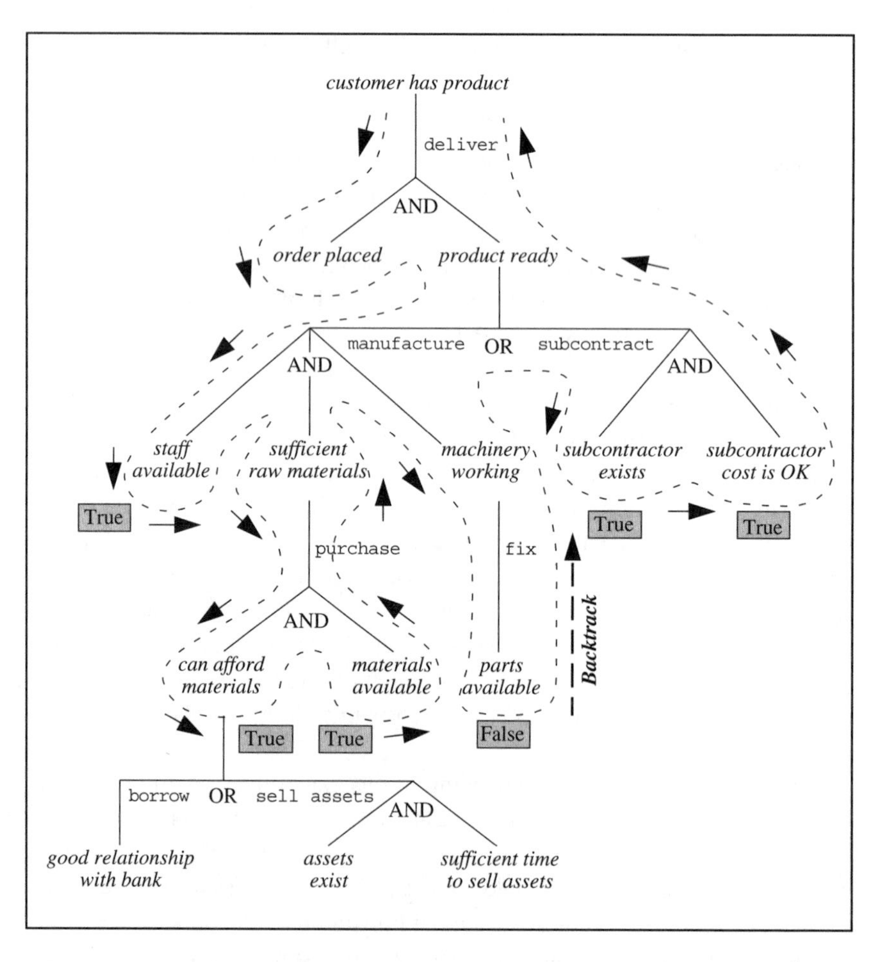

Figure 13.5 Inefficient search using STRIPS: the system backtracks on finding that there are no parts available for fixing the broken machinery

13.5.3 Hierarchical planning with ABSTRIPS

ABSTRIPS, i.e., *abstraction-based STRIPS* [10], is an extension of STRIPS that incorporates hierarchical planning. In ABSTRIPS, preconditions and operators are unchanged from STRIPS, except that some preconditions are considered more important than others. Before attempting to generate a plan, ABSTRIPS assigns an importance rating (or criticality) to each precondition. The highest criticality is ascribed to those preconditions that cannot be altered by the planner, and lower criticalities are given to preconditions that can be satisfied by generating a subplan. Planning proceeds initially by considering only the operators that have the highest criticality, thereby generating a skeleton plan. This is said to be a plan in the highest *abstraction space*. Details of the skeleton plan are filled by progressively considering lower criticality levels. In this way, subplans are generated to satisfy the preconditions in the higher level plans until all the preconditions in a plan have been achieved. The plan at any given level (save for the highest and lowest) is a refinement of the skeleton plan provided by the layer above, and is itself a skeleton plan for the level below.

ABSTRIPS adopts a semiautomated approach to the assigning of criticalities to preconditions. The user supplies a set of values, which are subsequently modified by ABSTRIPS using some simple heuristics. We will illustrate the process with reference to our example of supplying a product to a customer. The preconditions in our model include having something, something's being affordable, or something existing. The existence or otherwise of something is beyond our powers to alter and, thus, intuitively warrants the highest criticality value. On the other hand, there is a variety of different ways in which having something can be achieved, and so these preconditions might be given the lowest criticality. A sensible set of user-supplied criticality values might be as follows:

precondition:	*user-supplied criticality:*
we have an item	1
something is affordable	2
something exists or is available	3
other considerations	2

ABSTRIPS then applies heuristics for modifying the criticality values, given a particular world state. The preconditions are examined in order of decreasing user-supplied criticality, and modified as follows:

(a) Any preconditions that remain true or false, irrespective of the application of an operator, are given the maximum criticality. Let us call these *fundamental preconditions*.

(b) If a precondition can be readily established, assuming that all previously considered preconditions are satisfied (apart from unsatisfied fundamental preconditions), then the criticality is left unaltered.

(c) If a precondition cannot be readily established as described in (b), it is given a criticality value between that for category (a) and the highest in category (b).

The criticality values supplied by the user are dependent only on the nature of the preconditions themselves, whereas the modified values depend upon the starting world state and vary according to circumstances. Consider for instance the following world state:

- customer does not have turbine disk;
- customer has placed an order for a turbine disk;
- staff is available;
- we have no raw materials;
- we have no money in the bank;
- we have a good relationship with our bank;
- we do not have any assets, nor time to sell assets;
- the machinery is broken;
- spare parts are not available;
- a subcontractor exists;
- the subcontractor cost is reasonable.

The following preconditions are given a maximum criticality (say 5) because they are fundamental, and cannot be altered by any operators:

- order placed by customer;
- subcontractor exists;
- subcontractor cost is OK;
- staff available;
- raw materials available from supplier;
- machinery parts available;
- good relationship with bank;
- assets exist;
- sufficient time to sell assets.

The precondition *machinery working* falls into category (c) as it depends directly on *spare parts available*, a fundamental precondition that is false in the current world model. The remaining preconditions belong in category (b), and, therefore, their criticalities are unchanged. Although *can afford materials* is not immediately satisfied, it is readily achieved by the operator `borrow`, assuming that *good relationship with bank* is true. Therefore, *can afford materials* falls into category (b) rather than (c). Similar arguments apply to *product ready* and *sufficient raw materials*. Given the world model described, the following criticalities might be assigned:

precondition:	*initial criticality:*	*modified criticality:*
staff available	3	5
subcontractor exists	3	5
raw materials available	3	5
machinery parts available	3	5
assets exist	3	5
order placed by customer	2	5
machinery working	2	4
subcontractor cost OK	2	5
good relationship with bank	2	5
sufficient time to sell assets	2	5
can afford materials	2	2
product ready	1	1
sufficient raw materials	1	1

Once the criticalities have been assigned, the process of generating a plan can proceed as depicted by the flowchart in Figure 13.6. Planning at each abstraction level is treated as elaborating a skeleton plan generated at the level immediately higher. The main procedure is called recursively whenever a subplan is needed to satisfy the preconditions of an operator in the skeleton plan. Figure 13.6 is based on the ABSTRIPS procedure described by Sacerdoti [10], except that we have introduced a variable lower limit on the criticality in order to prevent a subplan from being considered at a lower criticality level than the precondition it aims to satisfy.

When we begin planning, a `dummy` operator is used to represent the skeleton plan. The precondition of `dummy` is the goal that we are trying to achieve. Consider planning to achieve the goal *customer has product*, beginning at abstraction level 5 (Figure 13.7). The precondition to `dummy` is *customer has product*. This precondition is satisfied by the operator `deliver`, which has two preconditions. One of them (*order placed*) is satisfied, and the other (*product ready*) has a criticality less than 5. Therefore `deliver` becomes the skeleton plan for a lower abstraction level.

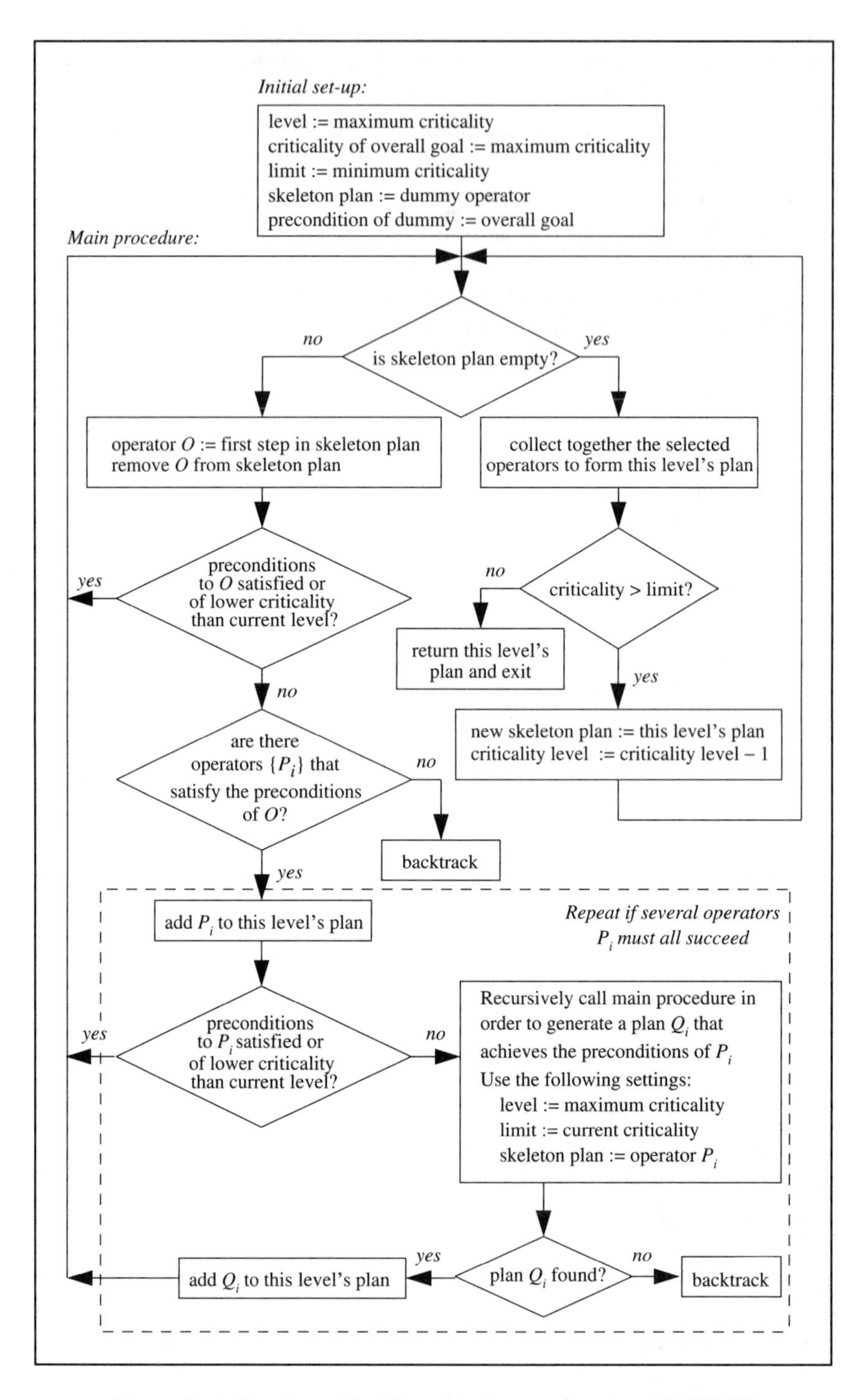

Figure 13.6 Planning with a hierarchical system based on ABSTRIPS

The skeleton plan cannot be elaborated in levels 4, 3, or 2, as the criticality of *product ready* is only 1. At level 1, operators that achieve *product ready* are sought and two are found, namely, `manufacture` and `subcontract`. Both of these operators have preconditions of the highest criticality, so there is no reason to give one priority over the other. Supposing that `manufacture` is selected, the main procedure is then called recursively, with the preconditions to `manufacture` as the new goal. The precondition of the highest criticality is *staff available*, and this is found to be satisfied. At the next level *machinery working* is examined, and the main procedure is called recursively to find a plan to satisfy this precondition. However, no such plan can be found as *parts*

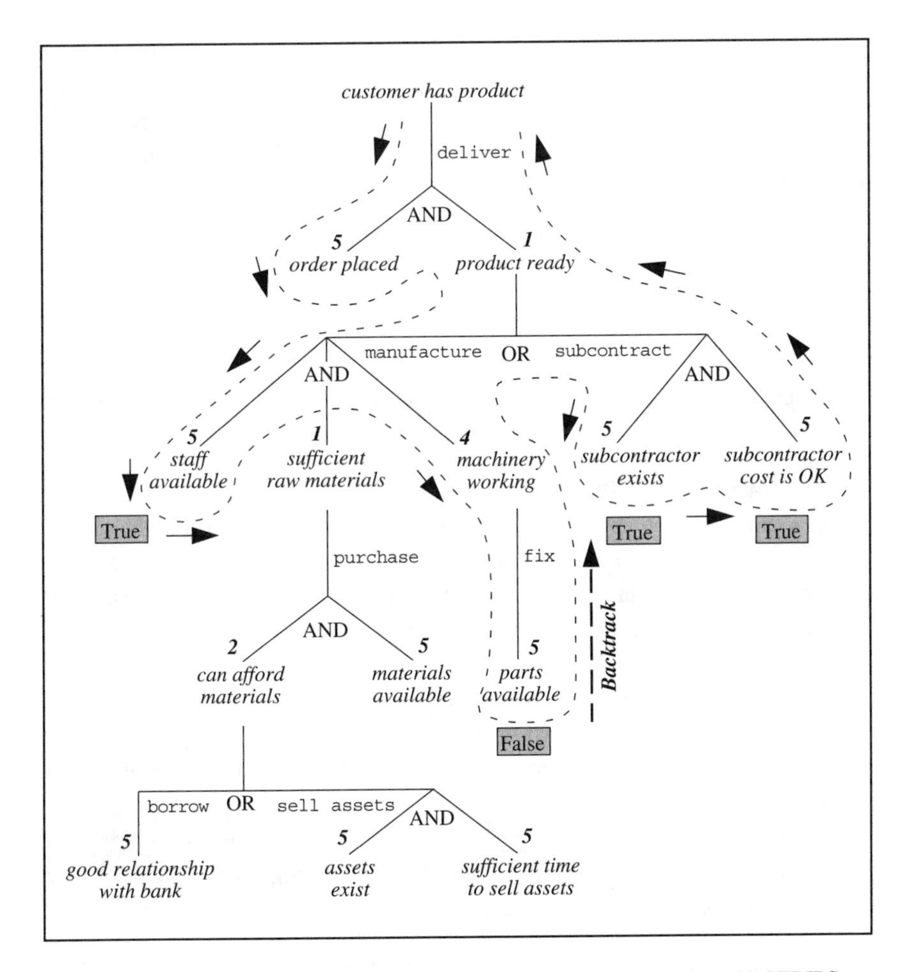

Figure 13.7 More efficient search using a hierarchical planner based on ABSTRIPS. The figures shown alongside the preconditions are the criticality levels

available is false. The plan to manufacture is abandoned at this stage and the planner backtracks to its alternative plan, `subcontract`. The preconditions to `subcontract` are satisfied and this becomes the plan.

A hierarchical planner can solve problems with less searching and backtracking than its nonhierarchical equivalent. The above example (shown in Figure 13.7) is more efficient than the STRIPS version (Figure 13.5), as the hierarchical planner did not consider the details of borrowing money and buying raw materials before abandoning the plan to manufacture. Because a complete plan is formulated at each level of abstraction before the next level is considered, the hierarchical planner can recognize dead ends early, as it did with the problem of fixing the machinery. If more complex plans are considered, involving many more operators, the saving becomes much greater still.

The planner described here, based on ABSTRIPS, is just one of many approaches to hierarchical planning. Others adopt different means of determining the hierarchical layers, since the assignment of criticalities in ABSTRIPS is rather *ad hoc*. Some of the other systems are less rigid in their use of a skeleton plan. For example, the Nonlin system [11] treats the abstraction levels as a guide to a skeleton solution, but is able to replan or consider alternatives at any level if a solution cannot be found or if a higher-level choice is faulty.

13.6 Postponement of commitment

13.6.1 Partial ordering of plans

We have already seen that an incentive for hierarchical planning is the notion that we are better off deferring detailed decisions until after more general decisions have been made. This is a part of the principle of *postponement of commitment*, or the principle of *least commitment*. In the same context, if the order of steps in a plan makes no difference to the plan, the planner should leave open the option of doing them in any order. A plan is said to be *partially ordered* if it contains actions that are unordered with respect to each other, i.e., actions for which the planner has not yet determined an order and which may possibly be in parallel.

If we refer back to the Prolog planner described in Section 13.3.3, we see that, given a particular world state and the goal of supplying a product to a customer, the following plan was generated:

```
[borrow_money, purchase_materials, fix_machine, manufacture,
deliver]
```

This plan contains a definite order of events. As can be inferred from the search tree in Figure 13.2, some events must occur before others. For example, the product cannot be delivered until after it has been manufactured. However, the operators `fix_machine` and `purchase_materials` have been placed in an arbitrary order. These operators are intended to satisfy two subgoals (*machinery working* and *sufficient raw materials*, respectively). As both subgoals need to be achieved, they are conjunctive goals. A planning system is said to be *linear* if it assumes that it does not matter in which order conjunctive goals are satisfied. This is the so-called *linear assumption*, which is not necessarily valid, and which can be expressed as follows:

> *according to the linear assumption, subgoals are independent and thus can be sequentially achieved in an arbitrary order.*

Nonlinear planners are those that do not rely upon this assumption. The generation of partially ordered plans is the most common form of nonlinear planning, but Hendler et al. [12] have pointed out that it is not the only form. The generation of a partially ordered plan avoids commitment to a particular order of actions until information for selecting one order in preference to another has been gathered. Thus a nonlinear planner might generate the following partially ordered plan:

$$\left[\begin{bmatrix}\texttt{borrow\_money, purchase\_materials}\\ \texttt{fix\_machine}\end{bmatrix}, \texttt{manufacture, deliver}\right]$$

If it is subsequently discovered that fixing the machine requires us to borrow money, this can be accommodated readily because we have not committed ourselves to fixing the machine before seeking a loan. Thus, a single loan can be organized for the purchase of raw materials and for fixing the machine.

The option to generate a partially ordered plan occurs every time a planner encounters a conjunctive node (i.e., AND) on the search tree. Linear planners are adequate when the branches are decoupled, so that it doesn't matter which action is performed first. Where the ordering is important, a nonlinear planner can avoid an exponential search of all possible plan orderings. To emphasize how enormous this saving can be, just ten actions have more than three million (i.e., 10!) possible orderings.

Some nonlinear planners (e.g., HACKER [13] and INTERPLAN [14]) adopt a different approach to limiting the number of orderings that need be considered. These systems start out by making the linearity assumption. When confronted with a conjunctive node in the search tree, they select an arbitrary order for the actions corresponding to the separate branches. If the selected order is subsequently found to create problems, the plan is fixed by reordering. Depending on the problem being addressed, this approach may be inefficient as it can involve a large amount of backtracking.

We have already seen that the actions of one branch of the search tree can interact with the actions of another. As a further example, a system might plan to purchase sufficient raw materials for manufacturing a single batch, but some of these materials might be used up in the alignment of machinery following its repair. Detecting and correcting these interactions is a problem that has been addressed by most of the more sophisticated planners. The problem is particularly difficult in the case of planners such as SIPE that allow actions to take place concurrently. SIPE tackles the problem by allocating a share of limited resources to each action and placing restrictions on concurrent actions that use the same resources. Modeling the process in this way has the advantage that resource conflicts are easier to detect than interactions between the effects of two actions.

13.6.2 The use of planning variables

The use of planning variables is another technique for postponing decisions until they have to be made. Planners with this capability could, for instance, plan to purchase `something`, where `something` is a variable that does not yet have a value assigned. Thus the planner can accumulate information before making a decision about what to purchase. The instantiation of `something` may be determined later, thus avoiding the need to produce and check a plan for every possible instantiation.

The use of planning variables becomes more powerful still if we can progressively limit the possible instantiations by applying constraints to the values that a variable can take. Rather than assuming that `something` is either unknown or has a specific value (e.g., `gearbox part #7934`), we could start by applying the constraint that it is a gearbox component. We might then progressively tighten the constraints and, thereby, reduce the number of possible instantiations.

13.7 Job-shop scheduling

13.7.1 The problem

As noted in Section 13.1, scheduling is a planning problem where time and resources must be allocated to operators that are known in advance. The term *scheduling* is sometimes applied to the internal scheduling of operations within a knowledge-based system. However, in this section we will be concerned with scheduling only in an engineering context.

Job-shop scheduling is a problem of great commercial importance. A job shop is either a factory or a manufacturing unit within a factory. Typically, the job shop consists of a number of machine tools connected by an automated palletized transportation system, as shown in Figure 13.8. The completion of a job may require the production of many different parts, grouped in lots. Flexible manufacturing is possible, because different machines can work on different part types simultaneously, allowing the job shop to adapt rapidly to changes in production mix and volume.

The planning task is to determine a schedule for the manufacturing of the parts that make up a job. As noted in Section 13.1, the operations are already known in this type of problem, but they still need to be organized in the most efficient way. The output that is required from a scheduling system is (typically) a Gantt chart like that shown in Figure 13.9. The decisions required are, therefore:

- the allocation of machines (or other resources) to each operation;
- the start and finish times of each operation; although it may be sufficient to specify only the order of operations, rather than their projected timings.

The output of the job shop should display *graceful degradation*, i.e., a reduced output should be maintained in the event of accidental or pre-programmed machine stops, rather than the whole job shop grinding to a halt.

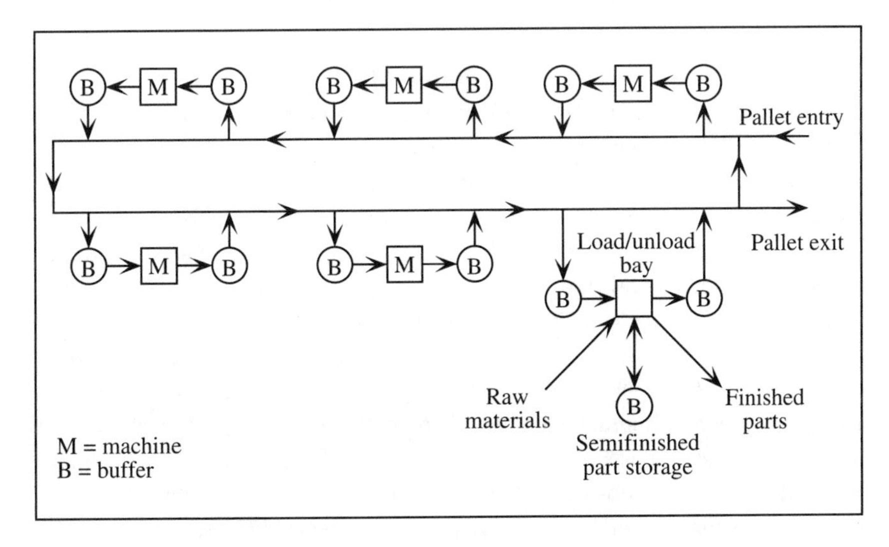

Figure 13.8 A possible job shop layout (adapted from [15])

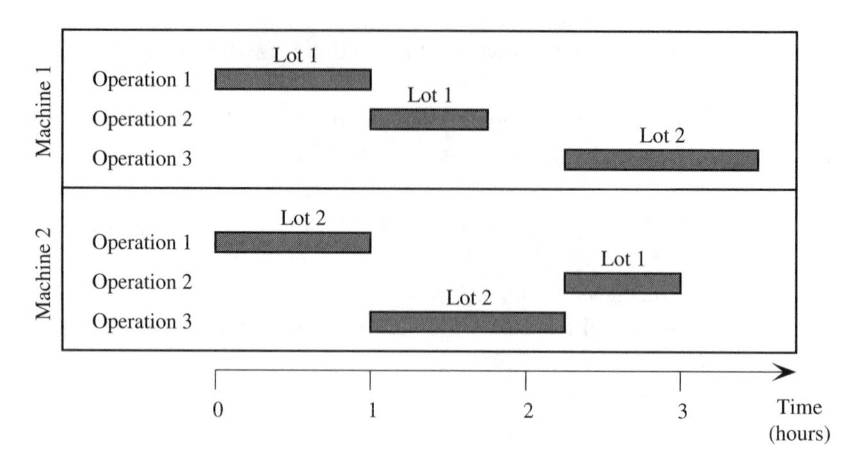

Figure 13.9 A Gantt chart:
giving a visual representation of a schedule

The schedule must ensure that all jobs are completed before their due dates, while taking account of related considerations such as minimizing machine idle times, queues at machines, work in progress, and allowing a safety margin in case of unexpected machine breakdown. Some of these considerations are *constraints* that *must* be satisfied, and others are *preferences* that we would *like* to satisfy to some degree. Several researchers (e.g., Bruno et al. [15] and Dattero et al. [16]) have pointed out that a satisfactory schedule is required, and that this may not necessarily be an optimum. A similar viewpoint is frequently adopted in the area of engineering design (Chapter 12).

13.7.2 Some approaches to scheduling

The approaches to automated scheduling that have been applied in the past can be categorized as either analytical, iterative, or heuristic [17]. None of these approaches has been particularly successful in its own right, but some of the more successful scheduling systems borrow techniques from all three approaches. The analytical approach requires the problem to be structured into a formal mathematical model. Achieving this normally requires several assumptions to be made, which can compromise the validity of the model in real-life situations. The iterative approach requires all possible schedules to be tested and the best one to be selected. The computational load of such an approach renders it impractical where there are large numbers of machines and lots. The heuristic approach relies on the use of rules to guide the scheduling. While this can save considerable computation, the rules are often specific to just one situation, and their expressive power may be inadequate.

Bruno et al. [15] have adopted a semiempirical approach to the scheduling problem. A discrete event simulation (similar in principle to the ultrasonic simulation described in Chapter 4) serves as a test bed for the effects of action sequences. If a particular sequence causes a constraint to be violated, the simulation can backtrack by one or more events and then test a new sequence. Objects are used to represent the key players in the simulation, such as lots, events, and goals. Rules are used to guide the selection of event sequences using priorities that are allocated to each lot by the simple expression:

$$\text{priority} = \frac{\text{remaining machining time}}{\text{due date} - \text{release time}}$$

Liu [17] has extended the ideas of hierarchical planning (Section 13.5) to include job-shop scheduling. He has pointed out that one of the greatest problems in scheduling is the interaction between events, so that fixing one problem (e.g., bringing forward a particular machining operation) can generate new problems (e.g., another lot might require the same operation, but its schedule cannot be moved forward). Liu therefore sees the problem as one of maintaining the integrity of a global plan, and dealing with the effects of local decisions on that global plan. He solves the problem by introducing planning levels. He starts by generating a rough utilization plan — typically based upon the one resource thought to be the most critical — that acts as a guideline for a more detailed schedule. The rough plan is not expanded into a more detailed plan, but rather a detailed plan is formed from scratch, with the rough plan acting as a guide.

Rather than attempt to describe all approaches to the scheduling problem, one particular approach will now be described in some detail. This approach involves constraint-based analysis (CBA) coupled with the application of preferences.

13.8 Constraint-based analysis

13.8.1 Constraints and preferences

There may be many factors to take into account when generating a schedule. As noted in Section 13.7.1, some of these are *constraints* that *must* be satisfied, and others are *preferences* that we would *like* to satisfy. Whether or not a constraint is satisfied is generally clear cut, e.g., a product is either ready on time or it is not. The satisfaction of preferences is sometimes clear cut, but often it is not. For instance, we might prefer to use a particular machine. This preference is clear-cut because it will either be met or it will not. On the other hand, a preference such as "minimize machine idle times" can be met to varying degrees.

13.8.2 Formalizing the constraints

Four types of scheduling constraints that apply to a flexible manufacturing system can be identified:

- *Production constraints*
  The specified quantity of goods must be ready before the due date, and quality standards must be maintained throughout. Each lot has an earliest start time and a latest finish time.

- *Technological coherence constraints*
  Work on a given lot cannot commence until it has entered the transportation system. Some operations must precede others within a given job, and sometimes a predetermined sequence of operations exists. Some stages of manufacturing may require specific machines.

- *Resource constraints*
  Each operation must have access to sufficient resources. The only resource that we will consider in this study is time at a machine, where the number of available machines is limited. Each machine can work on only one lot at a given time, and programmed maintenance periods for machines must be taken into account.

- *Capacity constraints*
  In order to avoid unacceptable congestion in the transportation system, machine use and queue lengths must not exceed predefined limits.

Our discussion of constraint-based analysis will be based upon the work of Bel et al. [2]. Their knowledge-based system, OPAL, solves the static (i.e., "snapshot") job-shop scheduling problem. It is, therefore, a classical planner (see Section 13.2). OPAL contains five modules (Figure 13.10):

- an object-oriented database for representing entities in the system such as lots, operations, and resources (including machines);

- a constraint-based analysis (CBA) module that calculates the effects of time constraints on the sequence of operations (the module generates a set of precedence relations between operations, thereby partially or fully defining those sequences that are viable);

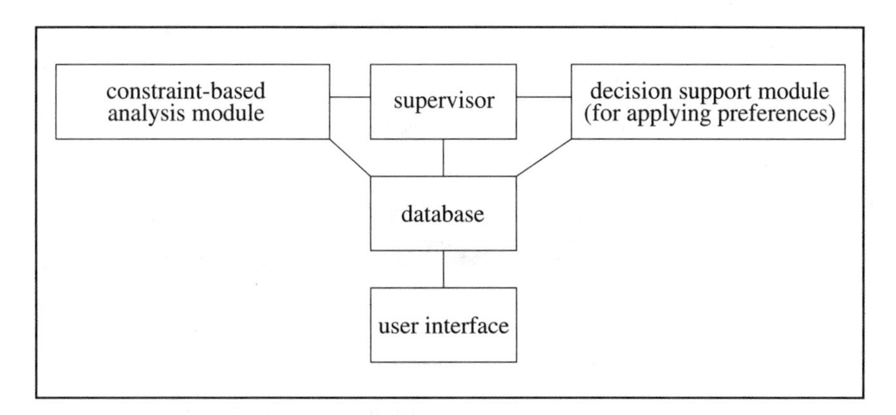

Figure 13.10 Principal modules in the OPAL scheduling system [2]

- a decision support module that contains rules for choosing a schedule, based upon practical or heuristic experience, from among those that the CBA module has found to be viable;

- a supervisor that controls communication between the CBA and decision support modules, and builds up the schedule for presentation to the user;

- a user interface module.

Job-shop scheduling can be viewed in terms of juggling operations and resources. Operations are the tasks that need to be performed in order to complete a job, and several jobs may need to be scheduled together. Operations are characterized by their start time and duration. Each operation normally uses resources, such as a length of time at a given machine. There are two types of decision, the timing or (sequencing) of operations and the allocation of resources. For the moment we will concentrate on the CBA module, which is based upon the following set of assumptions:

- there is a set of jobs J comprised of a set of operations O;
- there is a limited set of resources R;
- each operation has the following properties:
    - it cannot be interrupted;
    - it uses a subset r of the available resources;
    - it uses a quantity q_i of each resource r_i in the set r;
    - it has a fixed duration d_i.

A schedule is characterized by a set of operations, their start times, and their durations. We will assume that the operations that make up a job and their durations are predefined. Therefore, a schedule can be specified by just a set of

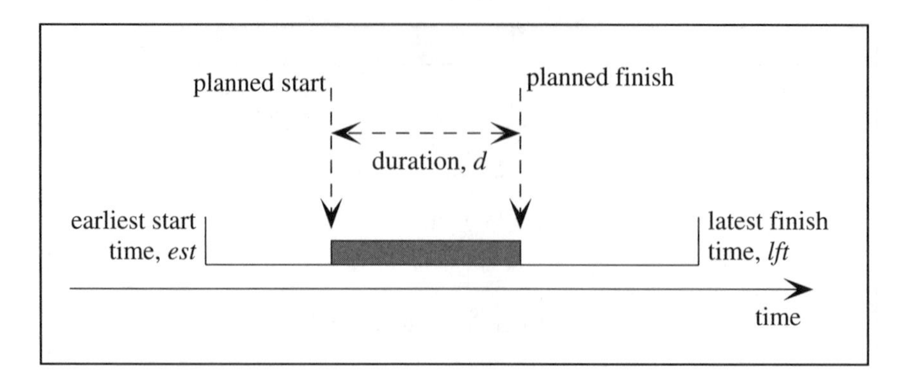

Figure 13.11 Scheduling an operation within its available time window

start times for the operations. For a given job, there is an earliest start time (est_i) and latest finish time (lft_i) for each operation O_i. Each operation has a time window which it *could* occupy and a duration within that window that it *will* occupy. The problem is then one of positioning the operation within the window as shown in Figure 13.11.

13.8.3 Identifying the critical sets of operations

The first step in the application of CBA is to determine if and where conflicts for resources arise. These conflicts are identified through the use of *critical sets*, a concept that is best described by example. Suppose that a small factory employs five workers who are suitably skilled for carrying out a set of four operations O_i. Each operation requires some number q_i of the workers, as follows:

	O_1	O_2	O_3	O_4
Number of workers required, q_i	2	3	5	2

A critical set of operations I_c is one that requires more resources than are available, but where this conflict would be resolved if *any* of the operations were removed from the set. In our example, the workers are the resource and the critical sets are $\{O_1,O_3\}$, $\{O_2,O_3\}$, $\{O_4,O_3\}$, $\{O_1,O_2,O_4\}$. Note that $\{O_1,O_2,O_3\}$ is not a critical set since it would still require more resources than are available if we removed O_1 or O_2. The critical sets define the conflicts for resources, because the operations that make up the critical sets cannot be carried out simultaneously. Therefore, the first sequencing rule is as follows:

> *one operation of each critical set must precede at least one other operation in the same critical set.*

Applying this rule to the above example produces the following conditions:

(i) either (O_1 precedes O_3) or (O_3 precedes O_1);

(ii) either (O_2 precedes O_3) or (O_3 precedes O_2);

(iii) either (O_4 precedes O_3) or (O_3 precedes O_4);

(iv) either (O_1 precedes O_2) or (O_2 precedes O_1) or
(O_1 precedes O_4) or (O_4 precedes O_1) or
(O_2 precedes O_4) or (O_4 precedes O_2).

These conditions have been deduced purely on the basis of the available resources, without consideration of time constraints. If we now introduce the known time constraints (i.e., the earliest start times and latest finish times for each operation) the schedule of operations can be refined further and, in some cases, defined completely. The schedule of operations is especially constrained in the case where each conflict set is a *pair* of operations. This is the *disjunctive* case, which we will consider first before moving on to consider the more general case.

13.8.4 Sequencing in the disjunctive case

As each conflict set is a pair of operations in the disjunctive case, no operations can be carried out simultaneously. Each operation has a defined duration (d_i), an earliest start time (est_i), and a latest finish time (lft_i). The scheduling task is one of determining the actual start time for each operation.

Consider the task of scheduling the three operations *A*, *B*, and *C* shown in Figure 13.12(a). If we try to schedule operation *A* first, we find that there is insufficient time for the remaining operations to be carried out before the last *lft*, irrespective of how the other operations are ordered (Figure 13.12(b)). However, there is a feasible schedule if operation *C* precedes *A*. This is an example of the general rule:

```
/* Rule 13.1 */
IF (latest lft - estA) < ? di
THEN at least one operation must precede A.
```

Similarly, there is no feasible schedule that has *A* as the last operation since there is insufficient time to perform all operations between the earliest *est* and the *lft* for *A* (Figure 13.12(c)). The general rule that describes this situation is:

```
/* Rule 13.2 */
IF (lftA - earliest est) < ? di
THEN at least one operation must follow A.
```

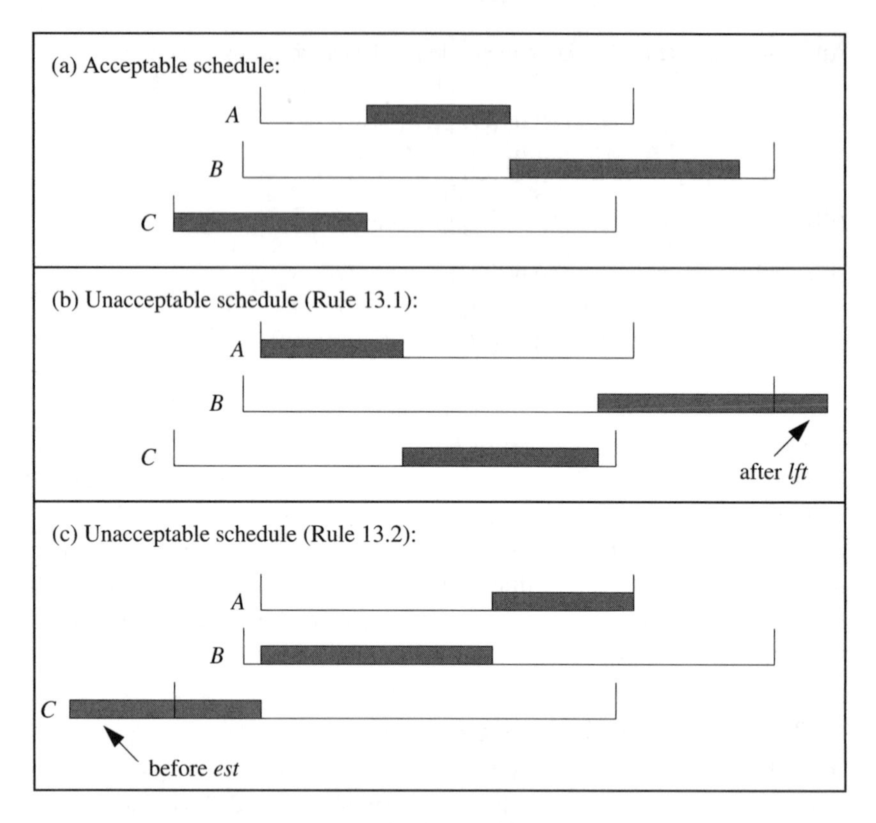

Figure 13.12 Sequencing in the disjunctive case:
(a) An acceptable schedule
(b) *A* cannot be the first operation (Rule 13.1)
(c) *A* cannot be the last operation (Rule 13.2)

13.8.5 Sequencing in the nondisjunctive case

In the nondisjunctive case, at least one critical set contains more than two operations. The precedence rules described above can be applied to those critical sets that contain only two operations. Let us consider the precedence constraints that apply to the operations O_i of a critical set having more than two elements. From our original definition of a critical set, it is not possible for all of the operations in the set to be carried out simultaneously. At any one time, at least one operation in the set must either have finished or be waiting to start. This provides the basis for some precedence relations. Let us denote the critical set by the symbol *S*, where *S* includes the operation *A*. Another set of operations that contains all elements of *S* apart from operation *A* will be denoted by the letter *W*. The two precedence rules that apply are as follows:

```
/* Rule 13.3 */
IF   for every pair {Oi, Oj} of operations in set W:
     lfti-estj < di + dj
     /* Oi cannot be performed after Oj has finished */
AND  for every operation (Oi) in set W:
     lfti-estA < dA + di
     /* Oi cannot be performed after OA has finished) */
THEN at least one operation in set W must have finished before
     A starts
```

```
/* Rule 13.4 */
IF   for every pair {Oi, Oj} of operations in set W:
     lfti-estj < di + dj
     /* Oi cannot be performed after Oj has finished */
AND  for every operation (Oi) in set W:
     lftA-esti < dA + di
    /* OA cannot be performed after Oi has finished */
THEN A must finish before at least one operation in set W
     starts
```

The application of Rule 13.3 is shown in Figure 13.13(a). Operations *B* and *C* have to overlap (the first condition) and it is not possible for *A* to finish before one of either *B* or *C* has started (the second condition). Therefore, operation *A must* be preceded by at least one of the other operations. Note that, if this is not possible either, there is no feasible schedule. (Overlap of all the operations is unavoidable in such a case, but since we are dealing with a critical set there is insufficient resource to support this.)

Rule 13.4 is similar and covers the situation depicted in Figure 13.13(b). Here operations *B* and *C* again have to overlap, and it is not possible to delay the start of *A* until after one of the other operations has finished. Under these circumstances operation *A must* precede at least one of the other operations.

13.8.6 Updating earliest start times and latest finish times

If Rule 13.1 or 13.3 has been fired, so we know that at least one operation must precede *A*, it may be possible to update the earliest start time of *A* to reflect this restriction, as shown in Figures 13.14a and 13.14b. The rule that describes this is:

```
/* Rule 13.5 */
IF some operations must precede A (by Rule 13.1 or 13.3)
AND [the earliest (est+d) of those operations] > estA
THEN the new estA is the earliest (est+d) of those operations
```

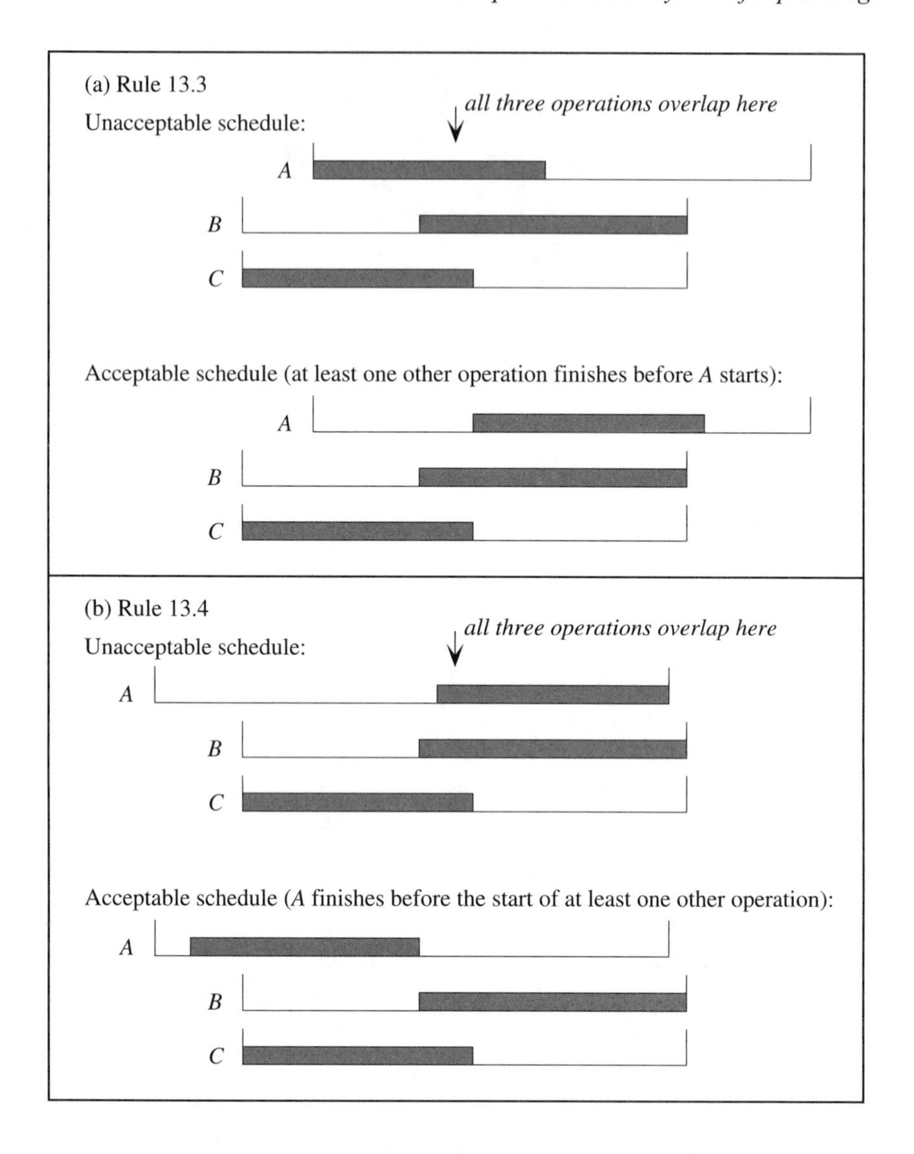

Figure 13.13 Sequencing in the nondisjunctive case

Similarly, if Rule 13.2 or 13.4 has been fired, so we know that at least one operation must follow operation *A*, it may be possible to modify the *lft* for *A*, as shown in Figures 13.14(a) and 13.14(c). The rule that describes this is:

```
/* Rule 13.6 */
IF some operations must follow A (by Rule 13.2 or 13.4)
AND [the latest (lft-d) of those operations] < lftA
THEN the new lftA is the latest (lft-d) of those operations
```

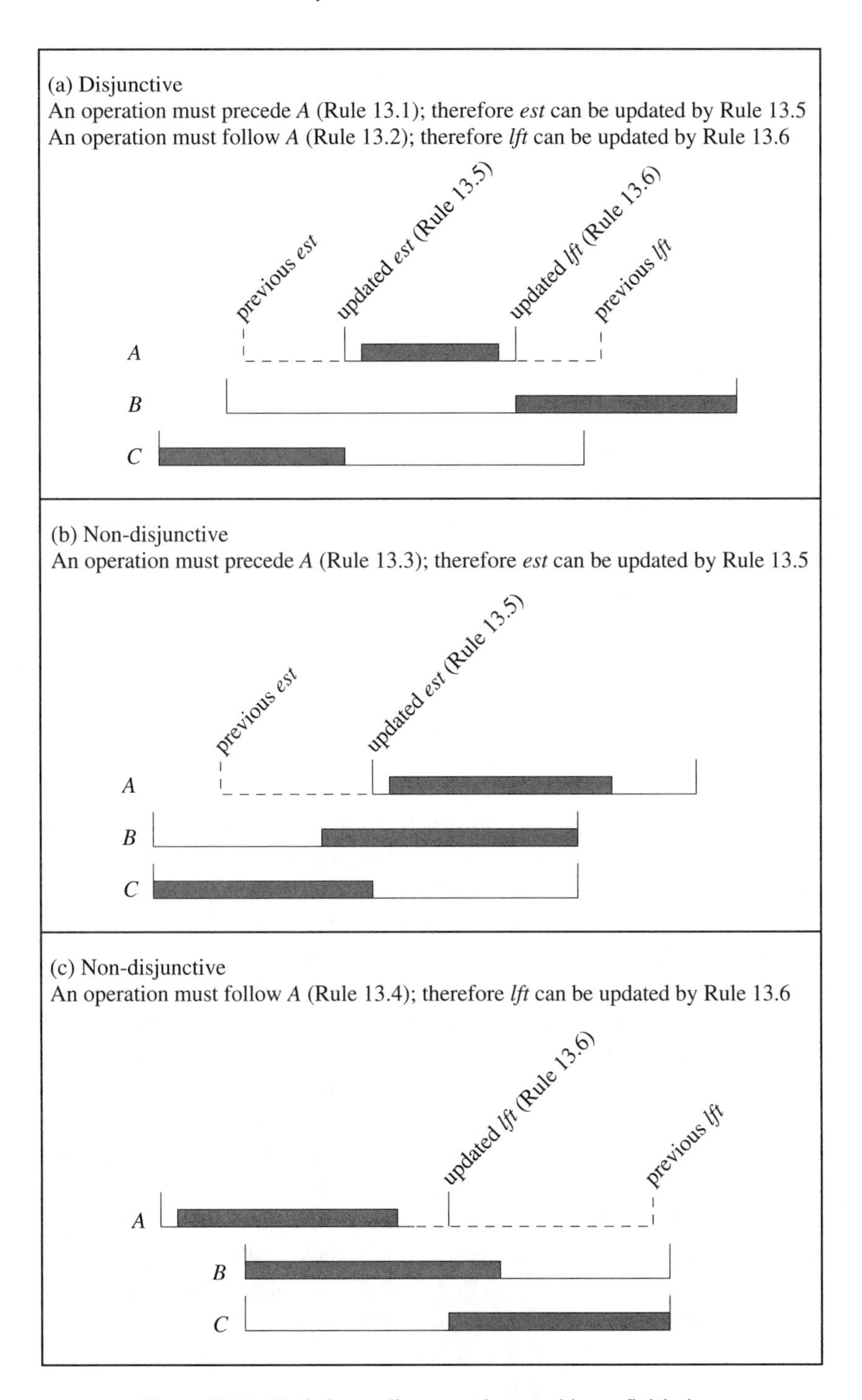

Figure 13.14 Updating earliest start times and latest finish times

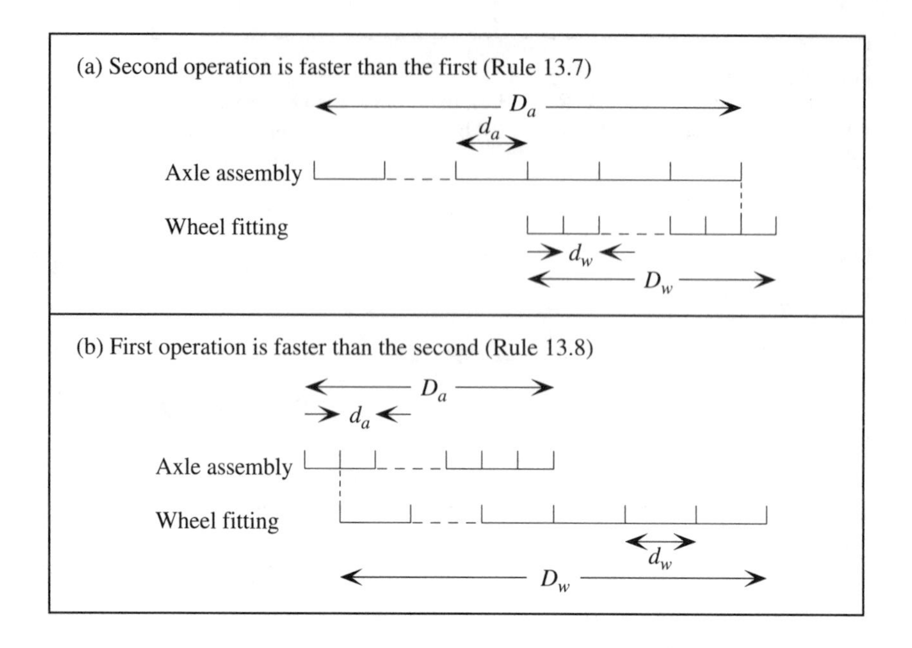

Figure 13.15 Overlapping operations in batch processing:
(a) Rule 13.7 applies if wheel fitting is faster than axle assembly
(b) Rule 13.8 applies if axle assembly is faster than wheel fitting

The new earliest start times and latest finish times can have an effect on subsequent operations. Consider the case of a factory that is assembling cars. Axle assembly must precede wheel fitting, regardless of any resource considerations. This sort of precedence relation, which is imposed by the nature of the task itself, is referred to as a *technological coherence constraint.* Suppose that, as a result of the arguments described above, the *est* for axle assembly is delayed to a new time est_a. If the axle assembly takes time d_a per car, then the new *est* for wheel-fitting (est_w) will be est_a+d_a (ignoring the time taken to move the vehicle between assembly stations).

However, a car plant is unlikely to be moving just one car through the assembly process, rather a whole batch of cars will need to be scheduled. These circumstances offer greater flexibility as the two assembly operations can overlap, provided that the order of assembly is maintained for individual cars. Two rules can be derived, depending on whether axle assembly or wheel fitting is the quicker operation. If wheel fitting is quicker, then a sufficient condition is for wheels to be fitted to the last car in the batch immediately after its axles have been assembled. This situation is shown in Figure 13.15(a), and is described by the following rule:

```
/* Rule 13.7 */
IF da > dw
THEN estw = esta + Da - (n-1)dw
AND lfta = lftw - dw
```

where n is the number of cars in the batch, D_a is nd_a, and D_w is nd_w. If axle assembly is the quicker operation, then wheel fitting can commence immediately after the first car has had its axle assembled (Figure 13.15(b)). The following rule applies:

```
/* Rule 13.8 */
IF da < dw
THEN estw = esta + da
AND lfta = lftw - Dw + (n-1)da
```

13.8.7 Applying preferences

Constraint-based analysis can produce one of three possible outcomes:

- the constraints cannot be satisfied, given the allocated resources;
- a unique schedule is produced that satisfies the constraints;
- more than one schedule is found that satisfies the constraints.

In the case of the first outcome, the problem can be solved only if more resources are made available or the time constraints are slackened. In the second case, the scheduling problem is solved. In the third case, pairs of operations exist that can be sequenced in either order without violating the constraints. The problem then becomes one of applying *preferences* so as to find the most suitable order. Preferences are features of the solution that are considered desirable, but unlike constraints they are not compulsory. Bel et al. [2] have attempted to apply preferences by using fuzzy rules (see Chapter 3 for a discussion of fuzzy logic). Their rule-based system constitutes the decision-support module in Figure 13.10.

The pairs of operations that need to be ordered are potentially large in number and broad in scope. It is, therefore, impractical to produce a rule base that covers specific pairs of operations. Instead, rules for applying preferences usually make use of variables (see Section 2.6), so that they can be applied to different pairs of operations. The rules may take the form:

```
IF
(attribute x of operation ?a) > (attribute x of operation ?b)
THEN
?a precedes ?b
```

where any pair of operations can be substituted for `a` and `b`, but the attribute `x` is specified in a given rule. A typical expression for `x` might be the duration of its available time window (i.e., *lft* – *est*). The rules are chosen so as to cover some general guiding principles or goals, such as:

- maximize overall slack time;
- perform operations with the least slack time first;
- give preference to schedules that minimize resource utilization;
- avoid tool changes.

In OPAL, each rule is assigned a degree of relevance R with respect to each goal. Given a goal or set of goals, some rules will tend to favor one ordering of a pair of operations, while others will favor the reverse. A consensus is arrived at by a rather complicated scoring (or "voting") procedure. The complete set of rules is applied to each pair of operations that need to be ordered, and the scoring procedure is as follows:

- The degree to which the condition part of each rule is satisfied is represented using fuzzy sets. Three fuzzy sets are defined (`true`, `maybe`, and `false`), and the degree of membership of each is designated μ_t, μ_m, and μ_f. Each membership is a number between 0 and 1, such that for every rule:

$$\mu_t + \mu_m + \mu_f = 1$$

- Every rule awards a score to each of the three possible outcomes (a precedes b, b precedes a, or no preference). These scores (V_{ab}, V_{ba}, and $V_{no\_preference}$) are determined as follows:

$$V_{ab} = \min(\mu_t, R)$$
$$V_{ba} = \min(\mu_f, R)$$
$$V_{no\_preference} = \min(\mu_m, R)$$

- The scores for each of the three possible outcomes are totaled across the whole rule set. The totals can also be normalized by dividing the sum of the scores by the sum of the R values for all rules. Thus:

$$\text{Total score for ab} = \Sigma(V_{ab}) / \Sigma(R)$$
$$\text{Total score for ba} = \Sigma(V_{ba}) / \Sigma(R)$$
$$\text{Total score for no preference} = \Sigma(V_{no\_preference}) / \Sigma(R)$$

The scoring procedure can have one of three possible outcomes:

- One ordering is favored over the other. This outcome is manifested by $\Sigma(V_{ab}) \gg \Sigma(V_{ba})$ or vice versa.

- Both orderings are roughly equally favored, but the scores for each are low compared with the score for the impartial outcome. This indicates that there is no strong reason for preferring one order over the other.

- Both orderings are roughly equally favored, but the scores for each are high compared with the score for the impartial outcome. Under these circumstances, there are strong reasons for preferring one order, but there are also strong reasons for preferring the reverse order. In other words there are strong conflicts between the rules.

13.8.8 Using constraints and preferences

OPAL makes use of a supervisor module (Figure 13.10), which controls the constraint-based analysis (CBA) and decision support (DS) modules. In OPAL and other scheduling systems, constraint-based analysis is initially applied in order to eliminate all those schedules that cannot meet the time and resource constraints. A preferred ordering of operations is then selected from the schedules that are left. As we have seen, the DS module achieves this by weighting the rules of optimization, applying these rules, and selecting the order of operations that obtains the highest overall score.

If the preferences applied by the DS module are not sufficient to produce a unique schedule, the supervising module calls upon the CBA and DS modules repeatedly until a unique solution is found. The supervising module stops the process when an acceptable schedule has been found, or if it cannot find a workable schedule.

There is a clear analogy between the two stages of scheduling (constraint-based analysis and the application of preferences) and the stages of materials selection (see Section 12.8). In the case of materials selection, constraints can be applied by ruling out all materials that fail to meet a numerical specification. The remaining materials can then be put into an order of preference, based upon some means of comparing their performance scores against various criteria, where the scores are weighted according to a measure of the perceived importance of the properties.

13.9 Replanning and reactive planning

The discussion so far has concentrated on predictive planning, that is, building a plan that is to be executed at some time in the future. Suppose now that while a plan is being executed something unexpected happens, such as a machine breakdown. In other words, the actual world state deviates from the expected world state. A powerful capability under these circumstances is to be able to replan, i.e., to modify the current plan to reflect the current circumstances. Systems that are capable of planning or replanning in real time, in a rapidly changing environment, are described as *reactive* planners. Such systems monitor the state of the world during execution of a plan and are capable of revising the plan in response to their observations. Since a reactive planner can alter the actions of machinery in response to real-time observations and measurements, the distinction between reactive planners and control systems (Chapter 14) is vague.

To illustrate the distinction between predictive planning, replanning, and reactive planning, let us consider an intelligent robot that is carrying out some gardening. It may have planned in advance to mow the lawn and then to prune the roses (predictive planning). If it finds that someone has borrowed the mower, it might decide to mow after pruning, by which time the mower may have become available (replanning). If a missile is thrown at the robot while it is pruning, it may choose to dive for cover (reactive planning).

Collinot et al. have devised a system called SONIA [18], that is capable of both predictive and reactive planning of factory activity. This system offers more than just a scheduling capability, as it has some facilities for selecting which operations will be scheduled. However, it assumes that a core of essential operations is preselected. SONIA brings together many of the techniques that we have seen previously. The operations are ordered by the application of constraints and preferences (Section 13.8) in order to form a predictive plan. During execution of the plan, the observed world state may deviate from the planned world state. Some possible causes of the deviation might be:

- machine failure;
- personnel absence;
- arrival of urgent new orders;
- the original plan may have been ill-conceived.

Under these circumstances, SONIA can modify its plan or, in an extreme case, generate a new plan from scratch. The particular type of modification chosen is largely dependent on the time available for SONIA to reason about the problem. Some possible plan modifications might be:

- cancel, postpone, or curtail some operations in order to bring some other operations back on schedule;
- reschedule to use any slack time in the original plan;
- reallocate resources between operations;
- reschedule a job — comprising a series of operations — to finish later than previously planned;
- delay the whole schedule.

A final point to note about SONIA is that it has been implemented as a blackboard system (see Chapter 9). The versatility of the blackboard architecture is demonstrated by the variety of applications in which it is used. An application in data interpretation was described in Chapter 11, whereas here it is used for predictive planning. SONIA uses the blackboard to represent both the planned and observed world states. Separate knowledge sources perform the various stages of predictive planning, monitoring, and reactive planning. In fact, SONIA uses two blackboards: one for representing information about the shop floor and a separate one for its own internal control information.

13.10 Summary

This chapter began by defining a classical planner as one that can derive a set of operations to take the world from an initial state to a goal state. The initial world state is a "snapshot," which is assumed not to alter except as a result of the execution of the plan. While being useful in a wide range of situations, classical planners are of limited use when dealing with continuous processes or a rapidly changing environment. In contrast, reactive planners can respond rapidly to unexpected events.

Scheduling is a special case of planning, where the operators are known in advance and the task is to allocate resources to each and to determine when they should be applied. Scheduling is particularly important in the planning of manufacturing processes.

A simple classical planner similar to STRIPS has been described. More sophisticated features that can be added to extend the capabilities of a planning system were then described and are summarized below.

(i) *World modeling*

Unlike STRIPS, the Prolog program shown in Section 13.3.3 does not explicitly update its world model as operators are selected. Proper maintenance of a world model ensures that any side effects of a plan are recorded along with intended effects.

(ii) *Deductive rules*

Deductive rules permit the deduction of effects that are additional to those explicitly included in the operator representation. Because they do not form part of the operator representation, they can allow different effects to be registered depending on the current context.

(iii) *Hierarchical planning*

STRIPS may commit itself to a particular problem-solving path too early, with the result that it must backtrack if it cannot complete the plan that it is pursuing. Hierarchical planners can plan at different levels of abstraction. An abstract (or "skeleton") plan is formed first, such as to build a house by digging foundations, then building walls, and finally putting on a roof. The detailed planning might include the precise shape, size, and placement of the timber. The abstract plan restricts the range of possibilities of the detailed planning.

(iv) *Nonlinearity*

Linear planners such as STRIPS make the assumption that it does not matter in which order the subgoals for a particular goal are satisfied. Partial ordering of operations is a form of nonlinear planning in which the ordering of operations is postponed until either more information becomes available or a decision is forced. In some cases is may be possible for operations to be scheduled to run in parallel.

(v) *Planning variables*

The use of variables also allows postponement of decisions. Plans can be generated using variables that have not yet been assigned a value. For instance, we might plan to go `somewhere` without specifying where. The instantiation of `somewhere` may become determined later, and we have saved the effort of considering all of the possibilities in the meantime.

(vi) *Constraints*

The use of planning variables is more powerful still if we can limit the possible instantiations by applying constraints on the values that a variable can take. Constraint-based analysis can be used to reduce the number of possible plans, or even to find a unique plan that meets the constraints.

(vii) *Preferences*
If constraint-based analysis yields more than one viable plan or schedule, preferences can be applied in order to select between the alternatives.

(viii) *Replanning*
The ability to modify a plan in the light of unexpected occurrences was discussed.

Hierarchical plans, partially ordered plans, and the use of planning variables are all means of postponing commitment to a particular plan. This is known as the *principle of least commitment.*

Systems that have only some of the above features are adequate in many situations. A few specific applications have been considered in this chapter, but much of the research effort in planning has been concerned with building general purpose planning systems that are domain-independent. The purpose of such systems is to allow knowledge relevant to any particular domain to be represented, rather like an expert system shell (see Chapters 1 and 10).

References

1. Charniak, E. and McDermott, D., *Introduction to Artificial Intelligence*, Addison-Wesley, 1985.
2. Bel, G., Bensana, E., Dubois, D., Erschler, J., and Esquirol, P., "A knowledge-based approach to industrial job-shop scheduling," in *Knowledge-Based Systems in Manufacturing*, Kusiak, A. (Ed.), pp. 207–246, Taylor and Francis, 1989.
3. Stuart, C. J., "An implementation of a multi-agent plan synchronizer," 9th International Joint Conference on Artificial Intelligence (IJCAI'85), Los Angeles, pp. 1031–1033, 1985.
4. Durfee, E. H., Lesser, V. R., and Corkhill, D. D., "Increasing coherence in a distributed problem solving network," 9th International Joint Conference on Artificial Intelligence (IJCAI'85), Los Angeles, pp. 1025–1030, 1985.
5. Fikes, R. E. and Nilsson, N. J., "STRIPS: a new approach to the application of theorem proving to problem solving," *Artificial Intelligence*, vol. 2, pp. 189–208, 1971.
6. Fikes, R. E., Hart, P. E., and Nilsson, N. J., "Learning and executing generalized robot plans," *Artificial Intelligence*, vol. 3, pp. 251–288, 1972.

7. Wilkins, D. E., "Representation in a domain-independent planner," 8th International Joint Conference on Artificial Intelligence (IJCAI'83), Karlsruhe, Germany, pp. 733–740, 1983.
8. Wilkins, D. E., "Domain-independent planning: representation and plan generation," *Artificial Intelligence*, vol. 22, pp. 269–301, 1984.
9. Wilkins, D. E., *Practical Planning: extending the classical AI planning paradigm*, Morgan Kaufmann, 1988.
10. Sacerdoti, E. D., "Planning in a hierarchy of abstraction spaces," *Artificial Intelligence*, vol. 5, pp. 115–135, 1974.
11. Tate, A., "Generating project networks," 5th International Joint Conference on Artificial Intelligence (IJCAI), Cambridge, MA, pp. 888–893, 1977.
12. Hendler, J., Tate, A., and Drummond, M., "AI planning: systems and techniques," *AI Magazine*, pp. 61–77, Summer 1990.
13. Sussman, G. J., *A Computer Model of Skill Acquisition*, Elsevier, 1975.
14. Tate, A., "Interacting goals and their use," 4th International Joint Conference on Artificial Intelligence (IJCAI'75), Tbilisi, Georgia, pp. 215–218, 1975.
15. Bruno, G., Elia, A., and Laface, P., "A rule-based system to schedule production," *IEEE Computer*, vol. 19, issue 7, pp. 32–40, July 1986.
16. Dattero, R., Kanet, J. J., and White, E. M., "Enhancing manufacturing planning and control systems with artificial intelligence techniques," in *Knowledge-based Systems in Manufacturing*, Kusiak, A. (Ed.), pp. 137–150, Taylor and Francis, 1989.
17. Liu, B., "Scheduling via reinforcement," *Artificial Intelligence in Engineering*, vol. 3, pp. 76–85, 1988.
18. Collinot, A., Le Pape, C., and Pinoteau, G., "SONIA: a knowledge-based scheduling system," *Artificial Intelligence in Engineering*, vol. 3, pp. 86–94, 1988.

Further reading

- Allen, J., Hendler, J., and Tate, A. (Eds.), *Readings in Planning*, Morgan Kaufmann, 1990.
- Campbell, S. and Fainstein, S. S. (Eds.), *Readings in Planning Theory*, Blackwell, 1996.

- Wilkins, D. E., *Practical Planning: extending the classical AI planning paradigm*, Morgan Kaufmann, 1989.
- Yang, Q., *Intelligent Planning: a decomposition and abstraction-based approach*, Springer Verlag, 1998.

Chapter fourteen

Systems for control

14.1 Introduction

The application of intelligent systems to control has far-reaching implications for manufacturing, robotics, and other areas of engineering. The control problem is closely allied with some of the other applications that have been discussed so far. For instance, a controller of manufacturing equipment will have as its aim the implementation of a manufacturing plan (see Chapter 13). It will need to interpret sensor data, recognize faults (see Chapter 11), and respond to them. Similarly, it will need to replan the manufacturing process in the event of breakdown or some other unexpected event, i.e., to plan reactively (Section 13.9). Indeed, Bennett [1] treats automated control as a loop of plan generation, monitoring, diagnosis, and replanning.

The systems described in Chapters 11 to 13 gather data describing their environment and make decisions and judgments about those data. Controllers are distinct in that they can go a step further by altering their environment. They may do this actively by sending commands to the hardware or passively by recommending to a human operator that certain actions be taken. The passive implementation assumes that the process decisions can be implemented relatively slowly.

Control problems appear in various guises, and different techniques may be appropriate in different cases. For example, a temperature controller for a furnace may modify the current flowing in the heating coils in response to the measured temperature, where the temperature may be registered as a potential difference across a thermocouple. This is *low-level* control, in which a rapid response is required but little intelligence is involved. In contrast, *high-level* or *supervisory* control takes a wider view of the process being controlled. For example, in the control of the manufacturing process for a steel component, a furnace temperature may be just one of many parameters that need to be adjusted. High-level control requires more intelligence, but there is often more time available in which to make the decisions.

The examples of control discussed above implicitly assumed the existence of a model of the system being controlled. In building a temperature controller, it is known that an increased current will raise the temperature, that this is registered by the thermocouple, and that there will be a time lag between the two. The controller is designed to exploit this model. There may be some circumstances where no such model exists or is too complex to represent. The process under control can then be thought of as a black box, whose input is determined by the controller, and whose output we wish to regulate. As it has no other information available to it, the controller must learn how to control the black box through trial and error. In other words it must construct a model of the system through experience. We will discuss two approaches, the BOXES algorithm (Section 14.7) and neural networks (Section 14.8).

As well as drawing a distinction between low-level and high-level control, we can distinguish between *adaptive* and *servo* control. The aim of an adaptive controller is to maintain a steady state. In a completely stable environment, an adaptive controller would need to do nothing. In the real world, an adaptive controller must adapt to changes in the environment that may be brought about by the controlled process itself or by external disturbances. A temperature controller for a furnace is an adaptive controller the task of which is to maintain a constant temperature. It must do this in spite of disturbances such as the furnace door opening, large thermal masses being inserted or removed, fluctuations in the power supply, and changes in the temperature of the surroundings. Typically, it will achieve this by using negative feedback (see Section 14.2, below).

A servo controller is designed to drive the output of the plant from a starting value to a desired value. Choosing a control action to achieve the desired output requires that a prediction be made about the future behavior of the controlled plant. This, again, requires a model of the controlled plant. Often a high-level controller is required to decide upon a series of servo control actions. This is known as *sequence* control. For instance, alloyed components are normally taken through a heat-treatment cycle. They are initially held at a temperature close to the melting temperature, then they are rapidly quenched to room temperature, and finally they are "aged" at an intermediate temperature.

14.2 Low-level control

14.2.1 Open-loop control

The open-loop control strategy is straightforward: given a control requirement, the controller simply sends a control action to the plant (Figure 14.1(a)). The controller must have a model of the relationship between the control action and

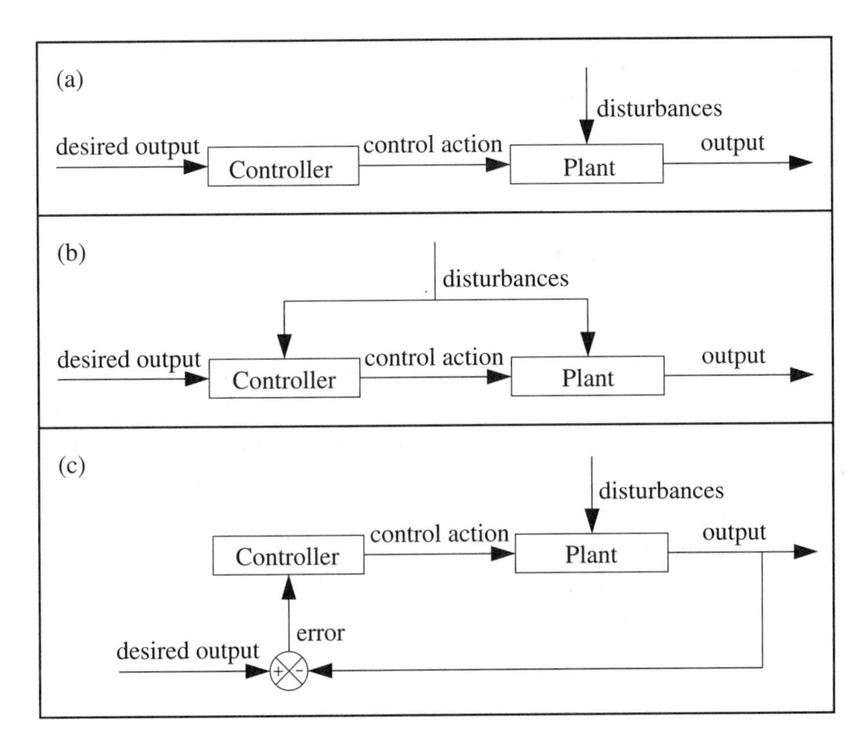

Figure 14.1 Three alternative strategies for control:
(a) open loop
(b) feedforward
(c) feedback (closed loop)

the behavior of the plant. The accuracy of control is solely dependent on the accuracy of the model, and no checks are made to ensure that the plant is behaving as intended. An open-loop temperature controller, for example, would send a set current to the heating coils in order to achieve an intended temperature. In order to choose an appropriate current, it would have an implicit model of the thermal capacity of the furnace and of the rate of heat loss. This strategy cannot correct for any disturbance to the plant (i.e., the furnace, in this example).

14.2.2 Feedforward control

The feedforward strategy takes account of disturbances to the plant by measuring the disturbances and altering the control action accordingly (Figure 14.1(b)). In the example of the temperature controller, fluctuations in the electrical power supply might be monitored. The nominal current sent to the coils may then be altered to compensate for fluctuations in the actual current. Note that the disturbance, and not the controlled variable itself (i.e., not the furnace temperature), is monitored. Any disturbances that are not measured are not taken into account. As with open-loop control, a model of the plant is needed in order to calculate the impacts of the control action and the measured disturbance.

14.2.3 Feedback control

This is the most common control strategy in both high-level and low-level control applications. The measured output from the plant is compared with the required value, and the difference or *error* is used by the controller to regulate its control action. As shown in Figure 14.1(c), the control action affects the plant and the plant output affects the controller, thereby forming a closed loop. Hence the strategy is also known as *closed-loop control.* This strategy takes account of all disturbances, regardless of how they are caused and without having to measure them directly. There is often a lag in the response of the controller, as corrections can only be made after a deviation in the plant output (e.g., the furnace temperature in the previous example) has been detected.

14.2.4 First- and second-order models

It has already been emphasized that a controller can be built only if we have a model, albeit a simple one, of the plant being controlled. For low-level control, it is often assumed that the plant can be adequately modeled on first- or second-order linear differential equations. Let us denote the input to the plant (which may also be the output of the controller) by the letter x, and the output of the plant by the letter y. In a furnace, x would represent the current flowing and y would represent the furnace temperature. The first-order differential equation would be:

$$\tau \frac{dy}{dt} + y = k_1 x \qquad (14.1)$$

and the second-order differential equation:

$$\frac{d^2 y}{dt^2} + 2\zeta\omega_n \frac{dy}{dt} + \omega_n^2 y = k_2 x \qquad (14.2)$$

where τ, k_1, k_2, ζ and ω_n are constants for a given plant, and t represents time. These equations can be used to tell us how the output y will respond to a change in input x.

Figures 14.2(a) and 14.2(b) show the response to a step change in x for a first- and second-order model, respectively. In the first-order model, the time for the controlled system to reach a new steady state is determined by τ, which is the *time constant* for the controlled system.

In the second-order model, the behavior of the controlled system is dependent on two characteristic constants. The *damping ratio* ζ determines the rate at which y will approach its intended value, and the *undamped natural angular frequency* ω_n determines the frequency of oscillation about the final value in the underdamped case (Figure 14.2(b)).

14.2.5 Algorithmic control: the PID controller

Control systems may be either analog or digital. In analog systems, the output from the controller varies continuously in response to continuous changes in the controlled variable. This book will concentrate on digital control, in which the data are sampled and discrete changes in the controller output are calculated accordingly. There are strong arguments to support the view that low-level digital control is best handled by algorithms, which can be implemented either in electronic hardware or by procedural coding. Such arguments are based on the observation that low-level control usually requires a rapid response, but little or no intelligence. This view would tend to preclude the use of intelligent systems. In this section we will look at a commonly used control algorithm, and in Section 14.6 we will examine the possibilities for improvement by using fuzzy logic.

It was noted in Section 14.2.3 that feedback controllers determine their control action on the basis of the error e, which is the difference between the measured output y from the plant at a given moment and the desired value, i.e., the reference r. The control action is often simply the assignment of a value to a variable, such as the furnace current. This is the *action variable*, and it is usually given the symbol u. The action variable is sometimes known as the *control* variable, although this can lead to confusion with the *controlled* variable y.

In a simple controller, u may be set in proportion to e. A more sophisticated approach is adopted in the PID (proportional + integral + derivative) controller. The value for u that is generated by a PID controller is the sum of three terms:

- P — a term proportional to the error e;
- I — a term proportional to the integral of e with respect to time;
- D — a term proportional to the derivative of e with respect to time.

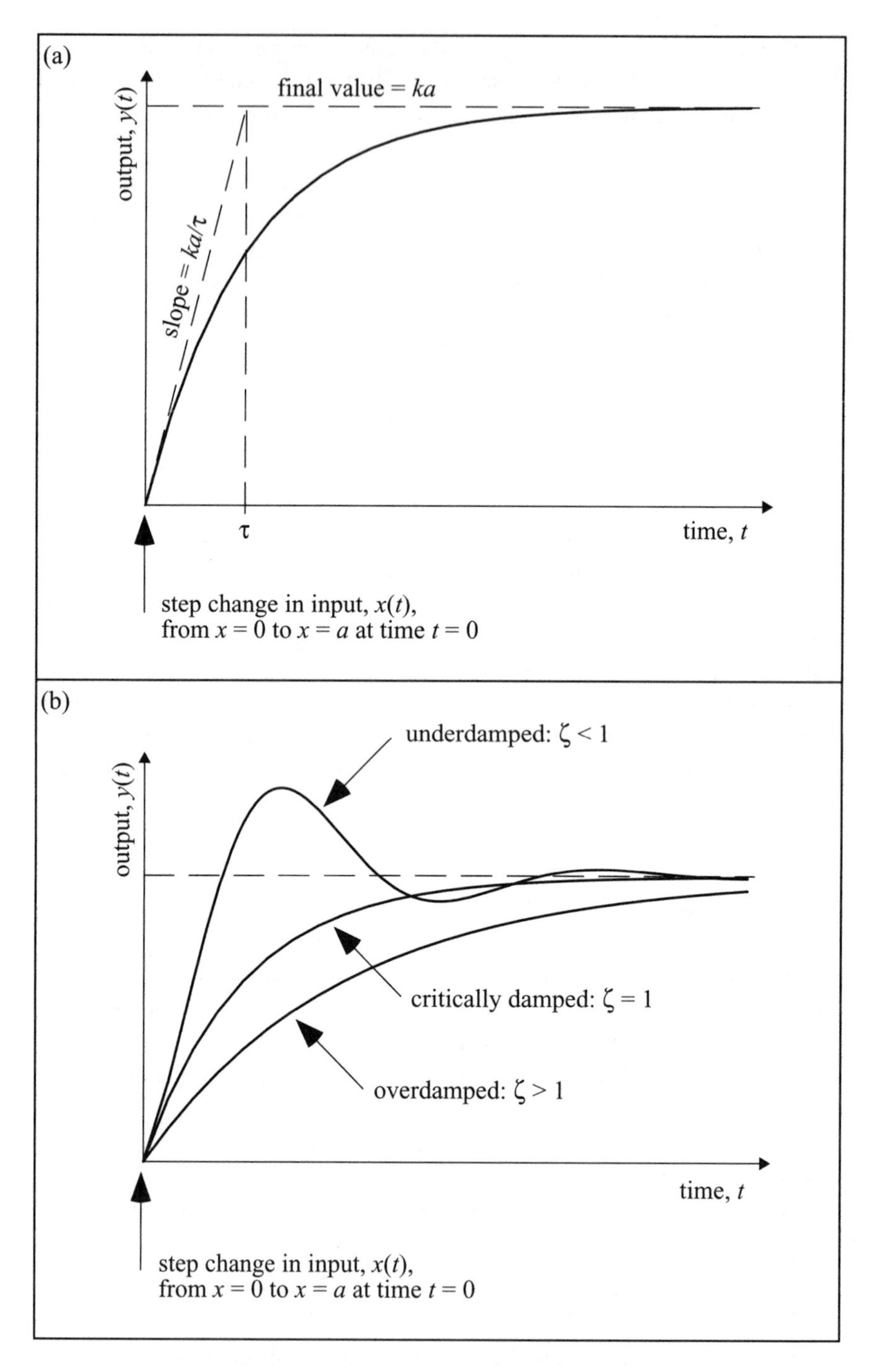

Figure 14.2 (a) first-order response to a step change
(b) second-order response to a step change

Thus, the value assigned to the action variable u by a PID controller would ideally be given by:

$$u = K_p \left(e + \frac{1}{\tau_i} \int e \, dt + \tau_d \frac{de}{dt} \right) \tag{14.3}$$

where K_p, τ_i, and τ_d are adjustable parameters of the controller that can be set to suit the characteristics of the plant being controlled. K_p is the *proportional gain,* τ_i is the *integral time,* and τ_d is the *derivative time*. Because values for e are samples at time intervals Δt, the integral and derivative terms must be approximated:

$$u = K_p \left(e(k) + \frac{\Delta t}{\tau_i} \sum_k e(k) + \tau_d \frac{e_{k+1} - e_k}{\Delta t} \right) \tag{14.4}$$

where k is the sample number, such that time $t = k\Delta t$. The role of the P term is intuitively obvious — the greater the error, the greater the control action that is needed. Through the I term, the controller output depends on the accumulated historical values of the error. This is used to counteract the effects of long-term disturbances on the controlled plant. The magnitude of the D term depends on the rate of change of the error. This term allows the controller to react quickly to sharp fluctuations in the error. The D term is low for slowly varying errors, and zero when the error is constant. In practice, tuning the parameters of a PID controller can be difficult. One commonly used technique is the Ziegler–Nichols method (see, for example, [2]).

14.2.6 Bang-bang control

Bang-bang controllers rely on switching the action variable between its upper and lower limits, with intermediate values disallowed. As previously noted, servo control involves forcing a system from one state to a new state. The fastest way of doing this, i.e., time optimal control, is by switching the action variable from one extreme to another at precalculated times, a method known as *bang-bang* control. Two extreme values are used, although Sripada et al. [3] also allow a final steady-state value for the action variable (Figure 14.3). Consider the control of an electric furnace. As soon as the new (increased) temperature requirement is known, the electric current is increased to the maximum value sustainable until the error in the temperature is less than a critical value e^*. The current is then dropped to its minimum value (i.e., zero) for time Δt, before being switched to its final steady-state value. There are, therefore, two parameters that determine the performance of the controller, e^* and Δt. Figure 14.3 shows the effects of errors in these two parameters.

Sripada et al. coded the three switchings of their bang-bang servo controller as a set of three rules. They acknowledged, however, that since the rules fired in sequence there was no reason why these could not have been procedurally coded. They had a separate set of rules for adjusting the parameters e^*, Δt, and K_p in the light of experience. The controller was

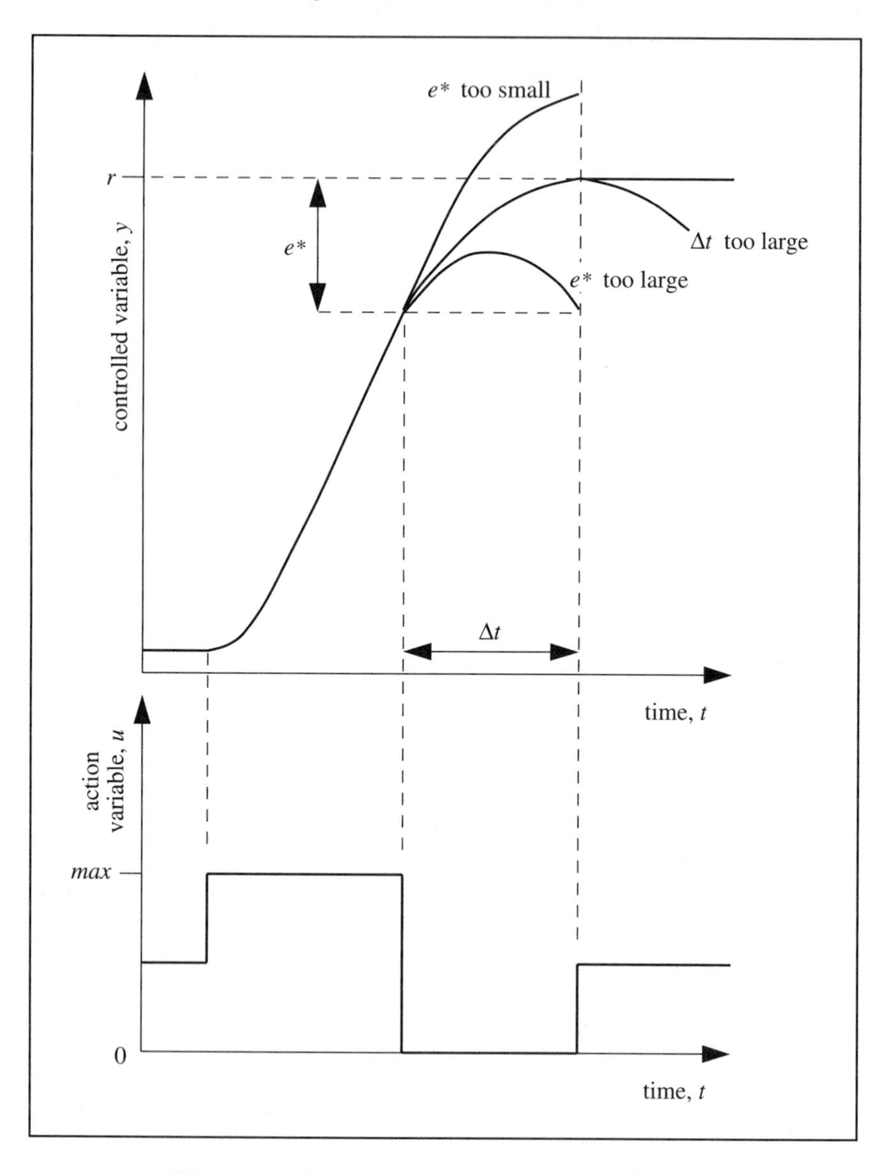

Figure 14.3 Bang-bang servo control (adapted from [3])

self-tuning, and the parameter adjustment can be thought of as controlling the controller. The extent to which the parameters are adjusted is in proportion to the degree of membership of the fuzzy sets `too large` and `too small`. Thus, a typical fuzzy rule might be:

```
IF y overshoots r
THEN e* is too large
AND e* becomes e* - Δe.
```

The proposition `y overshoots r` is fuzzy, with a degree of membership between 0 and 1, reflecting the amount of overshoot. This becomes the degree of membership for the fuzzy proposition `e* is too large`, and is also used to calculate the adjustment Δe. Techniques for scaling the adjustment (in this case Δe) to reflect the membership functions are discussed in detail in Section 14.6. Sripada et al. found that their self-tuning bang-bang controller outperformed a well-tuned PI controller, i.e., a PID controller with the *D* term set to zero.

14.3 Requirements of high-level (supervisory) control

Our discussion so far has concentrated on fairly straightforward control tasks, such as maintaining the temperature of a furnace or driving the temperature to a new set point. Both are examples of low-level control, where a rapid response is needed but scarcely any intelligence. In contrast, high-level control may be a complex problem concerning decisions about the actions to take at any given time. This may involve aspects of both adaptive and servo control. An important example of high-level control is the control of a manufacturing process. Control decisions at this level concern diverse factors such as choice of reactants and raw materials; conveyor belt speeds; rates of flow of solids, liquids, and gases; temperature and pressure cycling; and batch transport.

Leitch et al. [4] have identified six key requirements for a real-time supervisory controller:

- ability to make decisions and act upon them within a time constraint;
- handling asynchronous events — the system must be able to break out of its current set of operations to deal with unexpected occurrences;
- temporal reasoning, i.e., the ability to reason about time and sequences;
- reasoning with uncertainty;
- continuous operation;
- multiple sources of knowledge.

We have come across the third requirement in the context of planning (Chapter 13), and the last three requirements as part of monitoring and diagnosis (Chapter 11). Only the first two requirements are new.

14.4 Blackboard maintenance

One of the requirements listed above was for multiple sources of knowledge. These are required since high-level control may have many sources of input information and many subtasks to perform. It is no surprise, therefore, that the *blackboard model* is chosen for many control applications (see Chapter 9).

Leitch et al. [4] point out that, because continuous operation is required, a mechanism is needed for ensuring that the blackboard does not contain obsolete information. They achieve this by tagging blackboard information with the time of its posting. As time elapses, some information may remain relevant, some may gradually lose accuracy, and some may suddenly become obsolete. There are several ways of representing the lifetime of blackboard information:

(i) A default lifetime for all blackboard information may be assumed. Any information that is older than this may be removed. A drawback of this approach is that deductions made from information that is now obsolete may remain on the blackboard.

(ii) At the time of posting to the blackboard, individual items of information may be tagged with an expected lifetime. Suppose that item A is posted at time t_A with expected lifetime l_A. If item B is deduced from A at time t_B, where $t_B < t_A + l_A$, then the lifetime of B, l_B, would be $t_A + l_A - t_B$.

(iii) Links between blackboard items are recorded, showing the interdependencies between items. Figure 14.4 illustrates the application of this approach to control of a boiler, using rules borrowed from Chapter 2. There is no need to record the expected lifetimes of any items, as all pieces of blackboard information are ultimately dependent on sensor data, which are liable to change. When changes in sensor values occur, updates are rippled through the dependent items on the blackboard. In the example shown in Figure 14.4, as soon as the flow rate fell, `flow rate high` would be removed from the blackboard along with the inferences `steam escaping` and `steam outlet blockage` and the resulting control action. All other information on the blackboard would remain valid.

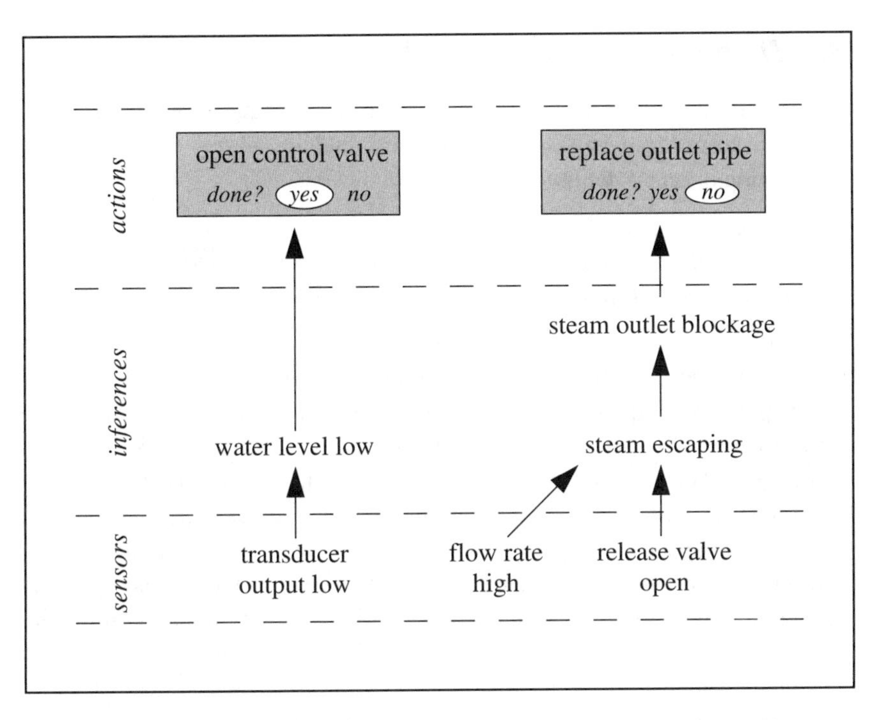

Figure 14.4 Storing dependencies between items of information on the blackboard

The third technique is a powerful way of maintaining the blackboard integrity, since it avoids unnecessary reevaluation of information. This is counterbalanced by the additional complexity of the blackboard and the computational load of maintaining the blackboard. These problems can be minimized by careful partitioning of the blackboard into levels of abstraction (from low-level statements about sensor values to high-level analysis of trends and policy) and subject groupings (low-level data about a conveyor belt are kept separate from low-level information about a boiler).

The control actions on the blackboard in Figure 14.4 are shown with a tag indicating whether or not the action has been carried out. The tag is necessary because there is a time lag between an action being taken and a resultant change in the sensor data. When the action is first added to the blackboard, the tag is set to "not yet done." This is changed as soon as the action is carried out. The action is not removed at this stage, as the knowledge source that generated it would simply generate it again. Instead, the control action remains on the blackboard until the supporting evidence is removed, when the sensor reading changes.

14.5 Time-constrained reasoning

An automatic control system must be capable of operating in real time. This does not necessarily mean fast, but merely fast enough. Laffey et al. [5] offer the following informal definition of real-time performance:

> *[real-time performance requires that] the system responds to incoming data at a rate as fast or faster than it is arriving.*

This conveys the idea that real-time performance is the ability to "keep up" with the physical world to which the system is interfaced. The RESCU process control system, applied to a chemical plant, receives data in blocks at five-minute intervals [4]. It, therefore, has five minutes in which to respond to the data in order to achieve real-time performance. This is an example of a system that is fast enough, but not particularly fast. The rate of arrival of data is one factor in real-time control, but there may be other reasons why decisions have to be made within a specified time frame. For instance, the minimum time of response to an overheated reactor is not dependent on the rate of updating the temperature data, but on the time over which overheating can be tolerated.

In order to ensure that a satisfactory (though not necessarily optimum) solution is achieved within the available time scale, an intelligent system needs to be able to schedule its activities appropriately. This is not straightforward, as the time taken to solve a problem cannot be judged accurately in advance. A conservative approach would be to select only those reasoning activities that are judged able to be completed comfortably within the available time. However, this may lead to an unsatisfactory solution while failing to use all the time available for reasoning.

Scheduling of knowledge sources is an internal control issue and would normally be kept apart from domain knowledge. On the other hand, the time constraints are domain-specific and should not be built into the inference engines or control software. A suitable compromise is to have a special knowledge source dedicated to the real-time scheduling of domain knowledge sources. This is the approach adopted in RESCU [4].

Some of the techniques that have been applied to ensure a satisfactory response within the time constraints are described below.

14.5.1 Prioritization of processes and knowledge sources

In the RESCU system, processes are partitioned on two levels. First, the processes of the overall system are divided into *operator*, *communications*, *log*, *monitor*, and *knowledge-based system*. The last is a blackboard system

(Chapter 9) and so it can be divided up into knowledge sources.* This is the second level of partitioning.

The separate processes at the overall system level are considered to be independent. Thus, they could run in parallel or, on a serial computer, a process can be temporarily suspended while a more important process is performed. Each of the five processes is assigned a priority number between 1 and 6. The communications and monitoring processes are given the highest priorities, so these can cause the temporary suspension of the other (lower priority) activities. The application of priorities to processes is similar to the application of priorities to rules in other systems (see Section 2.8.2).

The separate knowledge sources (KSs) at the knowledge-based system (KBS) level are not independent, as they all rely upon reading from, and writing to, the blackboard. In RESCU, once a KS has started, it will always run to completion before another KS starts. The whole KBS process, however, can be paused. A suitable size for a KS must be small enough to ensure that the reasoning does not continue when the data are out of date, and large enough to avoid overworking the scheduling software.

In order to provide adaptable and fairly intelligent scheduling of knowledge sources, RESCU's scheduling software is itself coded as a knowledge source. The scheduler checks the preconditions of KSs and any KSs that are applicable are assigned a priority rating. The one with the highest priority is chosen, and in the event of more than one having the highest priority, one is chosen at random. The applicable KSs that are not selected have their priority increased in the next round of KS selection, thereby ensuring that all applicable KSs are fired eventually.

14.5.2 Approximation

The use of priorities, described above, ensures that the *most important* tasks are completed within the time constraints, and less important tasks are completed only if time permits. An alternative approach, proposed by Lesser et al. [6], is to ensure that an *approximate* solution is obtained within the time constraints, and the solution is embellished only if time permits. Three aspects of a solution that might be sacrificed to some degree are:

- completeness,
- precision,
- certainty.

---

* RESCU actually schedules on the basis of groups of knowledge sources, referred to as *activities*.

Loss of completeness means that some aspects of the solution are not explored. Loss of precision means that some parameters are determined less precisely than they might have been. (The maximum precision is determined by the precision of the input data.) Loss of certainty means that some evidence in support of the conclusion has not been evaluated, or alternatives have been ignored. Thus, there is a trade-off between the quality of a solution and the time taken to derive it.

One approach to approximation would be to make a rough attempt at solving the problem initially and to use any time remaining to refine the solution incrementally. Provided that sufficient time was available to at least achieve the rough solution, a solution of some sort would always be guaranteed. In contrast, the approach adopted by Lesser et al. [6] is to plan the steps of solution generation so that there is just the right degree of approximation to meet the deadline.

Lesser et al. distinguish between *well-defined* and *ill-defined* approximations, although these names are possibly misleading. According to their definition, well-defined approximations have the following properties:

- a predictable effect on the quality of the solution and the time taken to obtain it;
- graceful degradation, i.e., the quality of the solution decreases smoothly as the amount of approximation is increased;
- loss of precision is not accompanied by loss of accuracy. If, for example, a reactor temperature is determined to lie within a certain temperature range, then this range should straddle the value that would be determined by more precise means.

Lesser et al. recommend that ill-defined approximations be used only as a last resort, when it is known that a well-defined approximation is not capable of achieving a solution within the available time. They consider six strategies for approximation, all of which they consider to be well defined. These strategies are classified into three groups: approximate search, data approximations, and knowledge approximations. These classifications are described below.

Approximate search

Two approaches to pruning the search tree, i.e., reducing the number of alternatives to be considered, are elimination of corroborating evidence and elimination of competing interpretations.

(i) *Eliminating corroborating evidence*

Once a hypothesis has been generated, and perhaps partially verified, time can be saved by dispensing with further corroborating evidence. This will have the effect of reducing the certainty of the solution. For corroborating data to be recognized as such, it must be analyzed to a limited extent before being discarded.

(ii) *Eliminating competing interpretations*
Elimination of corroborating evidence is a means of limiting the input data, whereas elimination of competing interpretations limits the output data. Solutions that have substantially lower certainties than their alternatives can be eliminated. If it is recognized that some solutions will have a low certainty regardless of the amount of processing that is carried out on them, then these solutions can be eliminated before they are fully evaluated. The net result is a reduced level of certainty of the final solution.

Data approximations

Time can be saved by cutting down the amount of data considered. Incomplete event processing and cluster processing are considered here, although elimination of corroborating evidence (see above) might also be considered in this category.

(i) *Incomplete event processing*
This approximation technique is really a combination of prioritization (see Section 14.5.1) and elimination of corroborating evidence. Suppose that a chemical reactor is being controlled, and data are needed regarding temperature and pressure. If temperature has the higher priority, then any data that support the estimation of the pressure can be ignored in order to save time. The result is a less complete solution.

(ii) *Cluster processing*
Time can be saved by grouping together data items that are related, and examining the overall properties of the group rather than the individual data items. For instance, the temperature sensors mounted on the walls of a reactor chamber might be clustered. Then, rather than using all of the readings, only the mean and standard deviation might be considered. This may lead to a loss of precision, but the certainty may be increased owing to the dilution of erroneous readings.

Knowledge approximations

Changes can be made to the knowledge base in order to speed up processing. Two possible approaches are:

(i) *Knowledge adaptation to suit data approximations*
This is not really a technique in its own right, but simply a recognition that data approximations (see above) require a modified knowledge base.

(ii) *Eliminating intermediate steps*
As noted in Section 14.3, shallow knowledge represents a means of by-passing the steps of the underlying deep knowledge. Therefore, it has potential for time saving. However, it may lead to a loss of certainty, as extra corroborating evidence for the intermediate steps might have been available. Shallow knowledge is also less adaptable to new situations.

14.5.3 Single and multiple instantiation

Single and multiple instantiation of variables in a rule-based system are described in Chapter 2. Under multiple instantiation, a single rule, fired only once, finds all sets of instantiations that satisfy the condition and then performs the conclusion on each. Under single instantiation, a separate rule firing is required for each set of instantiations. Depending on the strategy for conflict resolution, this may cause the conclusions to be drawn in a different order.

In time-constrained control applications, the choice between multiple instantiation and repeated single instantiation can be critical. Consider the case of an automatic controller for a telephone network [7]. Typically, the controller will receive statistics describing the network traffic at regular intervals Δt. Upon receiving the statistics, the controller must interpret the data, choose appropriate control actions, and carry out those actions before the next set of statistics arrives. Typically, there is a cycle of finding overloaded routes, planning alternative routes through the network, and executing the plan. The most efficient way of doing this is by multiple instantiation, as the total duration of the find-plan-execute cycle for all routes is smaller (Figure 14.5). However, the controller must finish within time Δt. The more overloaded links exist, the less likely it is that the controller will finish. In fact, it is feasible that the controller might have found all of the overloaded links and determined an alternative routing for each, but failed to perform *any* control actions (Figure 14.5(b)).

This problem is avoided by repeated use of single instantiation, shown in Figure 14.5(a), which is guaranteed to have performed *some* control actions within the available time. There is a crucial difference between the two approaches in a control application. Multiple instantiation involves the drawing up of a large plan, followed by execution of the plan. Repeated single instantiation, on the other hand, involves interleaving of planning and execution.

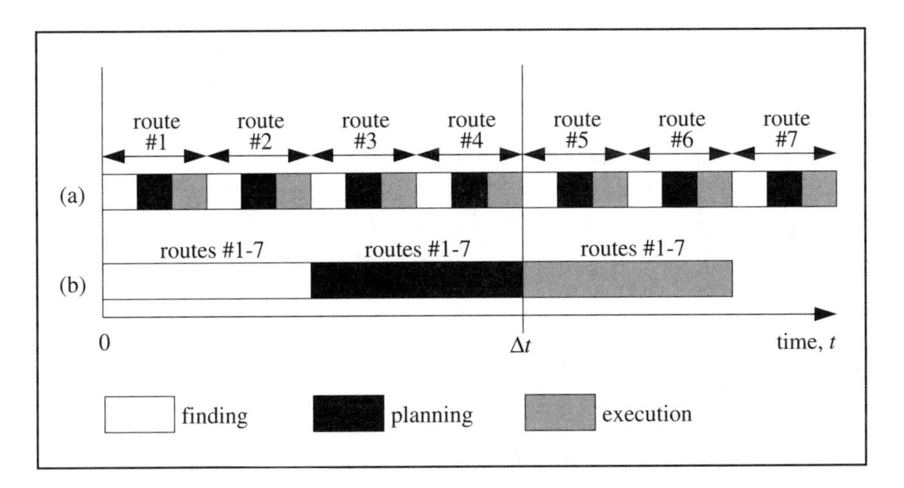

Figure 14.5 Controlling a telecommunications network within a time constraint (Δt):
(a) repeated single instantiation of variables
(b) multiple instantiation of variables

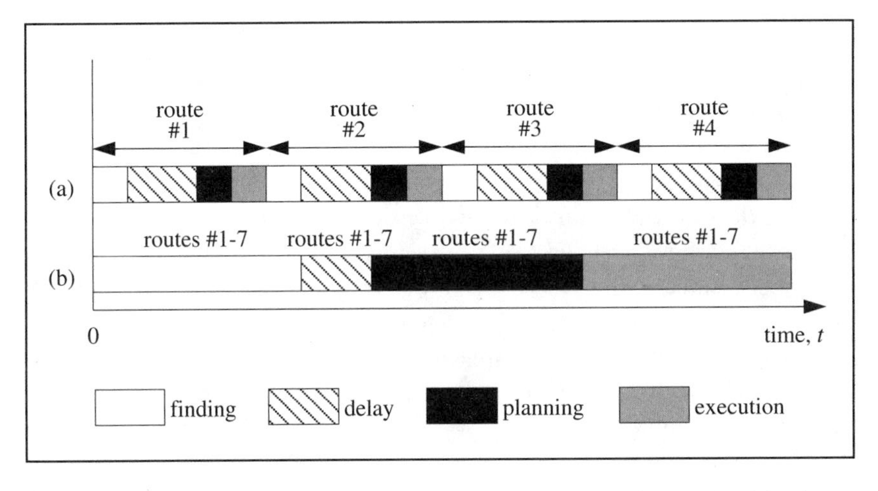

Figure 14.6 Controlling a telecommunications network where there is a delay in receiving information:
(a) repeated single instantiation of variables
(b) multiple instantiation of variables

Repeated single instantiation does not always represent the better choice. This depends on the specific application. Multiple instantiation is generally faster as it requires the rules to be selected and interpreted once only. Multiple instantiation can have other advantages as well. Consider the telecommunications example again, but now imagine that, in order to choose a suitable rerouting, the controller must request some detailed information from the local telephone exchanges. There will be a delay in receiving this information, thereby delaying the planning stage. If all the requests for information are sent at the same time (i.e., multiple instantiation, Figure 14.6(b)) then the delays overlap. Under single instantiation the total delay is much greater, since a separate delay is encountered for every route under consideration (Figure 14.6(a)).

14.6 Fuzzy control

14.6.1 Crisp and fuzzy control

Control decisions can be thought of as a transformation from state variables to action variables (Figure 14.7). *State variables* describe the current state of the physical plant and the desired state. *Action variables* are those that can be directly altered by the controller, such as the electrical current sent to a furnace, or the flow rate through a gas valve. In some circumstances, it may be possible to obtain values for the action variables by direct algebraic manipulation of the state variables. (This is the case for a PID controller; see Section 14.2.5.) Given suitably chosen functions, this approach causes values of action variables to change smoothly as values of state variables change. In high-level control, such functions are rarely available, and this is one reason for using rules instead to link state variables to action variables.

Crisp sets are conventional Boolean sets, where an item is either a member (degree of membership = 1) or it is not (degree of membership = 0). It follows that an item cannot belong to two contradictory sets, such as `large` and `small` as it can under fuzzy logic (see Chapter 3). Applying crisp sets to state and action variables corresponds to dividing up the range of allowable values into subranges, each of which forms a set. Suppose that a state variable such as temperature is divided into five crisp sets. A temperature reading can belong to only one of these sets, so only one rule will apply, resulting in a single control

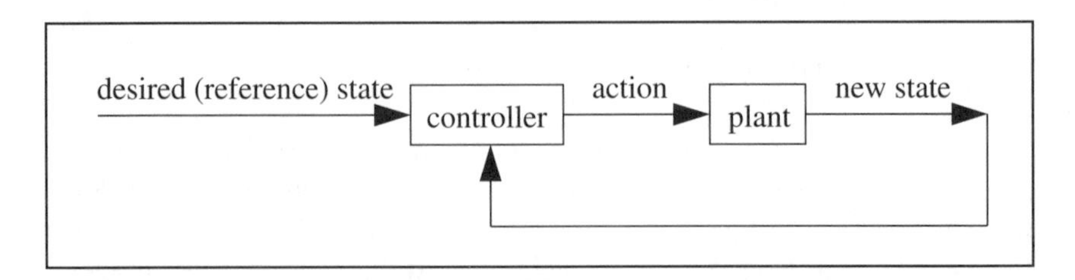

Figure 14.7 Control viewed as a transformation from state variables to action variables

action. Thus, the number of different control actions is limited to the number of rules, which in turn is limited by the number of crisp sets. The action variables are changed in abrupt steps as the state variables change.

Fuzzy logic provides a means of allowing a small number of rules to produce smooth changes in the action variables as the state variables change. The number of rules required is dependent on the number of state variables, the number of fuzzy sets, and the ways in which the state variables are combined in rule conditions. Numerical information is explicit in crisp rules, but in fuzzy rules it becomes implicit in the chosen shape of the fuzzy membership functions.

14.6.2 Firing fuzzy control rules

Some simple examples of fuzzy control rules are as follows:

```
/* Rule 14.1f */
IF temperature is high OR current is high
THEN reduce current

/* Rule 14.2f */
IF temperature is medium
THEN no change to current

/* Rule 14.3f */
IF temperature is low and current is high
THEN no change to current

/* Rule 14.4f */
IF temperature is low and current is low
THEN increase current
```

The rules and the fuzzy sets to which they refer are, in general, dependent on each other. Some possible fuzzy membership functions, μ, for the state variables `temperature` and `current`, and for the action variable `change in current`, are shown in Figure 14.8. Since the fuzzy sets overlap, a temperature and current may have some degree of membership of more than one fuzzy set. Suppose that the recorded temperature is 300°C and the measured current is 15 amperes. The temperature and current are each members of two fuzzy sets: `medium` and `high`. Rules 14.1f and 14.2f will fire, with the apparently contradictory conclusion that we should both reduce the electric current and leave it alone. Of course, what is actually required is *some* reduction in current.

Rule 14.1f contains a disjunction. Using Equation 3.36 (from Chapter 3), the possibility value for the composite condition is:

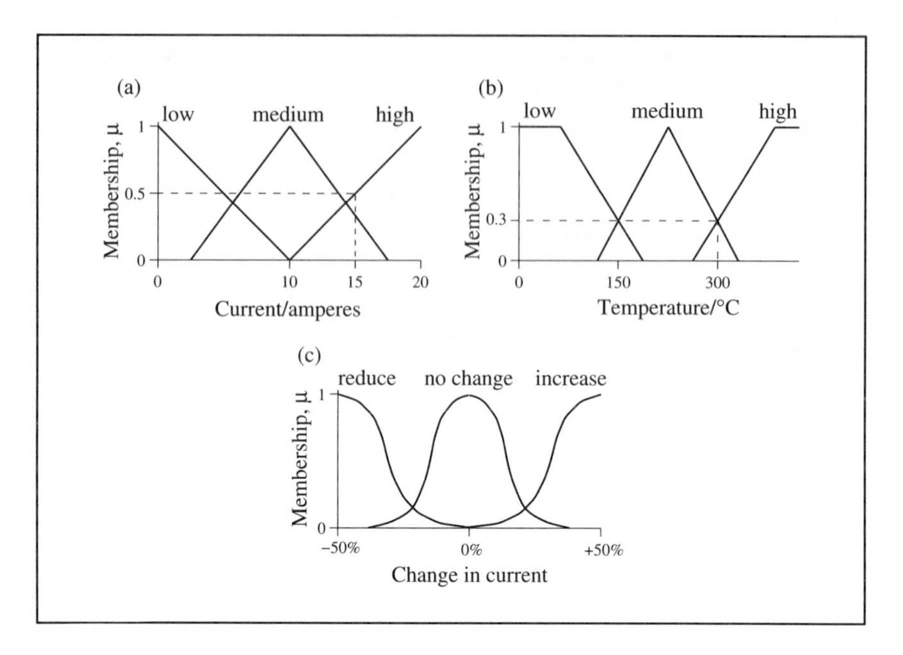

Figure 14.8 Fuzzy sets:
(a) electric current (state variable)
(b) temperature (state variable)
(c) change in electric current (action variable)

max{μ(`temperature is high`), μ(`current is high`)}.

At 300°C and 15 amps, μ(`temperature is high`) is 0.3 and μ(`current is high`) is 0.5. The composite possibility value is therefore 0.5, and μ(`reduce current`) becomes 0.5. Rule 14.2f is simpler, containing only a single condition. The possibility value μ(`temperature is medium`) is 0.3 and so μ(`no change in current`) becomes 0.3.

14.6.3 Defuzzification

After firing rules 14.1f and 14.2f, we have a degree of membership, or possibility, for `reduce current` and another for `no change in current`. These fuzzy actions must be converted into a single precise action to be of any practical use, i.e., they need to be *defuzzified.*

Assuming that we are using Larsen's Product Operation Rule (see Section 3.4.3), the membership functions for the control actions are compressed according to their degree of membership. Thus, the membership functions for `reduce current` and for `no change in current` are compressed so that their peak values become 0.5 and 0.3, respectively (Figure 14.9). Defuzzification

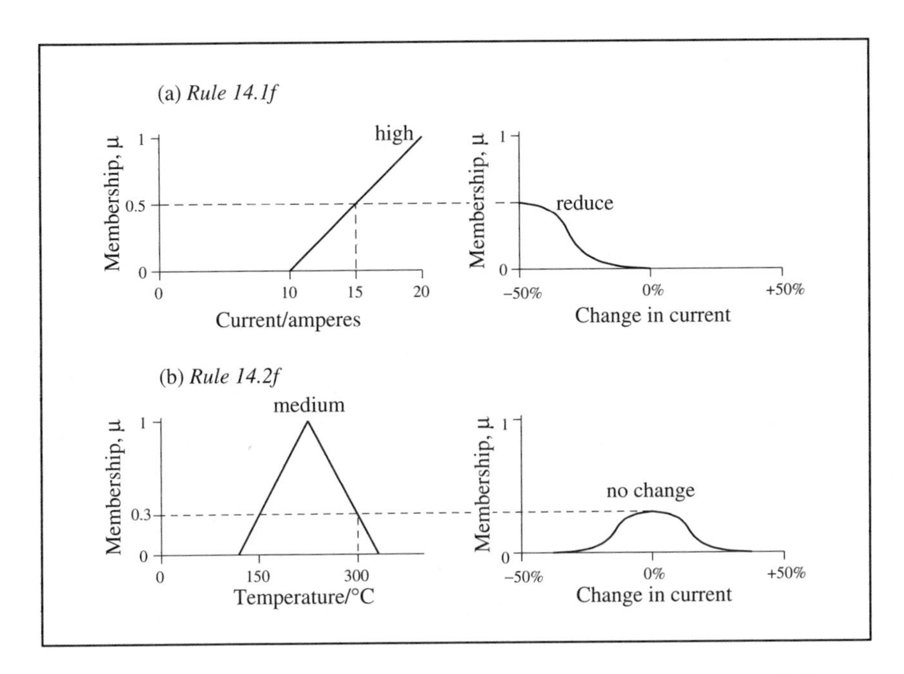

Figure 14.9 Firing fuzzy control rules using Larsen's Product Operator Rule

can then take place by finding the centroid of the combined membership functions. One of the membership functions, i.e., `reduce current`, covers an extremity of the fuzzy variable and, therefore, continues indefinitely toward $-\infty$. As discussed in Section 3.4.3, a method of handling this is required in order to find the centroid. Figure 14.10 shows the effect of applying the mirror rule, so that the membership function for `reduce current` is treated, for defuzzification purposes only, as though it were symmetrical around –50%. The centroid is then given by:

$$C = \frac{\sum_{i=1}^{N} a_i c_i}{\sum_{i=1}^{N} a_i} \tag{14.5}$$

Since we have used Larsen's Product Operation Rule, the values of c_i are unchanged from the centroids of the uncompressed shapes, C_i, and a_i is simply $\mu_i A_i$ where A_i is the area of the membership function prior to compression. In this particular case, all values of A_i are identical after the mirror rule is applied and thus the centroid is given by:

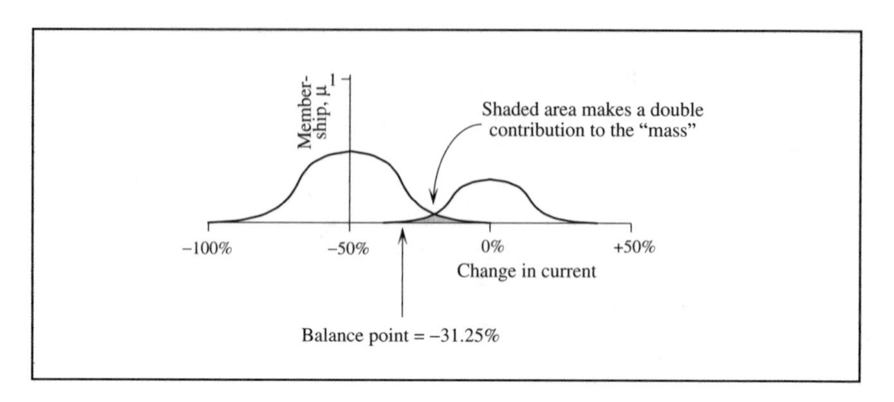

Figure 14.10 Defuzzifying a control action using the centroid method and mirror rule

$$C = \frac{(0.5 \times -50\%) + (0.3 \times 0\%)}{0.3 + 0.5} = -31.25\% \tag{14.6}$$

Thus the defuzzified control action is a 31.25% reduction in the current.

14.6.4 Some practical examples of fuzzy controllers

LINKman [8] is a fuzzy control system that has been applied to cement kiln control and other manufacturing processes. The amount of overlap of membership functions is deliberately restricted, thereby reducing the number of rule firings. This is claimed to simplify defuzzification and tuning of the membership functions, and to make control actions more transparent. All variables are normalized to lie within the range +1 to −1, representing the maximum and minimum extremes, where 0 represents the normal steady-state value or set point. It is claimed that, by working with normalized variables, the knowledge base can be more easily adapted to different plants.

ARBS — a blackboard system introduced in Section 11.5 and discussed further in Section 14.5 — has been applied to controlling plasma deposition, an important process in silicon chip manufacture [9]. This work has demonstrated the use of fuzzy rules that adjust more than one action variable simultaneously (i.e., *multivariable control*), in order to control a separate state variable that is measurable but not directly adjustable.

As well as looking at servo control (Section 14.2.6), Sripada et al. [3] have applied knowledge-based techniques to low-level adaptive control. They assert that, while a PID controller is adequate for reducing the drift in a typical plant output, it cannot cope with slight refinements to this basic requirement. In particular, they consider the application of constraints on the plant output, such that the output (y) must not be allowed to drift beyond $y_0 \pm y_C$, where y_0 is the set-point for y and y_C defines the constraints. Bang-bang control is required to move y rapidly toward y_0 if it is observed approaching $y_0 \pm y_C$. A further requirement is that control of the plant output be as smooth as possible close to the set point, precluding bang-bang control under these conditions. The controller was, therefore, required to behave differently in different circumstances. This is possible with a system based on heuristic rules, but not for a PID controller, which has a fixed predetermined behavior.

To determine the type and extent of control action required, the error e and the rate of change of the plant output dy/dt were classified with the following fuzzy sets:

e = zero;
e = small positive;
e = small negative;
e = large positive;
e = large negative;
e = close to constraint;
dy/dt = small (positive or negative);
dy/dt = large positive;
dy/dt = large negative.

According to the degree of membership of each of the nine fuzzy sets above, a rule base was used to determine the degree of membership for each of six fuzzy sets applied to control actions:

zero change;
small positive change;
small negative change;
large positive change;
large negative change;
drastic change (bang-bang).

Incorporating the last action as a fuzzy set enabled a smooth transition to bang-bang control as the plant output approached the constraints.

14.7 The BOXES controller

14.7.1 The conventional BOXES algorithm

It has already been emphasized (Section 14.1) that a controller can only function if it has a model for the system being controlled. Neural networks (see

Chapter 8 and Section 14.8, below) and the BOXES algorithm are techniques for generating such a model without any prior knowledge of the mechanisms occurring within the controlled system. Such an approach may be useful if:

- the system is too complicated to model accurately;
- insufficient information is known about the system to enable a model to be built;
- satisfactory control rules have not been found.

The BOXES algorithm may be applied to adaptive or servo control. Only the following information about the controlled system is required:

- its inputs (i.e., the possible control actions);
- its outputs (that define its state at any given time);
- the desired state (adaptive control) or the final state (servo control);
- constraints on the input and output variables.

Note that no information is needed about the relationships between the inputs and outputs.

As an example, consider a bioreactor [10, 11], which is a tank of water containing cells and nutrients (Figure 14.11). When the cells multiply, they consume nutrients. The rate at which the cells multiply is dependent only on the nutrient concentration C_n. The aim of the controller is to maintain the concentration of cells C_c at some desired value by altering the rate of flow u of nutrient-rich water through the tank. The state of the bioreactor at any time can be defined by the two variables C_n and C_c, and can be represented as a point in state space (Figure 14.12). For any position in state space there will be an

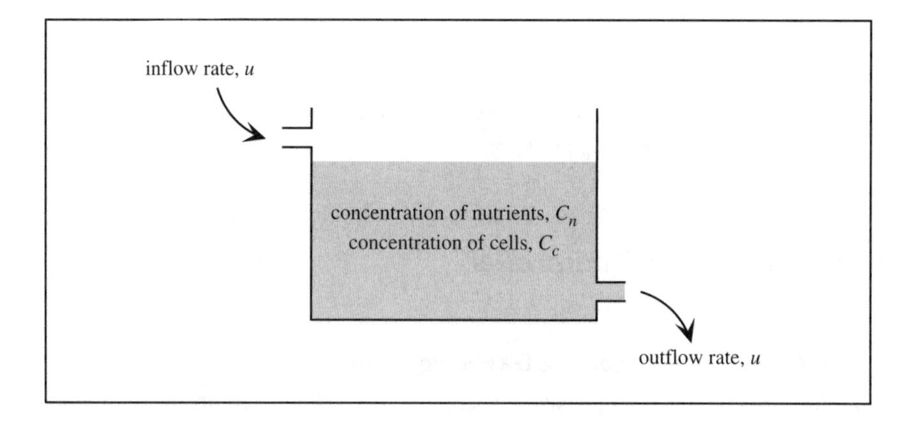

Figure 14.11 A bioreactor

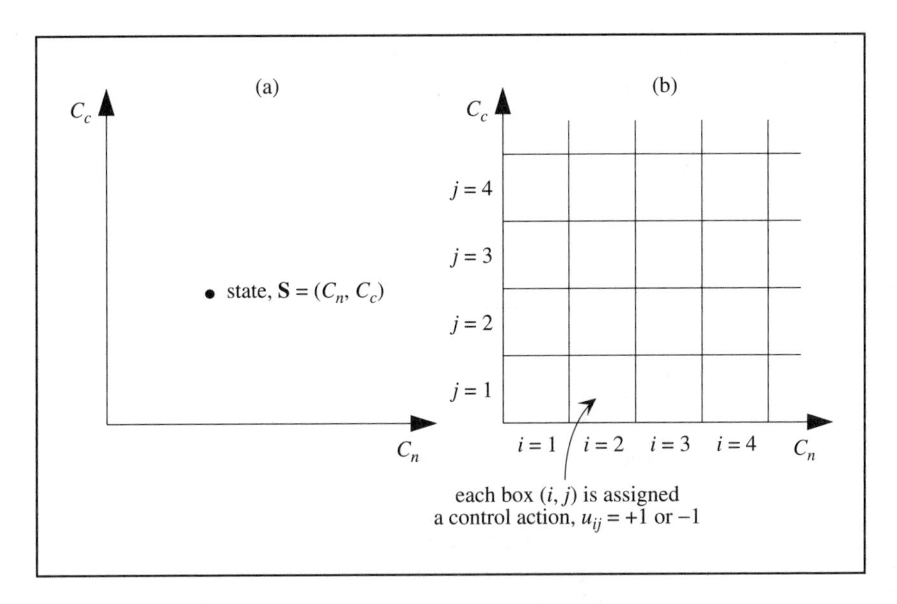

Figure 14.12 (a) state-space for a bioreactor
(b) state-space partitioned into boxes

appropriate control action u. By defining intervals in C_n and C_c we can create a finite number of boxes in state-space, where each box represents a collection of states that are similar to each other. A control action can then be associated with each box. A BOXES controller is completely defined by such a set of boxes and control actions, which together implicitly model the controlled system.

Control actions are performed at time $n\Delta t$, where n is an integer and Δt is the interval between control actions. At any such time, the system state will be in a particular box. That box is considered to be "visited," and its control action is performed. The system may be in the same box or a different one when the next control action is due.

The BOXES controller must be trained to associate appropriate control actions with each box. This is easiest using bang-bang control, where the control variables can take only their maximum or minimum value, denoted +1 and –1 respectively. In the case of the bioreactor, a valve controlling the flow would be fully open or fully shut. For each box, there is a recorded score for both the +1 and –1 action. When a box is visited, the selected control action is the one with the highest score. Learning is achieved by updating the scores.

In order to learn, the system must receive some measure of its performance, so that it can recognize beneficial or deleterious changes in its control strategy. This is achieved through use of a *critic*, which evaluates the controller's performance. In the case of the bioreactor, the controller's time to failure might be monitored, where "failure" occurs if the cell concentration C_c drifts beyond prescribed limits. The longer the time to failure, the better the performance of the controller. Woodcock et al. [11] consider this approach to be midway between supervised and unsupervised learning (Chapters 6 and 8), as the controller receives an indication of its performance but not a direct comparison between its output and the desired output.

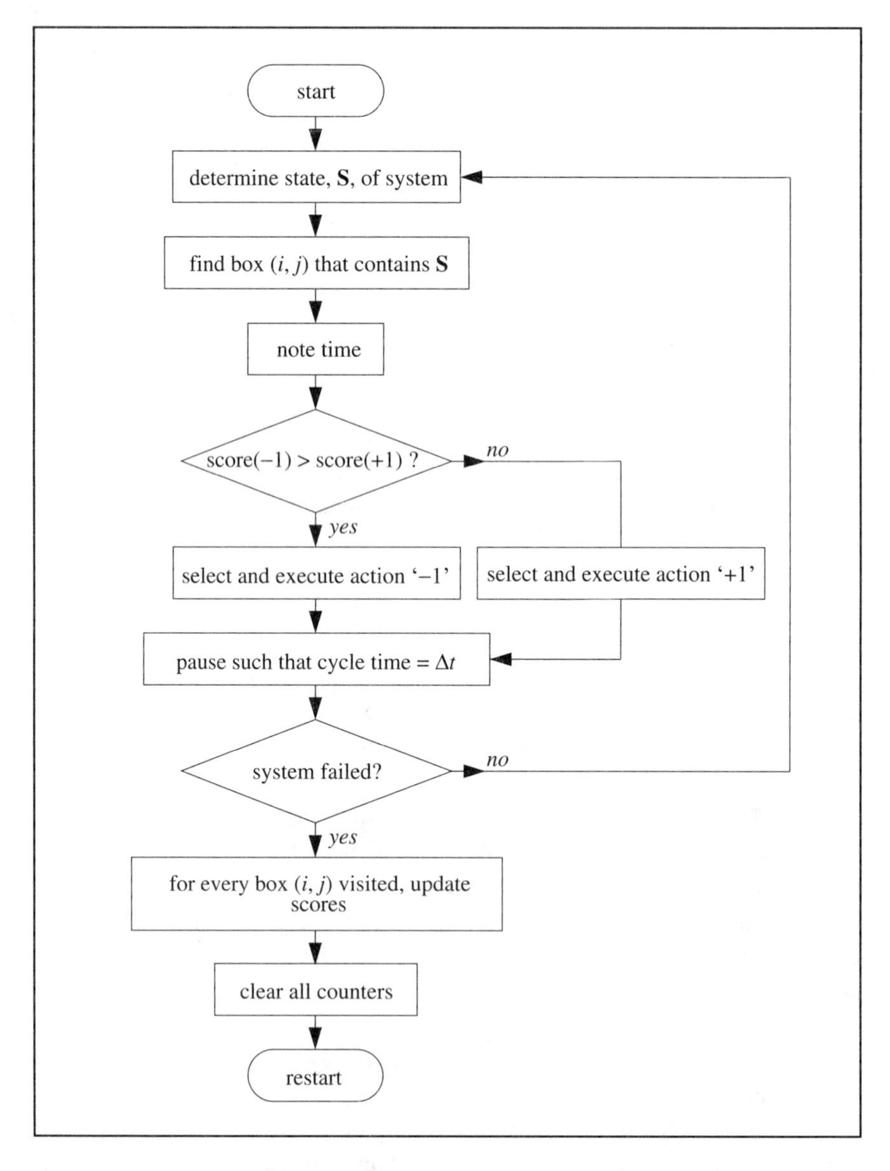

Figure 14.13 The BOXES learning algorithm, derived from [12]. In the example shown here, performance is gauged by time to failure

For each box, a score is stored for both the +1 action and the −1 action. These scores are a measure of "degree of appropriateness" and are based on the average time between selecting the control action in that particular box and the next failure.

The learning strategy of Michie and Chambers [12] for bang-bang control is shown in Figure 14.13. During a run, a single box may be visited N times. For each box, the times (t_1, ... t_i, ... t_N) at which it is visited are recorded. At the end of a run, i.e., after a failure, the time t_f is noted and the +1 and −1 scores for each visited box are updated. Each score is based on the average time to failure after that particular control action had been carried out, i.e., the lifetime l. The lifetimes are modified by a usage factor n, a decay factor α, a global lifetime l_g, a global usage factor n_g, and a constant β, thereby yielding a score. These modifications ensure that, for each box, both alternative actions have the chance to demonstrate their suitability during the learning process and that recent experience is weighted more heavily than old experience. The full updating procedure is as follows:

$$l_g = \alpha l_g + t_f \tag{14.7}$$

$$n_g = \alpha n_g + 1 \tag{14.8}$$

For each box where $score_{(+1)} > score_{(-1)}$:

$$l_{(+1)} = \alpha l_{(+1)} + \sum_{i=1}^{N} (t_f - t_i) \tag{14.9}$$

$$n_{(+1)} = \alpha n_{(+1)} + N \tag{14.10}$$

$$score_{(+1)} = \frac{l_{(+1)} + \beta \frac{l_g}{n_g}}{u_{(+1)} + \beta} \tag{14.11}$$

For each box where $score_{(-1)} > score_{(+1)}$:

$$l_{(-1)} = \alpha l_{(-1)} + \sum_{i=1}^{N} (t_f - t_i) \tag{14.12}$$

$$n_{(-1)} = \alpha n_{(-1)} + N \tag{14.13}$$

$$score_{(-1)} = \frac{l_{(-1)} + \beta \frac{l_g}{n_g}}{u_{(-1)} + \beta} \tag{14.14}$$

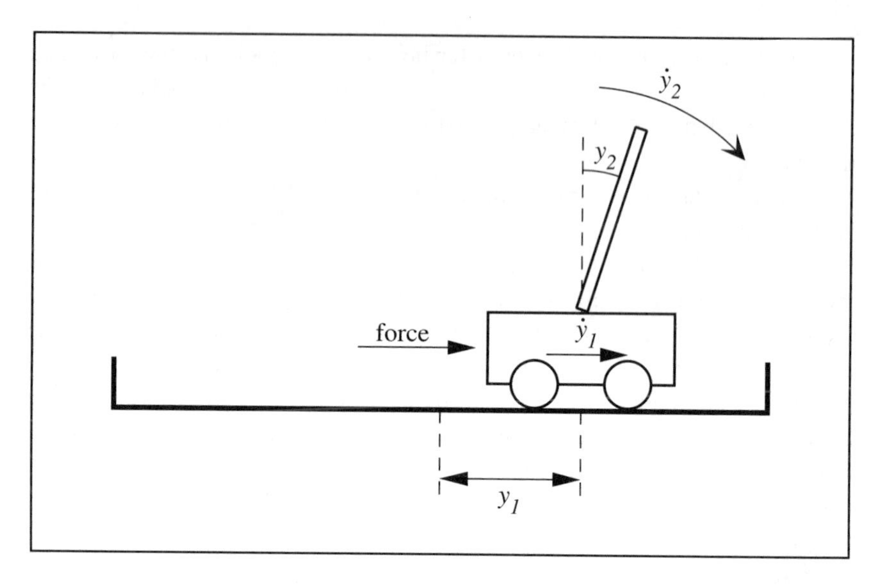

Figure 14.14 The cart-and-pole control problem

After a controller has been run to failure and the scores associated with the boxes have been updated, the controller becomes competent at balancing in only a limited part of the state-space. In order to become expert in all regions of state-space, the controller must be run to failure several times, starting from different regions in state-space.

The BOXES algorithm has been used for control of a bioreactor as described, and also for balancing a pole on a mobile cart (Figure 14.14). The latter is a similar problem to the bioreactor, but the state is described by four rather than two variables. The boxes are four-dimensional and difficult to represent graphically. In principle, the BOXES algorithm can be applied to state-space with any number of dimensions. The cart-and-pole problem, shown in Figure 14.14, has been used extensively as a benchmark for intelligent controllers. A pole is attached by means of a hinge to a cart that can move along a finite length of track. The cart and the pole are restricted to movement within a single plane. The controller attempts to balance the pole while keeping the cart on the length of track by applying a force to the left or right. If the force has a fixed magnitude in either direction, this is another example of bang-bang control. The four state variables are the cart's position y_1 and velocity $\dot{y}_1$ and the pole's angle y_2 and angular velocity $\dot{y}_2$. Failure occurs when y_1 or y_2 breach constraints placed upon them. The constraint on y_1 represents the limited length of the track.

Rather than use a BOXES system as an intelligent controller *per se*, Sammut and Michie [13] have used it as a means of eliciting rules for a rule-based controller. After running the BOXES algorithm on a cart-and-pole system, they found clear relationships between the learned control actions and the state variables. They expressed these relationships as rules and then proceeded to use analogous rules to control a different black box simulation, namely, a simulated spacecraft. The spacecraft was subjected to a number of unknown external forces, but the rule-based controller was tolerant of these. Similarly, Woodcock et al.'s BOXES controller [11] was virtually unaffected by random variations superimposed on the control variables.

One of the attractions of the BOXES controller is that it is a fairly simple technique, and so an effective controller can be built quite quickly. Woodcock et al. [11] rapidly built their controller and a variety of black box simulations using the Smalltalk object-oriented language (see Chapter 4). Although both the controller and simulation were developed in the same programming environment, the workings of the simulators were hidden from the controller. Sammut and Michie also report that they were able to build quickly their BOXES controller and the rule-based controller that it inspired [13].

14.7.2 Fuzzy BOXES

Woodcock et al. [11] have investigated the suggestion that the performance of a BOXES controller might be improved by using fuzzy logic to smooth the bang-bang control [14]. Where different control actions are associated with neighboring boxes, it was proposed that states lying between the centers of the boxes should be associated with intermediate actions. The controller was trained as described above in order to determine appropriate bang-bang actions. After training, the box boundaries were fuzzified using triangular fuzzy sets. The maximum and minimum control actions (bang-bang) were normalized to +1 and –1 respectively, and intermediate actions were assigned a number between these extremes.

Consider again the bioreactor, which is characterized by two-dimensional state-space. If a particular state **S** falls within the box (i, j), then the corresponding control action is u_{ij}. This can be stated as an explicit rule:

```
IF state S belongs in box (i,j)
THEN the control action is uij.
```

If we consider C_n and C_c separately, this rule can be rewritten:

```
IF Cn belongs in interval i AND Cc belongs in interval j
THEN the control action is uij.
```

The same rule can be applied in the case of fuzzy BOXES, except that now it is interpreted as a fuzzy rule. We know from Equation 3.36 (in Chapter 3) that:

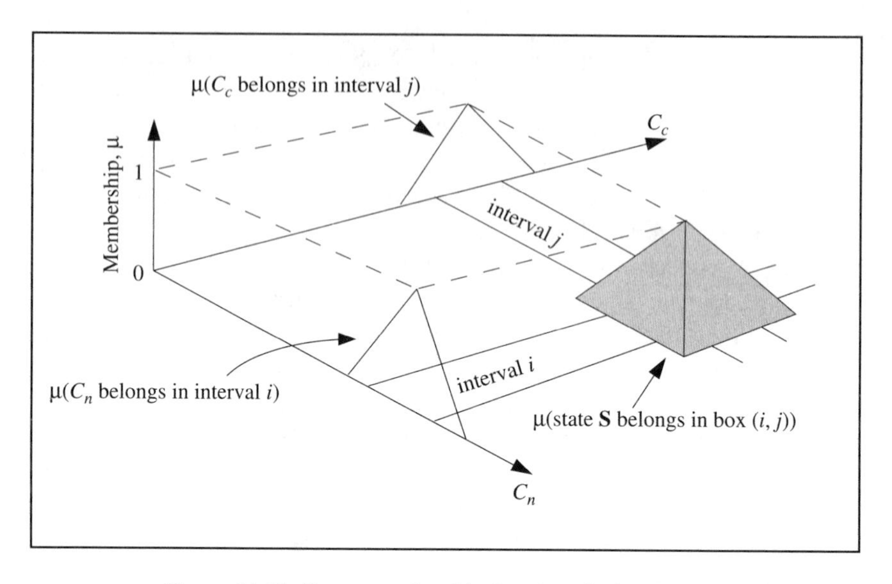

Figure 14.15 Fuzzy membership functions for boxes in the bioreactor state space (adapted from [11])

μ(`Cn belongs in interval i AND Cc belongs in interval j`) = min[μ(`Cn belongs in interval i`), μ(`Cc belongs in interval j`)]

Thus, if the membership functions for `Cn belongs in interval i` and `Cc belongs in interval j` are both triangular, then the membership function for `state S belongs in box (i,j)`, denoted by $\mu_{ij}(\mathbf{S})$, is a surface in state space in the shape of a pyramid (Figure 14.15). As the membership functions for neighboring pyramids overlap, a point in state space may be a member of more than one box. The control action u_{ij} for each box to which **S** belongs is scaled according to the degree of membership $\mu_{ij}(\mathbf{S})$. The normalized sum of these actions is then interpreted as the defuzzified action u_0:

$$u_0 = \frac{\sum_i \sum_j \mu_{ij}(\mathbf{S}) u_{ij}}{\sum_i \sum_j \mu_{ij}(\mathbf{S})}$$

This is equivalent to defuzzification using the centroid method (Chapter 3), if the membership functions for the control actions are assumed to be symmetrical about a vertical line through their balance points.

Woodcock et al. have tested their fuzzy BOXES controller against the cart-and-pole and bioreactor simulations (described above), both of which are adaptive control problems. They have also tested it in a servo control application, namely, reversing a tractor and trailer up to a loading bay. In none of these examples was there a clear winner between the nonfuzzy and the fuzzy boxes controllers. The comparison between them was dependent on the starting position in state-space. This was most clearly illustrated in the case of the tractor and trailer. If the starting position was such that the tractor could reverse the trailer in a smooth sweep, the fuzzy controller was able to perform best because it was able to steer smoothly. The nonfuzzy controller, on the other hand, was limited to using only full steering lock in either direction. If the starting condition was such that full steering lock was required, then the nonfuzzy controller outperformed the fuzzy one.

14.8 Neural network controllers

Neural network controllers tackle a similar problem to BOXES controllers, i.e., controlling a system using a model that is automatically generated during a learning phase. Two distinct approaches have been adopted by Valmiki et al. [15] and by Willis et al. [16]. Valmiki et al. have trained a neural network to associate directly particular sets of state variables with particular action variables, in an analogous fashion to the association of a box in state-space with a control action in a BOXES controller. Willis et al. adopted a less direct approach, using a neural network to estimate the values of those state variables that are critical to control but cannot be measured on-line. The estimated values are then fed to a PID controller as though they were real measurements. These two approaches are discussed separately below.

14.8.1 Direct association of state variables with action variables

Valmiki et al. have applied a neural network to a control problem that had previously been tackled using rules and objects, namely, the control of a glue dispenser [17]. As part of the manufacture of mixed technology circuit boards, surface-mounted components are held in place by a droplet of glue. The glue is dispensed from a syringe by means of compressed air. The size of the droplet is the state variable that must be controlled, and the change in the air pressure is the action variable.

Valmiki et al. have built a 6–6–5 multilayer perceptron (Figure 14.16), where specific meanings are attached to values of 0 and 1 on the input and output nodes. Five of the six input nodes represent ranges for the error in the size of the droplet. The node corresponding to the measured error is sent a 1, while the other four nodes are sent a 0. The sixth input node is set to 0 or 1 depending on whether the error is positive or negative. Three of the five output nodes are used to flag particular actions, while the other two are a coded

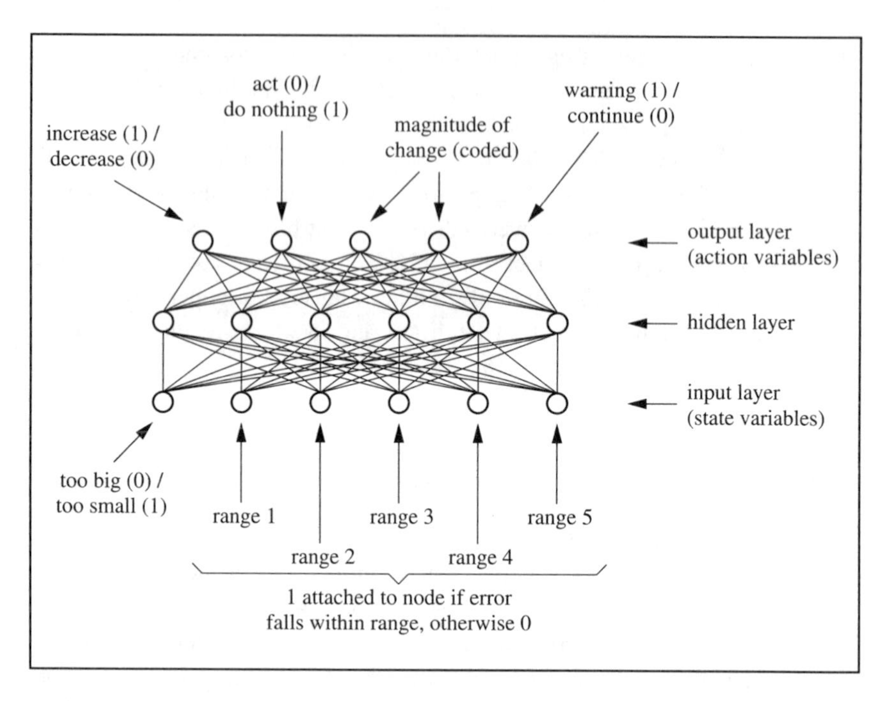

Figure 14.16 Using a neural network to map directly state variables to action variables (based on the glue-dispensing application of Valmiki et al. [15])

representation of the amount by which the air pressure should be changed, if a change is required. The three action flags on the output are:

- decrease (0) or increase (1) pressure;
- do something (0) or do nothing (1) (overrides the increase/decrease flag);
- no warning (0) or warning (1) if the error in the droplet size is large.

The training data were generated by hand. Since the required mapping of input states to outputs was known in advance — allowing training data to be drawn up — the problem could have been tackled using rules. One advantage of a neural network approach is that an interpolated meaning can be attached to output values that lie between 0 and 1. However, the same effect could also be achieved using fuzzy rules. This would have avoided the need to classify the state variables according to crisp sets. Nevertheless, Valmiki et al.'s experiment is important in demonstrating the feasibility of using a neural network to learn to associate state variables with control actions. This is useful where rules or functions that link the two are unavailable, although this was not the case in their experiment.

14.8.2 Estimation of critical state variables

Willis et al. [16] have demonstrated the application of neural network controllers to industrial continuous and batch-fed fermenters, and to a commercial-scale high purity distillation column. Each application is characterized by a delay in obtaining the critical state variable (i.e., the *controlled* variable), as it requires chemical or pathological analysis. The neural network allows comparatively rapid estimation of the critical state variable from secondary state variables. The use of a model for such a purpose is discussed in Section 11.4.4. The only difference here is that the controlled plant is modeled using a neural network. The estimated value for the critical state variable can be sent to a PID controller (see Section 14.2.5) to determine the action variable (Figure 14.17). As the critical variable can be measured off-line in each case, there is no difficulty in generating training sets of data. Each of the three applications demonstrates a different aspect to this problem. The chemotaxis learning algorithm was used in each case (see Section 8.4.3).

The continuous fermentation process is dynamic, i.e., the variables are constantly changing, and a *change* in the value of a variable may be just as significant as the absolute value. The role of a static neural network, on the other hand, is to perform a mapping of static input variables onto static output variables. One way around this weakness is to use the recent history of state variables as input nodes. In the continuous fermentation process, two secondary state variables were considered. Nevertheless, six input nodes were required, since the two previously measured values of the variables were used as well as the current values (Figure 14.18).

In contrast, the batch fermentation process should move smoothly and slowly through a series of phases, never reaching equilibrium. In this case, the time since the process began, rather than the time history of the secondary variable, was important. Thus, for this process there were only two input nodes, the current time and a secondary state variable.

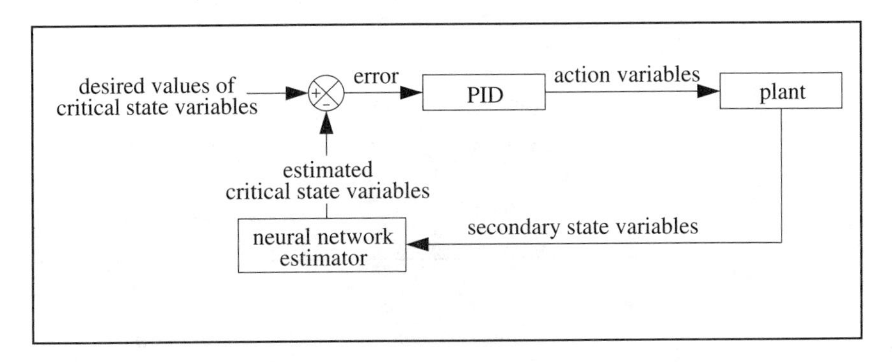

Figure 14.17 Using a neural network to estimate values for critical state variables

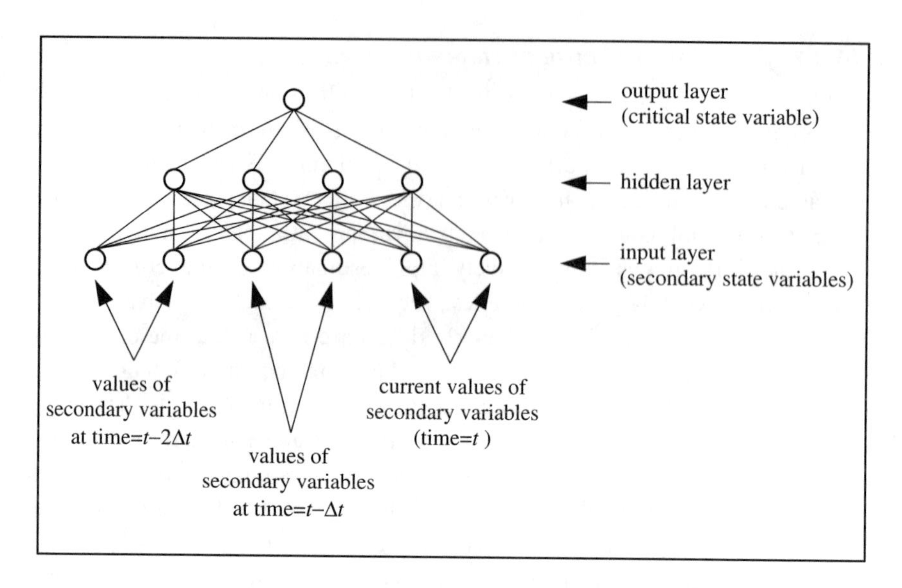

Figure 14.18 Using time histories of state variables in a neural network (based on the continuous fermentation application of Willis et al. [16])

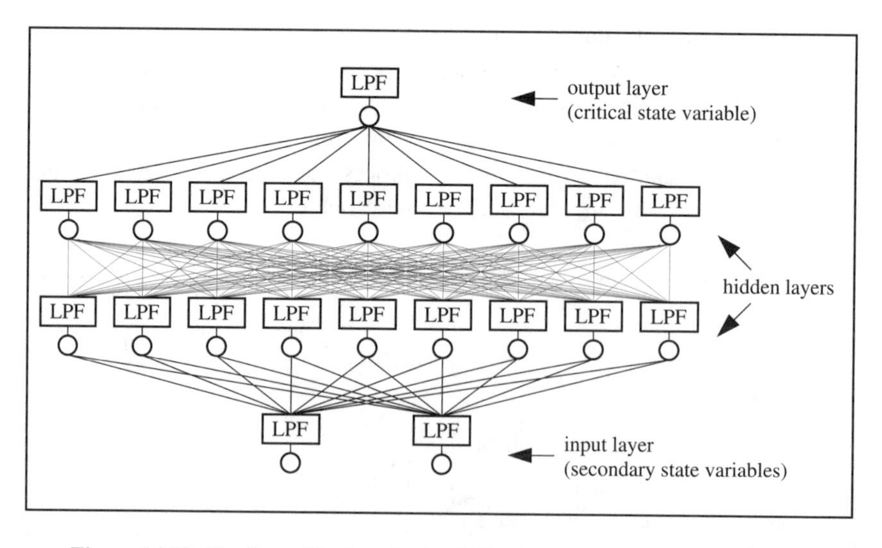

Figure 14.19 Dealing with changing variables by using low pass filters (LPF) (based on the industrial distillation application of Willis et al. [16])

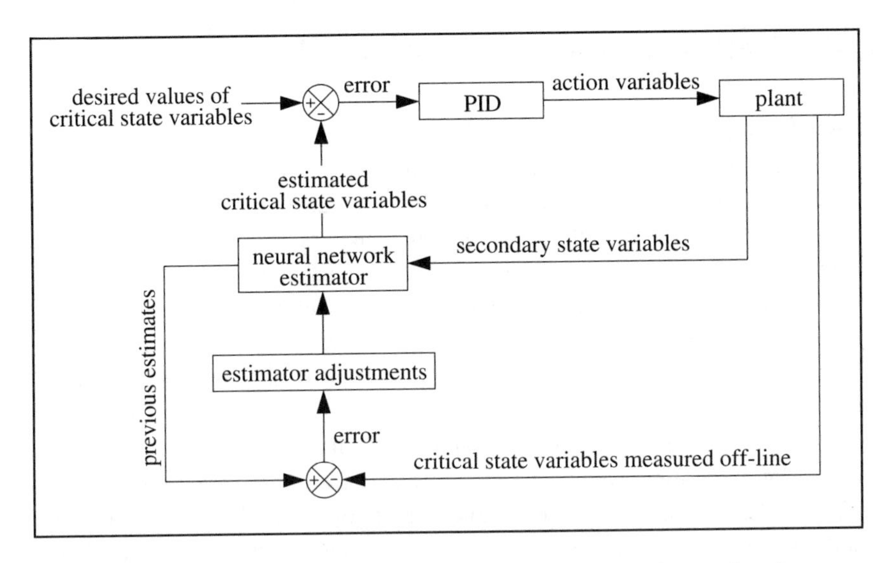

Figure 14.20 Feedback control of both the plant and the neural network estimator

In the methanol distillation process, an alternative approach was adopted to the problem of handling dynamic behavior. As it is known that the state variables must vary continuously, sudden sharp changes in any of the propagated values can be disallowed. This is achieved through a simple low-pass digital filter. There are several alternative forms of digital filter (see, for example, [18]), but Willis et al. used the following:

$$y(t) = \Omega\, y(t-1) + (1-\Omega)\, x(t) \qquad 0 \le \Omega \le 1 \tag{14.15}$$

where $x(t)$ and $y(t)$ are the input and output of the filter, respectively, at time t. The filter ensures that no value of $y(t)$ can be greatly different from its previous value, and so high-frequency fluctuations are eliminated. Such a filter was attached to the output side of each neuron (Figure 14.19), so that the unfiltered output from the neuron was represented by $x(t)$, and the filtered output was $y(t)$. Suitable values for the parameters Ω were learned along with the network weightings.

Willis et al. were able to show improved accuracy of estimation and tighter control by incorporating the digital filter into their neural network. Further improvements were possible by comparing the estimated critical state variable with the actual values, as these became known. The error was then used to adjust the output of the estimator. There were two feedback loops, one for the PID controller and one for the estimator (Figure 14.20).

14.9 Statistical process control (SPC)

14.9.1 Applications

Statistical process control (SPC) is a technique for monitoring the quality of products as they are manufactured. Critical parameters are monitored and adjustments are made to the manufacturing process *before* any products are manufactured that lie outside of their specifications. The appeal of SPC is that it minimizes the number of products that are rejected at the quality control stage, thereby improving productivity and efficiency. Since the emphasis of SPC lies in *monitoring* products, this section could equally belong in Chapter 11.

SPC involves inspecting a sample of the manufactured products, measuring the critical parameters, and inferring from these measurements any trends in the parameters for the whole population of products. The gathering and manipulation of the statistics is a procedural task, and some simple heuristics are used for spotting trends. The monitoring activities, therefore, lend themselves to automation through procedural and rule-based programming. Depending on the process, the control decisions might also be automated.

14.9.2 Collecting the data

Various statistics can be gathered, but we will concentrate on the mean and standard deviation* of the monitored parameters. Periodically, a sample of consecutively manufactured products is taken, and the critical parameter x is measured for each item in the sample. The sample size n is typically in the range 5–10. In the case of the manufacture of silicon wafers, thickness may be the critical parameter. The mean $\bar{x}$ and standard deviation σ for the sample are calculated. After several such samples have been taken, it is possible to arrive at a mean of means $\bar{\bar{x}}$ and a mean of standard deviations $\bar{\sigma}$. The values $\bar{\bar{x}}$ and $\bar{\sigma}$ represent the normal, or set-point, values for $\bar{x}$ and σ, respectively. Special set-up procedures exist for the manufacturing plant to ensure that $\bar{\bar{x}}$ corresponds to the set-point for the parameter x. Bounds called *control limits* are placed above and below these values (Figure 14.21). Inner and outer control limits, referred to as *warning limits* and *action limits,* respectively, may be set such that:

$$\text{warning limit for } \bar{x} = \bar{\bar{x}} \pm 2\frac{\bar{\sigma}}{\sqrt{n}}$$

---

* The *range* of sample values is often used instead of the standard deviation.

$$\text{action limit for } \bar{x} = \bar{\bar{x}} \pm 3\frac{\bar{\sigma}}{\sqrt{n}}$$

Action limits are also set on the values of σ such that:

$$\text{upper action limit for } \sigma = C_U \bar{\sigma}$$

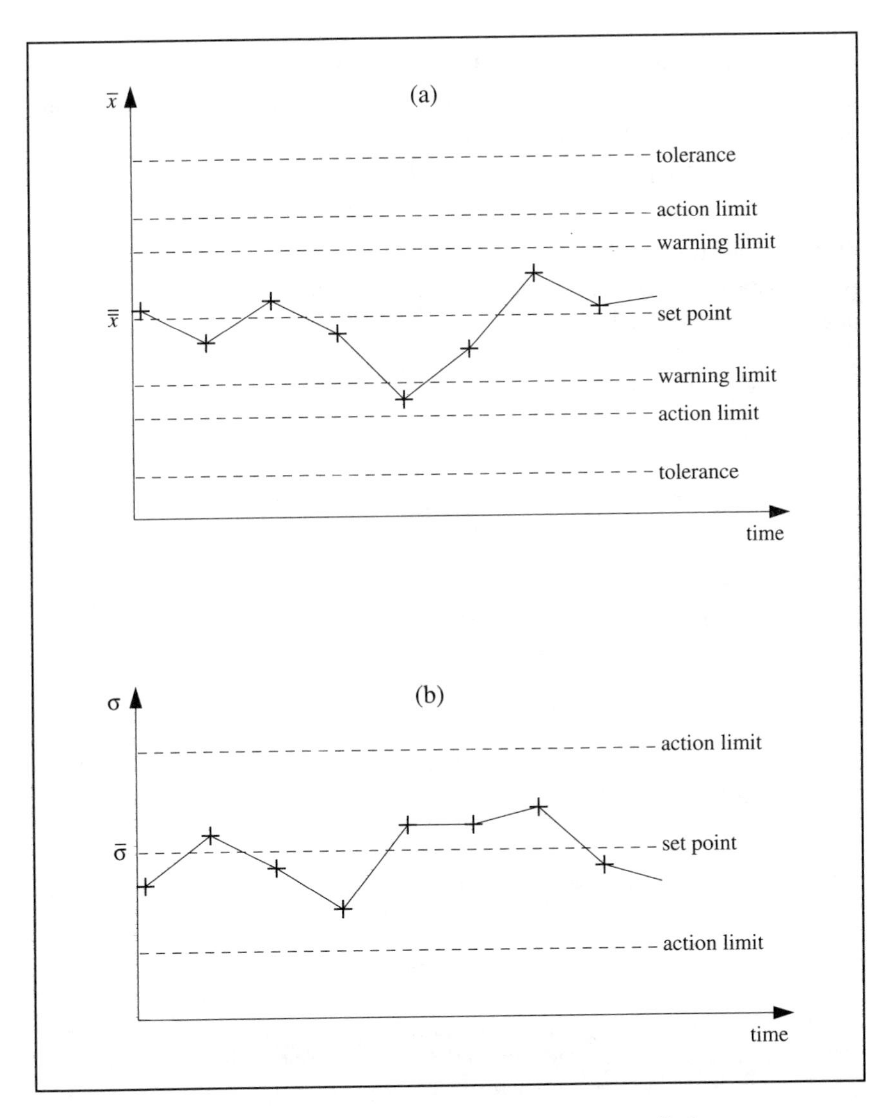

Figure 14.21 Control limits (action and warning) applied to:
(a) sample means ($\bar{x}$);
(b) sample standard deviations (σ)

$$\text{lower action limit for } \sigma = \frac{\overline{\sigma}}{C_L}$$

where suitable values for C_U and C_L can be obtained from standard tables for a given sample size n. Note that both C_U and C_L are greater than 1. The heuristics for interpreting the sample data with respect to the control limits are described in Section 14.9.3, below. Any values of $\overline{x}$ that lie beyond the action limits indicate that a control action is needed. The *tolerance* that is placed on a parameter is the limit beyond which the product must be rejected. It follows that if the tolerance is tighter than the action limits, then the manufacturing plant is unsuited to the product and attempts to use it will result in a large number of rejected products irrespective of SPC.

14.9.3 Using the data

As the data are gathered, a variety of heuristics can be applied. Some typical ones are reproduced below:

```
IF a single x̄ value lies beyond an action limit
THEN a special disturbance has occurred that must be
investigated and eliminated

IF there are x̄ values beyond both action limits
THEN the process may be deteriorating

IF two consecutive values of x̄ lie beyond a worrying limit
THEN the process mean may have moved

IF eight consecutive values of x̄ lie on an upward or downward
trend THEN the process mean may be moving

IF seven consecutive values of x̄ lie all above or all below x̿
THEN the process mean may have moved

IF there are σ values beyond the upper action limit
THEN the process may be deteriorating

IF eight consecutive values of σ lie on an upward trend
THEN the process may be deteriorating

IF there are σ values beyond the lower action limit
THEN the process may have improved and attempts should be made
to incorporate the improvement permanently
```

The conclusions of these rules indicate a high probability that a control action is needed. They cannot be definite conclusions, as the evidence is statistical.

Furthermore, it may be that the process itself has not changed at all, but instead some aspect of the measuring procedure has altered. Each of the above rules calls for investigation of the process to determine the cause of any changes, perhaps using case-based or model-based reasoning (Chapters 6 and 11).

14.10 Summary

Intelligent systems for control applications draw upon the techniques used for interpreting data (Chapter 11) and planning (Chapter 13). Frequently, the stages of planning are interleaved with execution of the plans, so that the controller can react to changes in the controlled plant as they occur. This contrasts with the classical planning systems described in Chapter 13, where the world is treated as a static "snapshot." As control systems must interact with a dynamic environment, time constraints are placed upon them. There is often a trade-off between the quality of a control decision and the time taken to derive it. In most circumstances it is preferable to perform a suboptimal control action than to fail to take any action within the time limits.

The control problem can be thought of as one of mapping a set of state variables onto a set of action variables. State variables describe the state of the controlled plant, and action variables, set by the controller, are used to modify the state of the plant. Adaptive controllers attempt to maintain one or more critical state parameters at a constant value, minimizing the effects of any disturbance. In contrast, servo controllers attempt to drive the plant to a new state, which may be substantially different from its previous state. The problems of adaptive and servo control are similar, as both involve minimizing the difference, or error, between the current values of the state variables and the desired values.

An approximate distinction can be drawn between low-level "reflex" control and high-level supervisory control. Low-level control often requires little intelligence and can be most effectively coded procedurally, for instance, as the sum of proportional, integral, and derivative (PID) terms. Improvements over PID control can be made by using fuzzy rules, which also allow some subtleties to be included in the control requirements, such as bounds on the values of some variables. Fuzzy rules offer a mixture of some of the benefits of procedures and crisp rules. Like crisp rules, fuzzy rules allow a linguistic description of the interaction between state and action variables. On the other hand, like an algebraic procedure, fuzzy rules allow smooth changes in the state variables to bring about smooth changes in the action variables. The nature of these smooth changes is determined by the membership functions that are used for the fuzzy sets.

Any controller requires a model of the controlled plant. Even a PID controller holds an implicit model in the form of its parameters, which can be tuned to specific applications. When a model of the controlled plant is not available, it is possible to build one automatically using the BOXES algorithm or a neural network. Both can be used to provide a mapping between state variables and action variables. They can also be used in a monitoring capacity, where critical state variables (which may be difficult to measure directly) are inferred from secondary measurements. The inferred values can then be used as feedback to a conventional controller. If a plant is modeled with sufficient accuracy, then predictive control becomes a possibility. A predictive controller has two goals, to tackle the immediate control needs and to minimize future deviations, based on the predicted behavior.

References

1. Bennett, M. E., "Real-time continuous AI," *IEE Proceedings–D*, vol. 134, pp. 272–277, 1987.
2. Franklin, G. F., Powell, J. D., and Emami-Naeini, A., *Feedback Control of Dynamic Systems*, 3rd ed., Addison-Wesley, 1994.
3. Sripada, N. R., Fisher, D. G., and Morris, A. J., "AI application for process regulation and process control," *IEE Proceedings–D*, vol. 134, pp. 251–259, 1987.
4. Leitch, R., Kraft, R., and Luntz, R., "RESCU: a real-time knowledge based system for process control," *IEE Proceedings–D*, vol. 138, pp. 217–227, 1991.
5. Laffey, T. J., Cox, P. A., Schmidt, J. L., Kao, S. M., and Read, J. Y., "Real-time knowledge-based systems," *AI Magazine*, pp. 27–45, Spring 1988.
6. Lesser, V. R., Pavlin, J., and Durfee, E., "Approximate processing in real-time problem solving," *AI Magazine*, pp. 49–61, Spring 1988.
7. Hopgood, A. A., "Rule-based control of a telecommunications network using the blackboard model," *Artificial Intelligence in Engineering*, vol. 9, pp. 29–38, 1994.
8. Taunton, J. C. and Haspel, D. W., "The application of expert system techniques in on-line process control," in *Expert Systems in Engineering*, Pham, D. T. (Ed.), IFS Publications / Springer-Verlag, 1988.
9. Hopgood, A. A., Phillips, H. J., Picton, P. D., and Braithwaite, N. S. J., "Fuzzy logic in a blackboard system for controlling plasma deposition

processes," *Artificial Intelligence in Engineering*, vol. 12, pp. 253–260, 1998.

10. Ungar, L. H., "A bioreactor benchmark for adaptive network-based process control," in *Neural Networks for Control*, Miller, W. T., Sutton, R. S., and Werbos, P. J. (Eds.), MIT Press, 1990.
11. Woodcock, N., Hallam, N. J., and Picton, P. D., "Fuzzy BOXES as an alternative to neural networks for difficult control problems," in *Applications of Artificial Intelligence in Engineering VI*, Rzevski, G. and Adey, R. A. (Eds.), pp. 903–919, Computational Mechanics / Elsevier, 1991.
12. Michie, D. and Chambers, R. A., "BOXES: an experiment in adaptive control," in *Machine Intelligence 2*, Dale, E. and Michie, D. (Eds.), pp. 137–152, Oliver and Boyd, 1968.
13. Sammut, C. and Michie, D., "Controlling a black-box simulation of a spacecraft," *AI Magazine*, pp. 56–63, Spring 1991.
14. Bernard, J. A., "Use of a rule-based system for process control," *IEEE Control Systems Magazine*, pp. 3–13, October 1988.
15. Valmiki, A. H., West, A. A., and Williams, D. J., "The evolution of a neural network controller for adhesive dispensing," IFAC Workshop on Computer Software Structures Integrating AI/KBS Systems in Process Control, Bergen, Norway, pp. 93–101, 1991.
16. Willis, M. J., Di Massimo, C., Montague, G. A., Tham, M. T., and Morris, A. J., "Artificial neural networks in process engineering," *IEE Proceedings–D*, vol. 138, pp. 256–266, 1991.
17. Chandraker, R., West, A. A., and Williams, D. J., "Intelligent control of adhesive dispensing," *Int. J. Computer Integrated Manufacturing*, vol. 3, pp. 24–34, 1990.
18. Lynn, P. A., Fuerst, W., and Thomas, B., *Introductory Digital Signal Processing with Computer Applications*, 2nd ed., Wiley, 1998.

Further reading

- Franklin, G. F., Powell, J. D., and Emami-Naeini, A., *Feedback Control of Dynamic Systems*, 3rd ed., Addison-Wesley, 1994.
- Franklin, G. F., Powell, J. D., and Workman, M. L., *Digital Control of Dynamic Systems*, 3rd ed., Addison-Wesley, 1998.
- Harris, C. J., *Advances in Intelligent Control*, Taylor and Francis, 1994.

- Lee, T. H., Harris, C. J., and Ge, S. S., *Adaptive Neural Network Control of Robotic Manipulators*, vol. 19, World Scientific, 1999.
- Miller, W. T., Sutton, R. S., and Werbos, P. J. (Eds.), *Neural Networks for Control*, MIT Press, 1991.
- Wang, H., Liu, G. P., Harris, C. J., and Brown, M., *Advanced Adaptive Control*, Pergamon, 1995.

Chapter fifteen

Concluding remarks

15.1 Benefits

This book has discussed a wide range of intelligent systems techniques and their applications. Whether any implemented intelligent system displays true intelligence — whatever that is assumed to mean — is likely to remain the subject of debate. Nevertheless, the following practical benefits have stemmed from the development of intelligent systems techniques.

Reliability and consistency
An intelligent system makes decisions that are consistent with its input data and its knowledge base (for a knowledge-based system) or numerical parameters (for a computational intelligence technique). It may, therefore, be more reliable than a person, particularly where repetitive mundane judgments have to be made.

Automation
In many applications, such as visual inspection on a production line, judgmental decision making has to be performed repeatedly. A well-designed intelligent system ought to be able to deal with the majority of such cases, while highlighting any that lie beyond the scope of its capabilities. Therefore, only the most difficult cases, which are normally the most interesting, are deferred to a person.

Speed
Intelligent systems are designed to automatically make decisions that would otherwise require human reasoning, judgment, expertise, or common sense. Any lack of true intelligence is compensated by the system's processing speed. An intelligent system can make decisions informed by a wealth of data and information that a person would have insufficient time to assimilate.

Improved domain understanding
The process of constructing a knowledge-based system requires the decision-making criteria to be clearly identified and assessed. This process frequently leads to a better understanding of the problem being tackled. Similar benefits can be obtained by investigating the decision-making criteria used by the computational intelligence techniques.

Knowledge archiving
The knowledge base is a repository for the knowledge of one or more people. When these people move on to new jobs, some of their expert knowledge is saved in the knowledge base, which continues to evolve after their departure.

15.2 Implementation

Since intelligent systems are supposed to be flexible and adaptable, development is usually based upon continuous refinements of an initial prototype. This is the *prototype–test–refine* cycle, which applies to both knowledge-based systems and computational intelligence techniques. The key stages in the development of a system are:

- decide the requirements;
- design and implement a prototype;
- continuously test and refine the prototype.

It is sometimes suggested that the first prototype and the first few revisions of a KBS be implemented using an expert system shell that allows rapid representation of the most important knowledge. When the prototype has demonstrated its viability, the system can be moved to a more sophisticated programming environment. This approach makes some sense if a mock-up is required in order to obtain financial backing for a project, but it also brings several disadvantages. Working in a shell that lacks flexibility and representational capabilities is frustrating and can lead to convoluted programming in order to force the desired behavior. Subsequent rewriting of the same knowledge in a different style is wasteful of resources. An arguably better approach is to work from the outset in a flexible programming environment that provides all the tools that are likely to be needed, or which allows extra modules to be added as required.

Software engineers, particularly those working on large projects, have traditionally been skeptical of the prototype–test–refine cycle. Instead, they have preferred meticulous specification, analysis, and design phases prior to implementation and testing. These attitudes have now changed, and rapid prototyping and iterative development have gained respectability across most areas of software engineering.

15.3 Trends

Intelligent systems are becoming increasingly distributed in terms of both their applications and their implementation. While large systems will remain important, e.g., for commerce and industry, smaller embedded intelligent systems have also started to appear in the home and workplace. Examples include washing machines that incorporate knowledge-based control systems, elevators that use fuzzy logic to decide at which floor to wait for the next passenger, and personal organizers that use neural networks to learn the characteristics of their owner's handwriting. Communication between embedded applications is likely to extend further their influence on our daily lives.

In addition to being distributed in their applications, intelligent systems are also becoming distributed in their implementation. Chapter 9 discussed the blackboard architecture for dividing problems into subtasks that can be shared among specialized modules. In this way, the right software tool can be used for each job. Similarly, Chapter 5 looked at the increasingly important technique of intelligent agents. The growth in the use of the Internet is likely to see increased communication between agents that reside on separate computers, and mobile agents that can travel over the net in search of information. Furthermore, Jennings argues that agent-based techniques are appropriate both for developing large complex systems and for mainstream software engineering [1].

Paradoxically, there is also a sense in which intelligent systems are becoming more integrated. Watson and Gardingen describe a sales support application that has become integrated by use of the World Wide Web, as a single definitive copy of the software accessible via the web has replaced distributed copies [2].

As a further aspect of integration, computers are required to assist in commercial decision making, based upon a wide view of the organization. For example, production decisions need to take into account and influence design, marketing, personnel, sales, materials stocks, and product stocks. These separate, distributed functions are becoming integrated by the need for communication between them. (Use of computers to support an integrated approach to manufacturing is termed *computer-integrated manufacturing*, or *CIM.*) However, smaller-scale systems are likely to remain at least as important. These include intelligent agents that serve as personal consultants to advise and inform us, and others that function silently and anonymously while performing tasks such as data interpretation, monitoring, and control.

References

1. Jennings, N. R., "On agent-based software engineering," *Artificial Intelligence*, vol. 117, pp. 277–296, 2000.
2. Watson, I. and Gardingen, D., "A distributed case-based reasoning application for engineering sales support," 16th International Joint Conference on Artificial Intelligence (IJCAI'99), Stockholm, Sweden, vol. 1, pp. 600–605, 1999.

Index